# THE AGE OF PROJECTS

# THE AGE OF PROJECTS

*Edited by Maximillian E. Novak*

Published by the University of Toronto Press in association with the UCLA Center for Seventeenth- and Eighteenth-Century Studies and the William Anderson Clark Memorial Library

Toronto Buffalo London
www.utppublishing.com
Printed in Canada

ISBN 978-0-8020-9873-3

Printed on acid-free paper

---

**Library and Archives Canada Cataloguing in Publication**

The age of projects / edited by Maximillian E. Novak.

(UCLA Center/Clark series ; 9)
Includes bibliographical references and index.
ISBN 978-0-8020-9873-3

1. Humanities. 2. Science and the humanities. 3. Intellectual life – History – 18th century. 4. Civilization, Western – 18th century. I. Novak, Maximillian E. II. William Andrews Clark Memorial Library III. University of California, Los Angeles. Center for 17th- & 18th-Century Studies IV. Series: UCLA Clark Memorial Library series ; 9

AZ341.A34 2008 001.309′033 C2008-901936-9

---

This book has been published with the help of a grant from the UCLA Center for Seventeenth- and Eighteenth-Century Studies.

University of Toronto Press acknowledges the financial support for its publishing activities of the Government of Canada through the Book Publishing Industry Development Program (BPIDP).

University of Toronto Press acknowledges the financial assistance to its publishing program of the Canada Council for the Arts and Ontario Arts Council.

# Contents

# Figures

**Boime Essay**



**O'Byrne Essay**

**Gierl Essay**

# Acknowledgments

The essays in this volume grew out of an extensive series of lectures presented at the William Andrews Clark Memorial Library, UCLA, between 2003 and 2004. I conceived of this lecture series while working with Joyce Kennedy and Michael Seidel on an edition of Daniel Defoe's *An Essay upon Projects* (New York: AMS Press, 1999). In writing the notes, I spent weeks going through the books and pamphlets on economic and social themes that are best preserved in the Kress Collection of microfilms. Although I had read a considerable number of these works decades earlier in writing my *Economics and the Fiction of Daniel Defoe* (Berkeley: University of California Press, 1962), in doing the detailed work of an editor, I was overwhelmed by the sheer number of works devoted to these subjects. It seemed to speak to what had formerly been called 'the spirit of the age' – a tremendous psychological and mental energy poured into the kind of enterprises that Defoe referred to as projects. It showed a side to the age that was inventive and expansive and often disorderly. During the Restoration – a time when an effort was made to discourage enthusiasm in matters of religion – Thomas Sprat waxed enthusiastic over the future of science in his *History of the Royal Society* (1668), and the enthusiasm for projects seemed to span the entirety of what is now called the 'Long Eighteenth Century.'

In titling this volume 'The Age of Projects' I hoped to reflect this aspect of the period. I am fully aware of some of the problems with using the 'age of' concept, but just as J.G.A Pocock has taught us to think in terms of various kinds of Enlightenments, so we may identify the spirit of projecting with much the same chronological period that I once called the 'Age of Disguise,' and which, to emphasize an important aspect of the time, might just as well be called the 'age of Newton and Newtonianism' or the 'Augustan Age.' If the latter designation suggests a devotion to unchanging concepts in literature, the arts, and society, my title stresses the belief that society may be shaped one way or another by inventions or various schemes, from those that attempt to establish traditions to those that endeavour to change or even revolutionize the present.

I want to thank all those who participated in the lecture series, including those whose essays do not appear in this volume: Paul Alkon, Jayne Lewis, Joseph Levine, Kirstie McClure, Claude Rawson, Peter Reill, Mona Scheuermann, Michael Seidel, Colleen Terrell, and Matthew Wickman. Along with the wonderful contributions by the audience, they helped to show how important the notion of projecting was to this period. I had some valuable suggestions about organizing the lecture series from Helen Deutsch and fresh ideas from the Fellows at the William Andrews Clark Memorial Library and at UCLA's Center for Seventeenth- and Eighteenth-Century Studies. I also received moral and financial support from the Center and from UCLA's Council on Research. The staff of the William Andrews Clark Library were extraordinarily helpful both during the lecture series and in helping me in the research necessary for editing this volume, and I want to thank them as well. In editing the essays, I had help from Sherry Rosenwein, Seaphina Goldfarb Tarrant and my wife, Estelle Gershgoren Novak, and most especially John St James.

MAXIMILLIAN E. NOVAK

# THE AGE OF PROJECTS

chapter one

# Introduction

MAXIMILLIAN E. NOVAK

> There is, 'tis true, a great difference between *New Inventions* and *Projects*, between Improvement of Manufactures or Lands, which tend to the immediate Benefit of the Publick, and Imploying of the Poor; and Projects fram'd by subtle Heads, with a sort of a *Deceptio Visus*, and *Legerdemain*, to bring People to run needless and unusual hazards: I grant it, and give a due preference to the first, and yet Success has so sanctifi'd some of those other sorts of Projects, that 'twould be a kind of Blasphemy against Fortune to disallow 'em.[1]

Daniel Defoe's description of the 1690s as '*The Projecting Age*' in his *Essay upon Projects* (1697) is the starting point for this collection of essays.[2] Whereas for us the word 'projects' has a relatively neutral meaning, in Defoe's time it had a distinctly unsavoury connotation, being associated with unscrupulous schemes for getting money. One contemporary definition described the projector as a 'speculator, a cheat,' and Defoe refers to '*the Despicable title of a* Projector,' in his dedication of his work to Dalby Thomas.[3] But Defoe attempts to transform the meaning of the word. For him, projectors are people who find a way out of their difficulties by coming up with novel ideas. Unlike the thieves or the schemers who have been placed in the same desperate condition, the projectors 'turn their thoughts to Honest Invention, founded upon the Platform of Ingenuity and Integrity.' Yet Defoe does not want projecting to be overly respectable, and his attempt, in the opening quotation, to merge the meanings of 'Inventions' and 'Projects' is entirely deliberate. After all, as his novels attest, he had a certain fascination with unscrupulous characters. Nevertheless he does want to suggest that projectors and projects make society change and move forward. And he proceeded to offer a number of ways in which improvements might be made in the England of 1697, from academies – for women, for refining the English language, and for improving the military – to new kinds of banks, pension schemes, and courts for bankrupts. And behind these wonderfully imaginative projects lay the heart of a projector, for at the time Defoe was in the brick-making business. Projects require public buildings, buildings require bricks, and who better to sell bricks than Daniel Defoe, who had already supplied some for one of Christopher Wren's public projects.[4]

Towards the end of his life, Defoe confessed unashamedly that at one time he was a projector,[5] and while he invested in what was supposed to be an improved diving machine for seeking treasure, most of his projects involved social engineering, such as his proposal to improve street lighting to prevent crime, or literary projects such as his planned history of trade. In collecting the essays for this volume, I have followed Defoe's lead in examining a broad range of projects in literature, history, the arts and engineering as well as the interest in exploration and in particular scientific discoveries between the Restoration and the early nineteenth century. During this period – what is now often called the 'Long Eighteenth Century' – there were certainly some major scientific discoveries. As Jenny Uglow has argued in her book *The Lunar Men*, there were groups of men entirely dedicated to scientific ideals, but even in this group, particularly with Matthew Boulton, there was a projecting spirit.[6]

This volume grew out of a series of seminars at the William Andrews Clark Memorial Library during the academic year of 2003–4 that I arranged for the Center for Seventeenth- and Eighteenth-Century Studies at UCLA. In ordering the essays, I have divided the book into three sections: Retrieving the Past, Improving the Present, and Envisioning the Future. There is a rough chronological pattern in this, though, as shall be seen, in areas such as music, the discovery of the past came relatively late. To some extent this order reflected a changing view of the world in Britain and Western Europe. For those steeped in traditional learning and in the belief that such learning contained all that was to be known, the most attractive aspects of such organizations as the Royal Society, given an official charter in 1662, lay in its promise to rediscover ancient wisdom. In the battle between the Ancients and the Moderns that smouldered through much of the seventeenth century and well into the next, writers such as Samuel Butler, John Dryden, and Jonathan Swift satirized what they considered to be a rage for novel ideas in life and literature. By the beginning of the nineteenth century, the advocates of progress and change in all forms of human endeavour were still embattled, but it was clear that they were winning their case.

Nevertheless, it has to be admitted that some of the best minds of the time were among the sceptics. Thomas Hobbes doubted that anything like genuine science lay behind Boyle's Air Pump;[7] Bishop Edward Stillingfleet thought he sensed heresy in John Locke's view of the human mind;[8] and Jonathan Swift refused to believe that Isaac Newton's system was anything more than another modern error.[9] Even at the end of the century, William Blake could voice dismay at the concern with the material world, when it was possible for the mind 'To see a World in a grain of sand / And a Heaven in a wild flower, / Hold Infinity in the palm of your Hand / And Eternity in an hour.'[10] In the early years of the 'Long Eighteenth Century,' it was easy enough to find absurdity in the numerous failed projects and in the opening volumes of the *Philosophical Transactions of the Royal Society* dealing with monstrous births and other oddities. In the third book of his *Travels*, Gulliver visits an academy of projectors, the Academy of Lagado, where he finds himself distinctly uncomfortable when embraced by an odorous 'scientist' working to convert excrement back into food. Offended by the smell caused through the discharge of a dog being used in an experiment on curing disease, yet somewhat impressed by an experiment in writing books by a random jumbling of letters, Gulliver laments the 'Ireconcilable Enemies to Science' among the common people who refuse to see the advantages of reducing all communication to the display of objects, thereby getting rid of the complexities of language and reducing words to things. Gulliver, the projector, even offers an addition to the scheme of one experimenter to detect insurrections against the government by studying the feces of those under suspicion, suggesting a method of reading ordinary language as containing coded messages of revolt. Swift's message is that a country ruled by projectors is doomed. For example, the country of Lagado is going to ruin as new experimental systems of agriculture have resulted in the destructions of the old system that had worked perfectly well.[11]

Although Swift's satire has often been seen as directed towards the scientific experiments of the Royal Society, it was more generally aimed at the spirit of change and novelty that had triumphed during the 1690s. During the Restoration there had been official disapproval of political and religious innovation. The Royal Society had to make its case for its essentially anti-radical designs. Thomas Sprat's and Robert Boyle's famous attacks upon obscure language and metaphor were only slightly disguised criticisms of the often inventive but uncontrolled use of imagery by some of the religious sects during the Interregnum.[12] Science was supposed to reveal what actually was, not to speculate on new possibilities. And as such it might seem harmless enough. Characters in Restoration comedy, such as Sir Nicholas Gimcrack in Shadwell's *The Virtuoso*, are distinguished by their foolishness in seeking new ways of doing things. To Swift, at least to Swift the satirist, Gimcrack and those like him failed to live a life within the 'common Forms' of existence.[13] Swift may have been on the wrong side of the flow of history, but there is a certain rightness in his criticisms that remains eternally true. To those who observed or lived through the five-year plans in the Soviet Union and China, Swift's satire has a special poignancy – simultaneously a sadness at the failure of the new systems of action that appeared to hold so much promise, accompanied by an uncomfortable recognition of some of their follies. Thus, given his suspicions about idealism, and his anxieties about all forms of embarrassment, it is hardly surprising that Swift proposed only limited goals in his *Project for the Advancement of Religion, and Reformation of Manners* (1709) – a work thought by many to be cynical, or even ironic, because it did not call for real reform, only for the appearance of reform.

Despite such sceptics, the period following the Glorious Revolution of 1688 in England saw a change – novelty was all the rage and schemes abounded. This was the time of the 'Financial Revolution,' and of a successful recoinage. John Dunton and his *Athenian Mercury* proclaimed that new ideas were good, and for a brief period, even Swift was impressed.[14] It was also a time of food riots, bankruptcies, and violent political squabbles. As an extension of the success of the *Athenian Mercury*, Dunton proposed the equivalent of modern Britain's 'Open University' for London, with classes on all kinds of exciting subjects.[15] As Defoe remarked, the projecting spirit is always urged forward by 'Necessity,' and these were difficult times.[16] Some of the proposals for new projects came directly from prisons. Moses Pitt, imprisoned for debt, published his *The Cry of the Oppress'd* (1691) calling for a reform of prisons and of the legal code just after he was released, and, as John Bender has shown in his *Imagining the Penitentiary*,[17] the impulse towards that kind of reform was an important inspiration for the eighteenth- and nineteenth-century novel, beginning with Defoe and continuing for most novels with social concerns, including those of Mary Wollstonecraft, William Godwin, and Mary Shelley.

From Pitt's proposals to James Oglethorpe's committee on prison reform, to John Howard's prison visitations, to Jeremy Bentham's Panopticon prison design, there is a sense that social institutions might be improved despite the seeming inertia of society. Yet over many of these proposals, particularly in the early years of our period, there hangs a poignant odour of failure, for, with the exception of those influenced by Jean-Jacques Rousseau at mid-century, most eighteenth-century thinkers believed that human nature was the same throughout earthly time and space and that since the institutions that needed change were a product of human nature, they must inevitably be flawed. In some instances, such as John Law's Mississippi scheme or the early steam engines, an intellectual grasp of possibilities exceeded the ability of practical performance. In Tom Stoppard's *Arcadia,* set towards the beginning of the nineteenth century, the young mathematical genius tries to explain why an engine proposed for the estate would be economically impractical. The steam engine would eventually transform Britain, but despite the continuing fascination for its possibilities, it would not do so until the Industrial Revolution was in full swing.

Defoe caught that poignancy very well in his *Essay upon Projects* – that mixture of hope and optimism with an expectation of failure. The *Essay* has a section on bankruptcy calling upon his personal experiences, and it is no accident that it is followed immediately by a section proposing a new hospital for the insane. A few projects were unquestionable successes. Defoe instances William Phipp's success in raising treasure from sunken Spanish ships and William Dockwra's development of the penny post.[18] Defoe's own investments in a diving engine and in a civet cat farm came to naught, but he was true believer. He was never more a projector than in the works of the last four years of his life, during which he proposed a wide range of reforms for the improvement of London and of the nation. In short, for him and for many of his contemporaries, projecting was the mode of the age, even if a great many of the schemes were unrealizable. Both at the beginning and at the end of the 'Long Eighteenth Century,' the spirit of the projector – the belief that human thought and action could transform society – was a vital force for change. To suggest that this collection of essays represents a complete history of this change would be both a distortion and an exaggeration. But what it does offer are particular moments in that larger history – moments devoted to organizing knowledge towards particular ends, sometime for the public good, sometimes for the individual aggrandizement of the projector.

## Retrieving the Past

In his *History of the Royal Society* (1667), Thomas Sprat drew a strong distinction between the projector, always pursuing his own self-interest, and the scientist, or Natural Philosopher, who is always engaged in an impartial search after truth. 'If they had speedily at first call'd for *mighty Treasures,*' Sprat wrote, 'and said aloud that their Enterprise requir'd the Exchequer of a Kingdom; they would onely have been contemn'd, as vain *Projectors,*' and suspected of the hoaxes usually associated with projects.[19] Despite his insistence on the disinterested nature of the scientific enterprise conducted by impartial gentlemen and aristocrats, Sprat himself, like any projector, insisted on the practical results that might accrue from experimentation. Yet projectors, as Daniel Defoe remarked in his *Essay upon Projects,* were sometimes far different from Sprat's disinterested scientists. They were often desperate beings whose schemes and inventions frequently arose out of desperation – out of a need to survive.

The first part of this book, 'Retrieving the Past,' will be devoted mainly to efforts at looking backward to discover in past knowledge – of science, languages, or the arts – matters thought to be of value for the present. Even Sprat, speaking for the Royal Society, believed that the new knowledge might be improved by attention to 'the Monuments of the Antients ... if the right use be made of them.'[20] English literature, especially, would benefit by greater understanding of the 'Antients,' and for all his advocacy of experimentation and new discoveries, he argued that the Royal Society might retrieve lost knowledge of the past. 'Nay,' he noted, 'even many of the lost *rarities* of *Antiquity* will be hereby restor'd,' and he hoped for a new form of archaeology that would recover matters 'overwhelm'd in the ruines of *Time.*'[21] Even a writer so little oriented toward the classical past as Defoe argued in his *Essay upon Projects* for a revival of the system of Roman roads throughout Britain. And a new breed of antiquarians searched in England's past, sometimes for the abstract purpose of understanding English history, sometimes, as J.G.A. Pocock has shown, for solutions to problems in contemporary politics.[22] At the same time, critics of music, art, and literature had to struggle to position their thought in relation to a respect for ancient learning and a growing confidence in the achievements of the 'Moderns.'

The first essay in this collection, Margery Kingsley's 'Family, Inheritance, and Clarendon's *History of the Rebellion,*' exemplifies the kind of relationship with the past that Sprat wanted. He specifically called for a history of the development of the opposition to Charles I and the eventual conflict between the King and Parliament.[23] Kingsley is particularly concerned with the methods of passing down history in the form of artefacts, as part of family inheritance, in an attempt to 'mould the future through its construction of the past.' Published in 1703 and dedicated to Queen

Anne, the prefatory material of Clarendon's *History* specifies the ways in which the *History* proposed the continuity of Stuart rule in England.

If Clarendon's *History* passed on a vision of the past in words, the legacies of the Restoration often included the kind of 'things' that Sprat believed had more permanent value, including medals, ribbons, and miniatures. But as Kingsley shows, the important matter for Clarendon was the inheritance of the past. His character portraits often contain an intricate web of family history and sometimes read more like genealogies than studies of personalities. And in willing his manuscripts and the *History* itself to his sons for publication, Clarendon was acting to transmit his own vision to his family and to the nation itself. Appearing, as it did, in a time that Kingsley describes as 'a new age of divided loyalties,' it presented a view of loyalty to the monarchy that was very much at odds with some of the new Whig theories of government. Kingsley sees this in terms of a historical project – a struggle for the present and future through a attempt to control a vision of the past.

For a brief time, John Dryden was a member of the Royal Society's 'committee for improving the English language,' when there was talk about an academy along the lines of the French Academy that would complement the Royal Society's scientific interests.[24] The committee disbanded with the plague of 1665 and never met thereafter, but Dryden's commitment to enriching the language continued throughout his life. In his essay on Dryden's layered language in his rendering of Chaucer's 'The Knight's Tale,' Paul Hammond demonstrates how Dryden succeeded in presenting three worlds: the classical setting about which Chaucer was writing, the medieval, reflecting Chaucer and his audience, and the Restoration world of Dryden and his contemporaries. As Hammond suggests, in recreating the classical world, Dryden added archaeological details not in Chaucer, and in recreating Chaucer's world, he added language drawn from heraldry, actually making Chaucer's world of jousts and pageantry even more vivid. But if his project in presenting these past worlds was linguistically complex, he had a somewhat different end in addressing the audience of his contemporaries.

Hammond suggests that 'The Knight's Tale' is very much a contemporary poem, using variations on Chaucer's themes to attack William III and his reign. Dryden succeeds in subtly attacking England's reigning monarch as a tyrant attempting to impose a standing army upon the nation, in revealing the Parliament as controlled by money ('Bought senates'), and in showing the clergy as hypocritical traitors. He also introduces broad philosophical discussions about chance and free will at a time when William III's victory over James II was often viewed as a Providential moment – a revelation of God's will. Hammond argues that Dryden's ideal prince in the poem is Theseus, and that by a complicated use of language, Dryden shows how grace may flow through a monarch to his people. Hammond sees in this a kind of utopian vision of a Catholic England, an ideal view of what England might have been like had not William III arrived in 1688 and brought with him a commitment to seemingly endless wars on the Continent.

Like Margery Kingsley's study of Clarendon's *History*, Elliott Visconsi's essay is about the ways in which the past might be retrieved for the English during the seventeenth and early eighteenth centuries. At the centre of Visconsi's discussion is the myth of Brutus fleeing the destruction of Troy to eventually arrive in England, where he founded London or the New Troy. As a historical project, it was a deliberate attempt at providing a respectable, classical myth of origin for a nation that too often seemed to exist on the periphery of great events occurring on the Continent. Although this myth of Trojan origins had lost its respectability as history long before the Restoration, it was widely known and accepted as something that might be used by poets and dramatists, provided that, as John Milton remarked, it were done 'judiciously.' And during the Exclusion Crisis (1678–82), it was used to establish political parallels between the fall of Troy and contemporary politics.

In focusing on John Dryden's *Troilus and Cressida* (1679), Visconsi shows how, in revising Shakespeare's satiric tragedy, Dryden tended to avoid some of the more obvious parallels between Trojan and English politics. It might be possible to read in the quarrels of Troilus and Hector a reflection of contemporary bickering among those leading the nation, but Visconsi sees Dryden reading the inheritance of the past in terms of an analysis of the 'soul' of each character as something that only great poets, such as Shakespeare and Dryden, might achieve. It is the poet who is capable of understanding the past and transmitting such understanding to the nation. At the end, Visconsi recapitulates his opening treatment of Pope's unfinished epic 'Brutus,' which was to deal with this theme in lofty terms. Visconsi sees Pope abandoning this mythic fiction for his *Essay upon Man*, where the grand vision of the poet that Dryden had put forward in his 'Grounds of Criticism in Tragedy,' attached to *Troilus and Cressida*, found its partial fulfilment in a poem that, in a projecting mode, attempted to embrace the psychology of the entire human race.

Unlike poetry and the drama of the Restoration and eighteenth century, musical performance had a somewhat different trajectory. Instead of revering the musical canon of the past, audiences throughout Europe expected to hear new compositions when they went to concerts. Britain was unique in gradually developing a canon of 'antient music' in the middle of the eighteenth century. In his essay in this volume, William Weber attempts to re-examine the development of the musical canon that he had set forth in his *The Rise of Musical Classics in Eighteenth-Century England* (1992). Although the pattern he had described involving the gradual triumph of 'classical music' between 1830 to 1850 still holds, here he wishes to emphasize a counter movement involving the persistence of older forms. Borrowing from the work of art historians such as Aby Warburg and Edgar Wind, he establishes a pattern of 'survival' in church music and in musical academies where the knowledge of Renaissance polyphony was a requirement.

Weber points to various survivals on the Continent – operas by Lully, court music in Frederick the Great's Prussia – but it was in Britain that there seemed to be a deeper memory. Some of the efforts at promoting English composers had a political basis in Tory and High Church opposition to Italian music, and there were occasional proposals, in the form of projects to this end. Various musical clubs and societies arose during the eighteenth century throughout the nation that continued to perform older English music. Purcell's *Te Deum* was preserved, and Handel continued to be performed long after his death in 1759 by the Academy of Antient Music. Although a canon of classical music eventually triumphed over the older musicians in the concert halls playing 'serious' music, the older music might be found here and there in popular concerts. In the end, Weber suggests, any canon is subject to the variability of taste.

## Improving the Present

When Sprat wrote of the benefits of 'retrieving the past,' it was, to some extent a sop to those who might have conceived of the ends of science as attempting to overthrow the established forms of knowledge. Nevertheless, his emphasis was always upon the ways in which science might result in new knowledge that would make life on earth better for his contemporaries. Although 'novelty' was satirized on the Restoration stage and in a number of literary works, Sprat asserted that he had no fear of '*alteration* and *Novelty*,' and he praised such discoveries as the microscope, the air pump, and pendulum clocks as representing genuine advances for mankind.[25] He advanced a theory of progress, suggesting that if Caesar and

Tacitus were to return to Britain of the Restoration, they would not recognize it.[26]

By the time Defoe came to write his *Essay upon Projects* in 1697, the age had become somewhat more sceptical. Projects seemed to be often the product of human minds forced by necessity and desperation to come up with something ingenious. Unlike Sprat's society of disinterested gentlemen, Defoe's projectors were often willing to be less than honest if actions of doubtful legality would rescue them. If Sprat had emphasized advances in science, Defoe was more interested in new institutions, and the way they might improve society. He sees the most evident advances made by science in the 'Art of War,' but he is not entirely happy with the new military tactics of the Nine Years War pursued by William III, even if he supports the notion of a standing army. Very much a child of the 1690s – the time of the Financial Revolution and the Society for Reformation of Manners, of economic, moral, social, and literary projects – he is excited by what is happening about him. A few years later, enthralled by the picture of Thomas Savery's steam engine, illustrated in Harris's *Lexicon Technicum* (1704), he imagines an engine that might enable mankind to journey to the moon, just as Byron was to do in *Don Juan* over a century later.

What seemed to inspire Defoe was the prospect of improvement. All of his proposals offered relatively simple ways of doing things better. Thus, towards the end of his life, he proposed improvements in lighting, a London university and academies for the training of musicians and artists. His way of seeing the world was the flip side of an age that continued to look to classical models taken from the art and institutions of ancient Greece and Rome as the ideals to follow. And the failure of many of the projects in the early part of the period covered by this volume – of the Mississippi Company's proposal for developing the vast Louisiana Territory, of John Law's scheme for using paper money, of the South Sea Company's attempt to trade with Spain's American colonies – provided ample material for the satirists. Yet Johnson's satire on the projector who tries to convince Rasselas, the eponymous hero of Johnson's philosophical romance, that it might be possible to fly out of the Happy Valley, does not negate the fact that Johnson was an admirer of projectors who sought to improve human life. And by Johnson's time, when members of the Lunar Society were making startling discoveries, when Josiah Wedgewood was proclaiming an 'Age of Miracles,'[27] there was no denying that inventions and new schemes were transforming Britain and parts of Europe. 'Improvements' were, in some ways, the essence of the Enlightenment project.[28]

In his essay, 'A Revolution in Political Economy,' Steven Pincus explores the background to the Glorious Revolution that brought William III to replace James II on the throne of England and, along with that event, the Financial Revolution that saw the creation of the Bank of England in 1694. Examining two theories about the period that produced the Glorious Revolution in 1688, J.G.A. Pocock's argument about the crucial importance of political ideas – specifically the theories of Niccolò Machiavelli and James Harrington – and the argument of C.B. Macpherson to the effect that the Glorious Revolution was based upon possessive individualism – the notion that the individual functioned only to protect his own property without any conception of the good of the nation – Pincus offers the view that political economists, arguing for the importance of trade, manufacturing, and labour for the national good, created the atmosphere that made the Revolution of 1688 possible. In this scheme of things, political events tended to follow in the wake of economic theories that posited the possibilities of infinite growth of wealth through trade. Even William III's invasion might be seen as a 'preemptive strike' against a possible English war against the Dutch possessions in India.

The villain of this drama was Josiah Child, head of the East India Company and proponent of a theory that wealth was finite and based exclusively upon land. As adviser to James II, he held to the notion that England could only gain wealth at the expense of other nations, hence his pursuit of conquest against the Dutch and Mogul in India. His faith in the importance of land was eventually to become the Tory position. The heroes on the Whig side of this ideological struggle are a mass of writers on economics, among them Carew Reynell, Richard Blome, William Carter, John Cary, John Locke, Dalby Thomas, and Daniel Defoe. They tended to believe in a mobile system of wealth through trade and the circulation of money, and they thought that the East India Company was bad for the nation. When William III needed an economy that would supply him with the money to fight his war against Louis XIV, he turned to a Whig theory of the economy and the possibility of large economic projects that would benefit the nation.

Kimberly Latta's essay complements Pincus's in that it examines the ambiguities in the writings of one political economist of the period of the Financial Revolution, Daniel Defoe. She attempts to read Defoe's ideas on projecting, against a broad historical background of economic ideas going back to Aristotle's discussion of 'natural' and 'unnatural' ways of getting wealth. Criticizing Solon's statement to the effect that there was no boundary to the accumulation of riches, Aristotle argued that there were indeed boundaries and that the notion of getting wealth with no foundation other than itself was 'both unethical and unreal.' Latta shows that Aristotle's ideas influenced the many attacks upon usury during the seventeenth century. On the other hand, Daniel Defoe wrote at a time when some economic writers were advocating the possibilities of endless growth. In his writings, Defoe distinguishes between 'honest' and 'dishonest' projecting, seemingly putting himself in the same camp with the followers of Aristotle and Tory proponents of Land. But as Latta suggests, Defoe's position, though complicated, was finally on the side of growth, circulation, and commerce.

Through a close reading of the many economic allegories in Defoe's journal, the *Review*, Latta reveals an admiration for 'Invention' and for 'Credit.' Sometimes the projecting mind is associated with the 'Whore,' who produces many worthless offspring, but Defoe called trade itself 'the Whore I really doated upon.' Latta does a remarkable job of treating the gendered meanings of Defoe's many allegories and their apparent contradictions. But it is in the latter part of her essay, where she discusses Defoe's attitude towards wit that she comes to some surprising conclusions. And in her view of 'Credit' (often depicted by Defoe as a lady) and the circulation of wealth there is a combination of respectability and the possibility of growth along with a certain amount of imagination and true inventiveness that allows Defoe to find a place, however uncomfortably, among the proponents of growth, change, and improvement.

Perhaps no project considered during the seventeenth and eighteenth centuries was more fascinating and perplexing than that involving the possibility of prolonging life. The examples provided by the Bible (accepted by almost everyone as historically valid) seemed to suggest that human beings could live far beyond the life expectancy of the period. What was the problem? Luxury? Bad diet? Bad air? Original Sin? The inequality of property and the environment created thereby? In his essay David Boyd Haycock sees Sir Francis Bacon offering this as a project for his age in *The Advancement of Learning* (1623) and in his essay of the same year 'The Historie of Life and Death,' and shows how the age continued to be fascinated by the possibility of extending life – and this at a time when the mortality rate was undoubtedly much higher than in the middle of the eighteenth century, when the Comte de Buffon estimated that half the human race perished before the age of eight.

Haycock traces the continuation of interest in this subject throughout the seventeenth and eighteenth centuries among those who thought they could attain the immortality that Adam and Eve had lost after departing from the Garden of Eden and those, such as Dr George Cheyne, who believed that a vegetarian diet might restore humanity to longer life. Even philosophers such as Thomas Hobbes, who lived to an advanced age and René Descartes, who, at forty-two, expected to live an additional hundred years but died at fifty-three, dabbled in theories of long life. Haycock suggests that there was another dimension to this quest. Bacon believed that human beings might have a kind of immortality through the institutions with which they were associated, such as his Salamon's House in *The New Atlantis*, and in some ways the Royal Society and the Society of Antiquaries fulfilled that ideal. Following another line of thought, Haycock explores David Hume's notion that, through the study of history, it might be possible to achieve a kind of immortality by understanding one's place in the flow of events. But for the most part, ignoring Swift's horrendous picture of the senile Struldbruggs in the third book of *Gulliver's Travels*, thinkers of the time continued to be fascinated by literal notions of immortality and by impostors, such as the Comte de St Germain, who claimed to be three hundred years old.

Perhaps the profoundest praise of projecting during this period was written by Samuel Johnson in *The Adventurer* of 16 October 1753. After dismissing such 'royal projectors' as Caesar and Peter the Great as harmful to the human race, Johnson defended those projectors 'who are searching out new powers of nature, or contriving new works of art.'[29] Such projectors may be responsible for 'the invention of those arts which are yet wanting to the felicity of life.'[30] He concludes with a ringing endorsement of those projectors whose efforts have benefited humanity:

> Those who have attempted much, have seldom failed to perform more than those who never deviate from the common roads of action: many valuable preparations of chemistry, are supposed to have risen from unsuccessful enquiries after the grand elixir: it is, therefore, just to encourage those, who endeavour to enlarge the power of art, since they often succeed beyond expectation; and when they fail, may sometimes benefit the world even by miscarriage.[31]

As the creator of the *Dictionary*, the edition of Shakespeare, the high moral tone of the *Rambler* essays, and what is usually called *The Lives of the Poets*, Johnson was the great literary projector of his period, and to a great extent his praise of projecting redounds upon himself.

In his 'Impossible Project,' Howard Weinbrot undertakes the task of overturning the accepted notion that the French have always regarded

Johnson's high reputation in his native country as an example of an eccentricity in British taste, never to be understood by French literary critics. What he demonstrates is that during the late eighteenth century, Johnson had a considerable reputation in France. And for the most part, French critics admired the *Dictionary*, the edition of Shakespeare, and even the often severe morality of the *Rambler*. Although Voltaire had declared himself an enemy of what he considered to be Shakespeare's violations of the 'rules' that, to his mind and the opinion of almost all critics on the Continent, all dramas should follow, Johnson's dismissal of such criticism as 'petty cavils of petty minds' gradually made headway. It was not only the Johnson of these major projects who was admired but also Johnson the biographer, the writer of fiction, travel literature, and poetry who received praise in the country that was at the very heart of the Enlightenment.

Perhaps no projects, during the 'Long Eighteenth Century,' have greater prestige at the start than those involving the establishment of academies, and none are more likely, over the years, to become rigid and stodgy. In this regard no type of academy has a worse reputation for being stultifying than the art academy. In today's popular culture it is sometimes known only because, according to the myth involving the creation of modern art, the innovating Impressionists held their own exhibition in defiance of the French Academy. In his essay, Albert Boime contextualizes the academies of art within the general aims of academies in Western Europe during this period. He focuses on the utopian ends of these academies – to conserve and propagate, to 'keep the faith and pass it on.' If the faith in tradition might seem retrogressive, those who believed in that tradition felt that maintaining it would lead to the improvement and perfection of both art and society. Ideally it was not to function as a kind of pedagogic institution like the university; rather it was to be more like a museum or gallery, exhibiting greatness in the form of statues of important scientists and artists or as a research centre attempting to regulate cultural production.

As Boime notes, almost all utopias had their ideal academies dedicated to perfection of the arts and knowledge and placed them in an ideal city. The same drive lay behind the notions of the perfect human nude and the perfect landscape in the art academies. Behind the architectural constructs of Étienne-Louis Boullé and Claude-Nicolas Ledoux there lies a utopian quest for the perfection of society, rendered in architectural terms, and at the centre of them is the ideal of the academy. Towards the end of his essay, Boime notes that 'perfection' often meant a Eurocentric and masculine ideal that often excluded women and minorities, raising the question whether these idyllic utopias and academies were not built on the notion of exclusion of the ugly and its equivalent in human terms – those seen as misfits.

## Envisioning the Future

The essays in part 3 treat the last half of the eighteenth century extending into the nineteenth and beyond. During this period, the strong consensus against all forms of religious and political 'enthusiasm,' which prevailed as a corrective after the civil wars of the seventeenth century, had almost evaporated. At the same time, the ideal of sensibility had begun producing a literature filled with ghosts, mystery, and heightened emotion. The political revolution in America seemed to foreshadow possible revolutions in the sciences and the arts. When the Montgolfier brothers, Joseph Michel and Jacques Étienne, managed a successful balloon flight on 5 June 1783 and fulfilled what had been only achieved in those imaginary voyages through space in the writings of Cyrano de Bergerac, Defoe, Swift, and Voltaire, enthusiasm seemed the only possible response. Utopia was in the air, both literally and figuratively, and the impulse to peer into the future was difficult to resist. William Godwin pronounced about the future, in his *Enquiry into Political Justice* (1793), 'There will be no war, no crimes, no administration of justice, as it is called, and no government. Beside this, there will be neither disease, anguish, melancholy, nor resentment.'[32]

In some ways, the very notion of projecting is always about the future, and several of the book's earlier essays considered matters such as the possibility of extending the span of life and future utopias. But the years we are considering in part 3 were particularly notable in this respect. In the hands of Constantin Volney's *Ruines*, history itself began assuming visionary status. And many of those convinced by Jean-Jacques Rousseau's arguments about the evils of society came to believe that by changing systems of education, property, and social relations, it might be possible to create a new kind of human being.[33] As mentioned previously, Jeremy Bentham argued that the architecture of prisons might be used to create solitude and introspection in the inmates and thereby improve their attitude towards the world.[34] Most of the projects dealt with in the papers in part 3 deal with more mundane subjects, but all are informed by this new intellectual environment.

Alison O'Byrne's essay on the construction of Westminster Bridge shows how the bridge was conceived as a project for giving London a kind of magnificence that it had previously lacked. She demonstrates how a specific piece of architecture might embody a kind of utopian ideal. Finally opened in 1750, Westminster Bridge continued to be the centre of debates over the identity of the city and the nation as a whole. It became a symbol of what O'Byrne calls 'London's politeness and modernity.' On its way to establishing itself as the most powerful nation in the world, Britain was prepared to show a London that was also civilized and aesthetically pleasing. Nevertheless, when the bridge was first proposed in 1722, it was attacked as a 'precarious Project,' in a renewal of the fears caused by the recent South Sea Bubble and the rash of projects that accompanied that time of wild investments in airy schemes. And when a wooden bridge was proposed, James Thomson's poem urged a public outcry over so base a scheme.

One aspect of building the stone bridge under the direction of the Swiss architect Charles Labelye was the intense public interest in the project – an interest that involved economics, politics, and aesthetics. *The Gentleman's Magazine* carried news of the progress of the bridge to a wide audience. Although there was some criticism of the bridge, particularly after one of its piers began sinking, for the most part it was considered a monument that most Englishmen, and particularly Londoners, might regard with pride, and the idealized society imagined in the view of the bridge presented by the painter Canaletto seemed to have pleased those who viewed the painting and purchased engraved versions. O'Byrne suggests at the end, after a close reading of William Wordsworth's sonnet 'Composed upon Westminster Bridge, 3 September 1802' that by the middle of the nineteenth century, the ideal of politeness and civil responsibility that lay behind the original Westminster Bridge was replaced by one of energetic industrialization.

If O'Byrne shows the way Westminster Bridge succeeded in transforming Britain's sense of itself, Manuel Schonhorn's essay is essentially about Laurence Sterne's literary project of transforming the concept of the heroic, as it had long existed in the worlds of religion, learning, and warfare, into a single ideal suitable for the second half of the eighteenth century and future centuries – that of the artist. The need for such a transformation, Schonhorn argues, lay in the crisis of the 1750s. Britain was on the brink of victory in the Seven Years War and a new period of power and confidence, but before 1759 it seemed to many as if decay, luxury, and corruption had sapped all sense of authority from the social fabric. John Brown's *Estimate of the Manners of the Times* (1757), after a scathing overview of the evils of contemporary society, called for some leader who would rescue the nation from moral collapse. Schonhorn argues that Tristram Shandy, embodying the ideals of his learned father, Walter, of the sensibility of the clergyman, Yorick, and the military honour of his Uncle Toby, was offered to the novel's readership as the reconciling artist figure needed to cure England of its ills.

After demonstrating the means by which Sterne identified himself with the three influences on Tristram's life, Schonhorn reveals the ways in which Tristram is embodied as the moral heir of Shandy Hall. He then points to the similarities between *The Life and Opinions of Tristram Shandy* and biographies and autobiographies of artists, from Giotto di Bondone to Pablo Picasso, in which the artist is seen as a prodigy with miraculous talents. Then, by an analysis of religious allusions contained in the dates of Tristram's birth and his 'circumcision,' Schonhorn suggests that such allusions add up to 'sacralizing the earthly legend of the artist.' Just as William III miraculously came to rescue England on the fifth of November 1688, so Sterne has his Tristram come to rescue his nation through his art.

Transforming the nation required not only new ideals, but new ways of organizing knowledge. It is a historical paradox that both Jonathan Swift's satire on projectors seeking to create a method of writing books through the random alignment of words in *Gulliver's Travels* and Gottfried Wilhelm Leibniz's genuine effort at creating a computer are both given some credit for contributing to the development of modern computers, particularly those used in translation or philology. As the master of irony, Swift might have appreciated the joke. Martin Gierl's essay is devoted to exploring a phenomenon he considers an important aspect of modernity – 'a potentially universal, fast growing linking of information and organization' – and some of the odd paths taken for arriving at this goal. Connected with this method of organizing knowledge are a variety of institutions, especially the academy, which Gierl sees as dedicated to 'the organization of organization.' And within the academy were the projectors attempting to connect utopian schemes with reality. The focal points of Gierl's investigation are Swift, Leibniz, and the historian of the Royal Society, Thomas Sprat. Sprat wanted to see the newly created Royal Society as a repository of information gathered from around the world – information that would be useful to all of humankind because it would be examined and tested.

Swift satirized the Royal Society and its communal quest for knowledge as a group of crazed scientists fixated on projects that could never succeed. And few of these projects were more absurd than the machine that, with help for a team of researchers consisting of a professor, students, and scribes, arbitrarily selected words to enable the most ignorant person to write books 'without the least Assistance from Genius or Study.' As Gierl suggests, in attacking the communal gathering of knowledge and putting forward a theory of individual learning and talent, Swift, for the purpose of his satire, was ignoring the degree to which, in composing *Gulliver's Travels*, he depended on his readings in Rabelais and Cervantes. As the founder of the Prussian Academy, Leibniz seems to have participated in as many impossible schemes as Swift's Lagadian Academy, including the idea for a computer that might have worked had the age had the capacity to put such a machine together. Yet, through its connection with the institution of the state, the Prussian Academy produced an extremely useful and widely used calendar. And in organizing knowledge through encyclopaedias, bibliographies, and book reviews, the institutions of the Enlightenment, particularly academies, did add remarkably to the sum of knowledge. Swift's contribution to cybernetics may not be so paradoxical after all. Imaginative creators of utopias and dystopias have always been notorious for prophesying the future one way or another.

If Martin Gierl's account of the gradual improvement in the accumulation of knowledge is relatively optimistic, Carole Fabricant's discussion of exploration during the latter half of the eighteenth century is both sceptical and pessimistic. She begins with Edmund Burke's optimistic pronouncement in 1777 about the West having 'the great Map of Mankind ... under our View.' But Fabricant sees the great explorers of the period – Captain James Cook, James Bruce, and Mungo Park – in the light of one of Defoe's definitions of a project: 'a vast Undertaking too big to be manag'd, and therefore likely enough to come to nothing.' All three explorers were burdened with the dual notion of Western explorations: on the one hand they were scientific projects intended to extend knowledge of the world, on the other they were colonial enterprises that aimed at the ownership of new lands and sovereignty over native peoples. This ambiguity was built into the 'Directions for Seamen bound for Voyages,' which included the keeping of diaries that were to be deposited with the Admiralty.[35] Thus, as Fabricant points out, Captain Cook's first voyage on the *Endeavour* was sponsored by both the Royal Society and the Admiralty.

Fabricant argues that, for the most part, the scientific aspects of these explorations were often useless. Cook was supposed to report on the transit of Venus across the sun from Tahiti, but his scientific instruments often failed and there were disagreements about the observations. Similarly, the heavy quadrant carried by Bruce to Abyssinia served for little except a test of manhood as he and Yasine the Moor succeeded in carrying it to the top of a hill where others failed. And Park's 'discovery' of the Niger, like Bruce's discovery of the source of the Blue Nile, was no discovery to the natives of these countries, one of whom wondered if there were not rivers enough in the explorers' native land. Though hailed as heroes by the British public, Fabricant finds evidence enough in the texts of their voyages to see them as mock-heroic figures. The failure of language hindered true discoveries as much as the often opaque nature of the societies these explorers encountered. And attempts at humane treatment of the natives often ended in seemingly accidental killings. Fabricant questions the scientific benefits that the West bestowed upon these native societies and finds something manic about the attempts to claim large territories already owned by the indigenous people. She concludes that the greed for land and the accompanying evils of colonialism are a sad achievement for these seemingly noble efforts at advancing knowledge.

Like Fabricant's study of exploration, the last two essays in this collection explore the ambiguous relationships between a world of genuine science and invention and the lingering effects of the projectors with their imaginative but unrealizable schemes. Sarah Kareem analyses the career of Rudolph Raspe, best known today as the author of *Baron Munchausen's Narrative of his Marvellous Travels* (1785), a work that exemplifies a modern meaning for projection as the act of 'projecting unconscious fears or fantasies onto others.' In some sense, Raspe represents the projecting spirit as Defoe sometimes defined it, as the creator of grandiose and imaginative schemes that are usually unrealizable. Munchausen, with his fantastic adventures, seems to express the far end of projecting – the imaginative (and hence amusing) lie tied somewhat weakly to reality. But another side of Raspe may be found in the fictionalized character of the fraudulent projector, Herman Dousterswivel, the alchemist and seeker after methods of transmuting base metals into gold in Sir Walter Scott's *The Antiquary* (1816). The real Raspe was supposed to have swindled a Scottish nobleman by planting pyrite, or fool's gold, on his estate and suggesting that much more might be discovered. Some of these accounts may belong to the myths surrounding his character, but as Kareem points out, Raspe wrote to Captain Cook in 1776 volunteering to go on the voyage to Tahiti, assuring Cook that he would be able to find gold and diamonds in the volcanic sands of the island.

Kareem situates Raspe in the kind of no-man's-land between genuine science and the arts. The description of him by one author as 'geologist, gem expert, probable spy and anonymous author of ... *Munchausen*' reflects a disparate career and a dissipation of energies. But playing down his need for patrons and the occasional mumbo-jumbo of alchemical terms used to impress them, Kareem suggests that Raspe's resort to mystification and literary allusions in the midst of what purported to be careful science was part of his character as well as being endemic to the transitional nature of the period. In his mineralogical works, he attempted to find a middle ground between the miner and the scientist, between the practical and the theoretical. Working with the concept of 'marvellous diligence,' Kareem sees this encompassing both the diligence of the Linneaean taxonomy that Raspe applied to a catalogue of James Tassie's study of gems and the matter-of-fact treatment of the marvellous in *Munchausen*. If, in the nineteenth century, 'marvelous diligence' became part of what was considered to be genuine creativity, Raspe was seldom able to join a sense of the marvellous and diligence together with complete success.

If Rudolph Raspe sometimes used his claims to being a scientist for dishonest ends, Larry Stewart shows how thoroughly such dishonesty was built into the conflation of business and industry from the eighteenth century even to this day. Like Carole Fabricant, he begins with Edmund Burke: not his view of geographical and social knowledge, but his famous vision of the ways in which the French Revolution would produce 'an ignoble oligarchy' composed of projectors ('money jobbers, speculators, and adventurers') rather than a state governed by the ideals of liberty, equality, and fraternity. Seeing in this a reflection of Burke's fears for Britain and the effect of the new powers of commerce, Stewart attempts to examine the relationship between science and business at the end of the century and, more particularly, the relationship between projects and calculation at a time when Isaac Newton and his followers had demonstrated how seemingly possible it was for someone of genius to master the powers of nature.

Both James Watt, who calculated the measurement of horsepower in engines, and Matthew Boulton, who famously told James Boswell that he sold 'what all the world desires to have – POWER,' were able to succeed by the mastery of measurement, but at a time when people were eager to invest in projects based on 'selling the future and not the present,' they were among the few who could claim such competence. Despite what Stewart calls 'the momentum of calculation,' most of the patents granted by the Crown during these years show little in the way of knowledge other than the desire to fleece investors. Patents were given for perpetual-motion machines, for seemingly magical methods of reducing friction, and for any invention that used the formula 'than by any means at present practiced.' Stewart instances one patent granted for an 'Engine in place of the steam-engine, to be worked without fire, wind, or water, and with or without a horse.' He concludes that Burke was right about a world governed by projectors without principles – that, despite the achievements of the Industrial Enlightenment, the rage for invention created an atmosphere that was barely cooled by the Bubble Act that, at the time of the South Sea Bubble, attempted to control investment in hare-brained schemes. Too often what appeared, and still appears in our present day, to be industry governed by the exploitation of technology turned out to be a congame directed at consumers, workers, investors, and society itself.

With this final essay, *The Age of Projects* concludes appropriately on a somewhat negative note. Thomas Sprat was not entirely wrong. Great discoveries were to be made, and Isaac Newton gave the age a model of the universe that satisfied the period's desire for order. On a practical level, it was hardly easy but eventually possible to build a bridge over the Thames at Westminster. Bridges had been built before. This is why those projects conceived in the spirit of retrieving the past were the easiest and the most successful. Only convention, taste, and a lack of curiosity impeded the creation of a balanced musical canon, though this appears to be a work in progress. Only the labour and genius of some great poets transformed the language of ancient poets into modern English. And only a new understanding of the ways in which tradition might be used was needed to create a new feeling of English nationality. But the projects of the age were more often than not the product of the mind imagining possibilities rather than being able to bring them to success. The academy, that seemingly most desirable of institutions, always projected unrealizable and utopian ideals. Leibniz was able to conceive of the possibilities of a computer, but it took well over two centuries to make one that worked. Even if this gap between the imagined and the achievable was too often not bridged by successful projects, the excitement in projecting was an essential and vitalizing part of the age.

## NOTES

1 Daniel Defoe, *An Essay upon Projects*, ed. Joyce Kennedy, Michael Seidel, and Maximillian Novak (New York: AMS Press, 1999), 11.

2 *Essay upon Projects*, 7. The title of this volume, 'The Age of Projects,' appears in Larry Stewart, *The Rise of Public Science: Rhetoric, Technology and Natural Philosophy in Newtonian Britain, 1660–1750* (Cambridge: Cambridge University Press, 1992), 184.
3 See *Essay upon Projects*, 1. For contemporary definitions see *The Oxford English Dictionary*.
4 See Maximillian Novak, *Daniel Defoe Master of Fictions* (Oxford: Oxford University Press, 2001), 118–19.
5 Daniel Defoe, *The Political History of the Devil* (1726), ed. Irving Rothman and R. Michael Bowerman (New York: AMS Press, 2003), 206.
6 New York: Farrar, Straus and Giroux, 2003, 125.
7 See Steve Shapin, *Leviathan and the Air-Pump: Hobbes, Boyle and the Experimental Life* (Princeton: Princeton University Press, 1985).
8 See, for example, Edward Stillingfleet, *Works*, 3 vols. (London, 1710), 3:563–4, 571–7.
9 *Prose Works*, ed. Herbert Davis, Louis Landa, et al., 14 vols. (Oxford: Blackwell, 1941–68), 4:123–4.
10 'Auguries of Innocence,' *Poetical Works*, ed. John Sampson (London: Oxford University Press, 1934), 171.
11 Swift, *Prose Works*, 11:162–72.
12 See, for example, Sprat, *The History of the Royal Society* (London, 1667), 113.
13 See *Prose Works*, 1:108.
14 See Swift's 'Ode to the Athenian Society,' in *Poems*, ed. Harold Williams, 2nd ed. (Oxford: Clarendon Press, 1958), 1:14–25.
15 The proposal appeared as a supplement in the *Athenian Mercury*, 13 July 1695.
16 'Necessity' is the opening word of Defoe"s introduction. See *Essay*, 7.
17 Chicago: University of Chicago Press, 1987.
18 See Defoe, *Essay*, 11–12, 15–16.
19 Sprat, *History of the Royal Society*, 77.
20 Ibid., 24.
21 Ibid., 436.
22 See Pocock, *The Ancient Constitution and the Feudal Law: A Study of English Historical Thought in the Seventeenth Century* (Cambridge: Cambridge University Press, 1957).
23 Sprat, *History*, 43–4.
24 James Winn, *Dryden and His World* (New Haven: Yale University Press, 1987), 129.
25 Sprat, *History*, 321–2.
26 Ibid., 384–9.
27 Uglow, *The Lunar Men*, 157.

28 See Carew Reynel, *The True English Interest* (London, 1679), 22, for an early example of this impulse.

29 No. 99, in *The Idler and The Adventurer*, ed. W.J. Bate, John Bullitt, and L.F. Powell, *The Yale Edition of the Works of Samuel Johnson* (New Haven: Yale University Press, 1958–), 2:431.

30 Ibid., 2:434.

31 Ibid., 2:435.

32 Ed. Jonathan Kramnick (Harmondsworth: Penguin, 1976), 777.

33 See the discussion of Thomas Day's educational experiments in Uglow's *The Lunar Men*, 182–8.

34 See Bender, *Imagining the Penitentiary*. For a more sceptical view of the purpose of such social engineering, see Michel Foucault, *Discipline and Punish: The Birth of the Prison*, trans. Alan Sheridan (New York: Vintage Books, 1996).

35 See *Philosophical Transactions of the Royal Society*, 70 vols. (reprint edition; New York: Johnson Reprints, 1963), 8 January 1666, 1:140–3.

# PART I

# RETRIEVING THE PAST

chapter two

# Family, Inheritance, and Clarendon's *History of the Rebellion*

MARGERY KINGSLEY

In the opening pages of his *History of the Rebellion*, the earl of Clarendon attempts to justify the seemingly vain exercise of writing historical manuscripts in an age of violence and printed polemic, asserting that 'we may not yet find the cure [of England's ills] so desperate, but that, by God's mercy, the wounds may again be bound up; though no question many must first bleed to death ... And I have the more willingly induced myself to this unequal task, out of the hope of contributing somewhat to that end; and though a piece of this nature (wherein the infirmities of some, and the malice of others, both things and persons, must be boldly looked upon and mentioned) is not likely to be published at least in the age in which it is writ, yet it may serve to inform myself, and some others, what we are to do, as well as to comfort us in what we have done.'[1] Reflecting a late Renaissance understanding of history as a genre that improves the future by educating mankind in the lessons of the past and placing the historian second only to God in his social efficacy, the *History* affirms its central role in a crucial social project, the reconstruction of English society in the wake of twenty years of civil conflict.[2]

Yet in the end neither Clarendon's self-defence nor the monumental efforts of his sons and the Oxford editors who finally brought the *History* to press seem to have saved Clarendon's project from a somewhat dubious status as the best of a bad lot in the eyes of modern readers, who are conditioned to see in the voluminous and sometimes rambling histories produced in the years after the civil wars a dull and shapeless outlet for political resentment rather than a blueprint for social progress. Overshadowed and certainly out-sexed by more dramatic and more innovative plans for social improvement such as the institution of charity schools, the founding of a national bank, and the improvement of English highways, historiography in the seventy or so years following the English civil wars has been generally dismissed as boring and predictably biased, abandoning ideals of historical objectivity in favour of a more utilitarian polemic in the early days of party politics.[3] As a result, even the most generous studies of Clarendon's *History* feel compelled to come to terms with Clarendon's politics and the more personal, apologetic function of his writing – issues not unrelated – before gesturing towards the more comforting narrative art of his much praised character portraits.[4]

Certainly it is true that Clarendon's *History*, though often politely cited as the most meaningful example of historical writing in an otherwise moribund age, exhibits many of the very characteristics that trouble those unimpressed by late-seventeenth-century historiography. The very composition of the *History*, mixing as it did Clarendon's early partisan defence of Charles I with the more personal self-exoneration implicit in the *Life*, raises hints of polemic and autobiography hardly exorcized by the narrative art of the text and, if anything, exacerbated by Clarendon's own insistence that his text be used to clear his name and enhance the political position of his sons. When the *History* was first published, moreover, it was packaged and even read in part as a personal family artefact with significant implications for partisan politics, an understanding to some extent created by the introductory material in the first two volumes. The preface to the first volume ends with an assessment of the impact of Clarendon's life and writing upon his sons, who 'have found themselves as well the better Christians, as the better Men, for the afflicted, as well as prosperous parts of their Fathers Life.'[5] At the same time, the dedication to Clarendon's granddaughter, Queen Anne, that opens the second volume constructs the *History* as a vital link between the generations of a single family, asking of the queen, 'to whom so naturally can the works of this Author, treating of the times of Your Royal Grandfather, be address'd, as to Your self; now wearing, with Lustre and Glory, that Crown, which, in those unhappy days, was treated with so much contempt and barbarity, and laid low even to the Dust.'[6] Even the frontispiece of the *History*, the full-page picture of Charles I common in earlier histories of the civil wars replaced with a bust of Clarendon surrounded by the symbols of royal power, reminds the reader that the text is a family affair. It is the legacy not only of a monarch's deeds but also of the author himself, passed down through his sons to his granddaughter and recounting the history of the royal family to which the author was connected through his daughter's marriage, and focusing most particularly on the unhappy ending of the queen's other, royal, grandfather.

This positioning of the *History* was quickly perpetuated both by its defenders, who often printed and reprinted copies of Clarendon's will as proof of the text's authenticity, and by its detractors, who accused Clarendon's sons of seeking undo influence over their royal niece. Over time, so firmly ingrained was the notion of the *History* as a family document that a 1732 defence of Clarendon's text was actually entitled *The Clarendon-family Vindicated.*[7] And thus in a sense Clarendon's text has from the beginning exemplified the political and personal uses of history traditionally so problematic for serious study of late-seventeenth-century historiography.

And yet ultimately, I would suggest, the involvement of Clarendon's text in both national and family politics does not so much obfuscate the history of the civil wars as it reveals a great deal about the often complicated cultural construction of history as a social project in the years following 1660. If anything, in fact, the very implication of Clarendon's text in the intertwined politics of nation and family reminds us that even as Clarendon, like other royalist writers of the Restoration, sought to make sense of the experience of civil war and to use that experience to build a new and stronger state, the conditions of the production of history in the late seventeenth century also challenged the authority of the historian – his or her ability to collect, to frame, to construct the past – in ways that affected the nature of English historical thought. More than any period before it, the late seventeenth century understood history as a product of the material remains of the past, and in the middle of the seventeenth century new modes of production and expanding markets made historical and political material more available within the domestic sphere. As a result, attempts to write national history, to repair the divisions of the English civil wars by creating a common interpretation of the past in which readers could share and on which they could base a unified course of action, had to compete with the multiple pasts generated by the material signs of what were often highly politicized family loyalties and identities: for most English men and women of the period, the history of the English civil wars was most immediately available not through the voluminous textual histories of the day, but through the material and ideological inheritances passed down to them by family and friends. The late-seventeenth-century historian, then, was forced to cope with the relatively new but persistent evidence of multiple family-generated histories that intruded upon and constructed the dynamics of present social and political divisions and threatened to expose narratives that sought to impose a common past as fiction, or polemic, or both. Thus, the family politics of Clarendon's text, replete as they were with both personal and national implications, do not so much reduce the accomplishment of the *History* as they provide access to the challenges facing those who sought to pursue the social project of history in the years after the civil wars. Family, that is, played a significant role in the construction of the past in the late seventeenth century, and in many ways Clarendon's narrative epitomizes the negotiation among histories forced by the changing production of the past within family groups – a negotiation that ultimately led to Clarendon's particular sense of the changing nature of history and the role of the historian at the end of the seventeenth century.

## I

As early as the 1590s, Sir Philip Sidney had disparaged historians as writers and thinkers unable to escape their material conditions; in the late seventeenth century, such assumptions about the materialist essence of the historical enterprise were arguably more fully realized than in any previous period. The rise of antiquarianism in the late sixteenth and early seventeenth centuries had heightened English consciousness of the extent to which history is mediated by its material evidence and imbued the educated English mind with the sense that it was indeed surrounded by the remains of the past, even in the very landscape. It was during this period, as well, that the British royal family began to appreciate the utility of material culture as a means of propagating authority and constructing a sense of national history. Prince Henry, eldest son of James I, was an avid collector of ancient and modern medals, many of which he passed on to his younger brother Charles, who as Charles I became the first British monarch to use medals as an effective means of celebrating important political events. At the same time, both changing techniques for the production of certain domestic goods and a growing market for luxury items made the dissemination and consumption of historical material far more widespread than in previous periods. The growth and expansion of the print industry in England made possible the publication of images of royalty, nobility, and historical events in prints and books, new pottery techniques enabled the production of more life-like images on slipware, and expanded incomes that permitted the purchase of commemorative prints, pottery, and jewellery put ownership of the material evidence of history within reach of those of relatively moderate means.

For those who had been children during the wars and interregnum (people who would have been in their fifties and sixties when Clarendon's history was finally published), this new abundance and awareness of political objects with historical significance meant that the history of the civil wars was partly the memory of the material signs and symbols of family allegiance. Wealthy families throughout England politicized their private space with pottery, prints, jewellery, and even embroideries and hangings.[8] One jug dating from the 1650s served as a daily reminder of radical religious identity, imploring the user to 'Fast and Pray,' while immediately after the Restoration, supporters of Charles II produced and bought ornamental plates and mugs decorated with pictures of the king and important events of the civil wars.[9] During the wars, the queen presented commemorative jewellery to those who contributed to the king's cash-strapped cause, even as others found more individualized ways of commemorating their own actions; one Major Carlos, involved in the escape of the Prince of Wales after the battle of Worcester, owned a gold case inscribed with the royal oak and some verses written in praise of himself.[10] Prints of Charles I, Charles II, and Oliver Cromwell were commonly available from printers and booksellers. Embroidery, too, was a popular means of representing one's political beliefs: after the death of Charles I, one common embroidery pattern in England was a reproduction of the cover of *Eikon Basilica*, while in a picture of Solomon receiving the Queen of Sheba, a politically inspired stitcher quite carefully replaced the features of Solomon with those of Charles I.[11] As a result of such active production and consumption of political goods, individuals living in England in the second half of the seventeenth century dwelt amidst the highly self-conscious material signs of their own recent history to an extent never before possible in that divided country.

There is, moreover, substantial evidence to suggest that during and after the civil wars individuals used the ritual transmission of such politicized objects within families to construct their own version of history and so reproduce family loyalties in the next generation. In 1653 Lewis Darcy received from his father, Thomas Darcy of York, a 'sword, inlayed and damasked with silver, which was given [to the elder Darcy] by the late RT Honble the Earl of Mulgrave,' apparently in commemoration of service during wartime.[12] Mulgrave was one of the few peers who remained sitting with parliament after the king fled to Oxford in January of 1642, and the language of the bequest subtly but movingly suggests both the pride of the elder Darcy in his connection with the peer and his political cause as well as a desire to impress that same sense of dignity and loyalty upon his son. Anthropologist Annette Weiner has shown that objects such as this, retained in families and inherited through generations, not only serve to forge a family identity but also create a sense of history for that group, a dynamic seemingly operative in the Darcys' situation, as the history of the civil wars is filtered for the younger Darcy through his father's gift.[13]

Likewise, in a will written in 1648 and proved in 1649, Sir Edmund Bacon left to his 'nephew,' Nicholas Bacon of Gislingham, a 'little black stacked piece inlayed with silver,' a 'case of redd stocked pistolls, and [the] flaske to them, inlayed with silver,' an 'achate with Queene Elizabeth's picture in it, and the chaine of achate it hangs at of thirty beads,' an 'Iseland comelian stonne sette in gold, with [an] antike figure in it,' and his 'meddall of the synode at Dort.'[14] In the same will he bequeathed 'towards the maintenance of the lecture upon the markett day in the towne of Botesdale, three pounds and ten shillings a yeare ... to be paid quarterly to the churchwardens of Botesdale for the time being, soe long as the Protestant religion continueth that is now professed in the church of England.'[15] This particular combination of gifts and conditions set by a man who moved in royalist circles during the civil wars implies an effort to transmit to his nephew the values of monarchy and the Church of England. His specific bequest of sentimental and symbolic objects, including the miniature of Queen Elizabeth, suggests both a special ideological tie to the younger Nicholas and a faith that the transfer of goods to the younger man makes the inheritance a vehicle for family identity and for the elder Bacon's own construction of a broader history of English politics and religion. As in the case of the Darcys' sword, Bacon's gifts help to construct an interpretation of the recent past that he seeks to impose upon his heir.

For the late seventeenth century, then, history was in part the product of ideologically loaded family inheritance, a material legacy that attempted to mould the future through its construction of the past. And thus, as it quite self-consciously styled itself as a family legacy, the prefatory material to the first edition of Clarendon's *History* did not simply attempt to manipulate the queen's emotional loyalties; it also sought to establish that text as the kind of material stuff of history so easily recognizable to wealthy readers of its day. As it did so, moreover, it suggested that family legacies were important in part because they provided the means by which belief systems could be replicated in the face of repression. Asserting that they want to 'make [Clarendon's] work public, in an age when so many memoirs, narratives, and pieces of history come out, as it were on purpose to justify the taking up arms against that king, and to blacken, revile, and ridicule the sacred majesty of an anointed head in distress,' the editors positioned their work as the family legacy of a man who suffered for the crown, partly as a means of distinguishing it from polemic and narrative history and suggesting its greater authenticity.[16]

In fact, in the uncertain political climate during and after the civil wars, the transmission of family history and identity through the ritual of inheritance seemed to many a more secure means of preserving social order as they understood it than the successions, declarations, and alterations that supposedly determined national politics, an issue important for royalist historians who sought in part to justify the Restoration by depicting its continuity with earlier Stuart rule and its seemingly providential preservation despite attempts to eradicate it. During the wars, royalists had worked hard to preserve the memory of the king and the promise of restoration within the private sphere through material goods such as mourning rings, medals, and prints. Families and family groups may even have been central to efforts to preserve the king's own belongings and the symbols of monarchy in preparation for a desired restoration. One Philip Kinnersley, who had been a Yeoman of the Removing Wardrobe to Charles I, seems to have purchased the anointing spoon and perhaps retained several of the king's personal swords, while records concerning the royal tapestries sold after the king's execution suggest that in 1654 several hangings, tapestries, and even carpets were in the hands of one Clement Kinnersley, Keeper of the Wardrobe under Oliver Cromwell.[17] Members of the dead king's household also preserved several ceremonial swords from the destruction ordered by parliament, constructing a promise of the future by preserving the history of English monarchy and the extended family that surrounded it.[18]

Even after the Restoration, the family and the transfer of property that helped to define family history seemed crucial to creating a sense of historical continuity and preserving existing institutions. In a will written in 1668, Sir John Denham, who had been active on the part of the king during the civil wars, left all his money and property to his daughters, provided that they 'observe and obey the [Church of England], they having beene soe well educated and exercised in their duty therein, and that they transferre the like charge to their posterity, that whysoever change or temptation may hereafter happen they may not change with it contrary to their owne understanding and consciousnesses.'[19] Here the gift of money rather than some object of personal or political significance in exchange for his daughters' explicitly demanded adherence to the

Church verges uncomfortably on blackmail. Yet rather than differing in kind from the seemingly less imperial transmission of swords and medals, Denham's conditions merely expose the extent to which the generational transfer of property in general is deeply interwoven with the mechanisms by which belief systems are preserved and reproduced in subsequent generations, as the will itself becomes a locus of social reproduction, inextricably linked to his money. Through his bequest, Denham's faith in the Anglican Church just as surely descends to female children who, as the agents of reproduction, will ensure the survival of his belief through future generations, whatever public pressures to do otherwise might arise. For Denham, that is, as for many royalists who had lived through the interregnum, the ritual transmission of family possessions and thereby family values was perceived as vital to historical continuity and national stability, constructing both a sense of the past and a viable future.

Clarendon too, himself a royalist writer concerned with the problem of the continuity of Stuart rule, seems to have construed the legacy of material inheritance as a crucial part of the successful transfer of value systems from one generation to the next and so as essential to the fate of the nation. One of the greatest acts of betrayal in the *History*, the alienation of the succession from Charles II, is framed not in terms of the murder of the king his father, as it is in so many other texts, or even in terms of rebellion per se. It is, rather, put in terms of the disruption of the young king's inheritance and the dispersal of his father's property. In a fit of absolute indignation, Clarendon recounts that after the death of Charles I, 'Cardinal Mazarine [of France] ... sent now to be admitted as a merchant to traffic in the purchase of the rich goods and jewels of the rifled crown, of which he purchased the rich beds, hangings, and carpets, which furnished his palace at Paris,' partly as a means of reinforcing diplomatic ties with Oliver Cromwell.[20] Likewise, the ambassador of the king of Spain, who Clarendon reports

> had always a great malignity towards the king, bought as many pictures, and other precious goods appertaining to the crown, as, being sent in ships to the Corunna in Spain, were carried from thence to Madrid upon eighteen mules. Christina, Queen of Sweden, purchased the choice of all the medals, and jewels, and some pictures of a great price, and received Cromwell's ambassador with great joy and pomp, and made an alliance with them. Archduke Leopold, who was governor of Flanders, disbursed a great sum of money for many of the best pictures, which adorned the several palaces of

> the king; which were all brought him to Brussels, and from thence carried by him into Germany. In this manner did the neighbour princes join to assist Cromwell with very great sums of money, whereby he was enabled to prosecute and finish his wicked victory over what yet remained unconquered, and to extinguish monarchy in this renowned kingdom; whilst they enriched and adorned themselves with the ruins and spoils of the surviving heir, without applying any part thereof to his relief.[21]

In this passage Clarendon indicts European leaders for their fickle behaviour and their lack of support for Charles II, as he ironically depicts them financing a rebellion that could endanger their own crowns while ignoring the dispossessed monarch in their midst. Yet significantly, he also suggests that Cromwell will eventually 'extinguish monarchy' not through brute force or a shift in ideology, but by disrupting the transfer of material possessions from one generation to the next. By denying him his father's 'things,' Cromwell has more effectively dispossessed the young heir than he did by defeating his father in battle, destroying the history and continuity of the royal family and thereby justifying his own usurpation.

Thus, for Clarendon, as for others in the period, family inheritance was important because it provided a mechanism by which the social structures and institutions of the past could be replicated, shaping a stable and comprehensible future; to interfere with the process of inheritance was, in effect, to jeopardize the future of the nation. And such privileging of family history is perhaps particularly understandable given the Restoration need for public compromise. As both commonwealth's men and royalists were forced after 1660 to make concessions in the interest of national unity and prosperity, the maintenance of family history and of a domestic space where beliefs and values could be preserved against the accommodations of the public sphere undoubtedly seemed all the more important. At the same time, however, it was this very capacity to enable compromise by offering a mechanism for shielding individual and family systems of belief that ultimately made family inheritance and family history problematic for royalist historians, encoding difference and multiplicity in a past where royalists sought desperately to find a common ground.

## II

Even as the transmission and inheritance of family legacy were understood and often used to generate a sense of identity, history, and continuity in the face of disorienting change, the very mechanisms by which they reproduced a threatened system of belief also frequently emphasized loyalty to family codes and family values over adherence to a changing national identity, undermining efforts to construct a common past, and thus threatening to create a divided future. When in 1654 Henry Marwood of York left to his nephew 'one deble spurr ryall with a ribbin to it to weare about him as a legacie so long as he pleaseth,' he in effect transformed a national legacy into a personal bond.[22] Even before the civil wars, spur royals were objects with significant political value, more like medals than like coins, having never, in all probability, been intended for common circulation but rather being granted as tokens to those who were touched for the king's evil or otherwise granted a ceremonial presence.[23] After the death of Charles I, parliament replaced existing coinage with the new coin of the commonwealth, and spur royals, which featured a picture of the monarch under whom they were coined, generally either Elizabeth or James I, took on additional political significance. In leaving the coin to his nephew as a 'legacie,' a reminder of the dead, however, Marwood attaches the political and symbolic value of the coin to his own personal history, transforming the political value of the coin into the value of the gift, not merely creating an affective bond with his nephew, but also invoking a web of responsibilities and reciprocities that bind the heir more effectively to the benefactor than to the nation.

Of course, it is uncertain to what extent the children and heirs exposed to such ideological inheritance and the materials that symbolized it actually adopted their parents' vision of history or their politics. One historian has estimated that during the civil wars about 20 per cent of families experienced some backlash as children rejected and even took arms against the beliefs of their parents, and doubtless after the wars many children followed the time-honoured tradition of rebelling against the value systems of their elders.[24] Yet the material evidence of the seventeenth century would certainly seem to suggest that while national history, the intertwined histories of England, Scotland, and Ireland, was the dominant public mode of representing the civil wars in print during the period, it was also true that for many individuals, their sense of history (and thus their sense of social reconstruction) was fractured in a way that forced them to negotiate between efforts to create a nationalist hegemonic vision of the past that would help England avoid future conflict and a family inheritance/history that often demanded loyalty to the very beliefs and values that had precipitated conflict in the first place. This fractured sense of history is at times evident in the reading habits of those who paradoxically sought in printed accounts of the civil wars both an understanding of a British past and a vindication of personal interests and loyalties. One copy of Sanderson's *The Life and Reign of Charles I* contains extensive marginal annotations, all of which refer exclusively to the events in Scotland and to the Earl of Montrose. Many, moreover, are cross-referenced to another, unidentified text in which Montrose appears to figure prominently. In this case the annotator is unknown, making it impossible to fully understand the particular interest in Montrose. Yet these annotations serve as a useful reminder that the readers of these massive histories did not necessarily read them disinterestedly from cover to cover, but dipped, sampled, and responded based on their own individual interests, which were often conditioned by family loyalties.[25]

Writers of the period, moreover, frequently depicted families as the vehicles for the transmission of divisive loyalties, fostering chaos precisely because of their ability to preserve and reproduce competing histories. Royalist writers in particular, eager to depict the civil wars as over and done with, expressed a great deal of anxiety about the role the family and family history played in reiterating the conflicts of the recent past. Historical ghost stories were common in the years around 1660, and in pamphlet after pamphlet, the 'bad boys' of the interregnum – Oliver Cromwell, John Bradshaw, Hugh Peters, and others – haunted the pages of the Restoration with historical accounts of their actions during the wars and with threats of renewed violence. Often they appear to family and friends, either seeking revenge, as ghosts so often do, or merely trying to pull their relations into some new hare-brained but potentially volatile scheme. In *The World in a Maize, Or, Olivers Ghost* (1659), Oliver Cromwell appears to his son Richard to upbraid him for his failures; the older Cromwell rehearses all of his own bloody actions, providing a yardstick by which the shame of the younger can be measured. At the same time, the frustrated ghost of Cromwell, exasperated with the weakness of his eldest son, pleads with him to take up the mantle of injustice, saying of his compatriots,

> I made them my slaves when they were at the best,
> And truly *Overton* I made him beleeve,
> He should not have any cause for to greeve,
> And *Bradshaw* was my Instrument to rize,
> By cutting one off we did win the prize,
> And prithee Dick canst not thou act these things?[26]

Dick, as the reader knows, could not, but the ghost's visitation is a reminder that the child of flesh and blood, a living soul with the power of revenge, cannot simply be dismissed. He or she is the literal product of the reproduction of past generations, heir to a fragmented history, reminding the reader of the basic biological fact of continuity between past and present and thus also inherently containing the lingering threat of revived conflict. Similarly, in *The Case is Altered. Or, Dreadful News from Hell*, the ghost of Cromwell describes to his wife plans to conquer new territory, explaining that he has come back to the world of the living in order 'to steak out and appoint a place where *Haselrigs* bones, *Scot*, and *Vanes* shall be intered; for absolutely if they would dispatch and come away, with their advice and my own, we could usurp a power from the Devil, and live in a corner by ourselves, without interruption.'[27] Even though he is dead, Cromwell's imperial aspirations live on, to be realized primarily through an infusion of energy from his friends and relations. And thus the history produced by the inheritance that is metaphorized in these pamphlets as the lingering corpse/spirit of the dead parent or protector repeatedly threatens to create new chaos and renewed civil strife.

So concerned, in fact, were Restoration royalists with the power of family inheritance to create both a divisive history and a divided future that the mock-will became a common genre, as popular pamphlets represented prominent members of the commonwealth attempting to reproduce their beliefs and institutions in the next generation through a material inheritance, much as Lewis Darcy and Sir Edmund Bacon had done. Thus, in *Hugh Peters Last Will and Testament*, Peters, often regarded by royalists as the religious mouthpiece of the commonwealth, is put to death and his executor must distribute his goods: 'Unto his Sisters he gave his Bable / Which stood in their services long as 'twas able, / And unto the Kirk his brains and eyes, / And unto the Queen of Morocco his thighs.'[28] Frequently these pamphlets do try to ease the anxiety about the power of a material inheritance by suggesting that the goods and ideologies conveyed in these documents are worthless. Yet the mere mention of Peter's 'gift' to the Queen of Morocco, in addition to reproducing popular stereotypes concerning sexual promiscuity among radical religious groups, suggests a certain potency and an ability to reproduce that potentially belies a seeming confidence that material inheritance and the divisive history it represents can easily be contained.

That reality was not lost on the historians of the civil wars, who were themselves both private individuals and the tellers of a national tale. And thus even while he implies the crucial role that family inheritance played both in the construction of history and the preservation of institutions and ideologies (particularly monarchy) in the face of repression, Clarendon too suggests that family inheritance can also be disruptive, creating and perpetuating conflict, and thereby endangering the very institutions it sought to preserve. In the *History*, while he laments the disruption of the royal succession in terms of the denial of Charles II's material inheritance, Clarendon also represents the social problems and divisions that can result from family history and the inheritance that transmits it. Despite claims that the *History*'s much-touted character portraits indicate his faith in the ability of the individual to make decisions and choose actions that affect the fate of the nation, that text in fact repeatedly depicts young men in particular who are significantly constrained by family history and inheritance.[29] Clarendon says of the second earl of Manchester, Lord Kimbolton, who sided with parliament in the early years of the wars, that

> the death of his lady, and the murder of that great favourite [Buckingham], his second marriage with the daughter of the earl of Warwick, and the very narrow and restrained maintenance, which he received from his father, and which would in no degree defray the expenses of the court, forced him too soon to retire to a country life, and totally to abandon both the court and London ... And in the end, even his piety administered some excuse to him [for abandoning the king]; for his father's infirmities and transgressions had so far exposed him to the inquisition of justice, that he found it necessary to procure the assistance and protection of those who were strong enough to violate justice itself; and so he adhered to those who were best able to defend his father's honour, and thereby to secure his own fortune.[30]

Manchester's actions, or lack of them, can thus be explained, according to Clarendon, not as a matter of personal choice, but as a path to which the finances, behaviours, and loyalties of his entire family have constrained him. He is not an individual operating within a political hegemony, but part of a group or clan whose decisions inevitably direct his own actions. Likewise the weaknesses of Edward Sackville, 4th earl of Dorset and a member of James I's privy council, are explained in terms of the extravagant spending habits of his older brother, which left the estate that descended to him too small to support his title, exposing him to 'many difficulties, and inconveniences.'[31] Though loyal to the king and of 'very good general reputation,' Sackville was never, in Clarendon's eyes, able to reach his full potential as a statesman because of the 'uneasy and strait fortune' that was his brother's legacy.

While Clarendon clearly recognized, moreover, that the actions of individuals are frequently dictated by family, the *History* also articulates concerns about the extent to which inheritance and the sense of family history that it created could lead to divided and divisive social loyalties that superseded a sense of common good and shared history. Writing of the lord Savile, an on-again-off-again adherent of the king who was eventually imprisoned by Charles I because of his political manoeuvring, Clarendon attributes his fickle politics to his 'particular malice to the earl of Strafford, which he had sucked in with his milk, (there having always been an immortal feud between the families; and the earl had shrewdly overborne his father)'; that feud, Clarendon says, 'had engaged him with all persons who were willing, and like to be able, to do him [Strafford] mischief.'[32] In this case the ideological inheritance of a family feud, symbolized by the material transfer of milk from mother to child, determines the actions of the individual and thus creates national as well as personal tensions. Likewise, the lack of support given to the king by the House of Lords is attributed in part to values and beliefs inherited from a family past: the lord Say, leader of the Lords opposition to the king, is said to have 'with his milk, sucked in an implacable malice against the government of the church.'[33] The resulting image of family loyalties and divisions being transferred like antibodies to the nursing infant symbolizes for Clarendon the extent to which inherited loyalties, and the personal and family histories that they create, can foster and perpetuate social divisions and undermine political hegemony. And thus in the *History* too close an adherence to family history and too much respect for one's patrimony can lead, ultimately, to overly narrow vision and to unnecessary conflict. In his famous portrait of Thomas Howard, Lord Arundel, Clarendon sneers 'he was willing to be thought a scholar, and to understand the most mysterious parts of antiquity, because he made a wonderful and costly purchase of excellent statues ... and had a rare collection of the most curious medals; whereas in truth he was only able to buy them, never to understand them; and as to all parts of learning he was almost illiterate, and thought no other part of history considerable, but what related to his own family; in which, no doubt, there had been some very memorable persons,' an appraisal that can perhaps best be understood as an unfriendly understatement. Arundel's father had died committed to the Tower of London, and Clarendon, who disliked Arundel intensely, was more than happy to suggest that the family studies of the son might lead him to the same supposedly treasonous and divisive end as his progenitor.[34]

Thus, the royalist writer of contemporary history after the civil wars was faced with the dilemma of family inheritance, a material legacy at once crucial to the royalist construction of history yet potentially fractious and divisive, endangering efforts to construct a national history and thus a national future. Providing the very sense of continuity and stability that enabled royalist versions of history, family history could not be laid aside lightly by royalist historians who based their version of the past on fictions of the continuity of Stuart rule. Yet, as we have seen, it was the very mechanisms of family inheritance that enabled, in royalist eyes, the preservation of dissent as well as loyalty, resistance as well as restoration. Thus, the challenge for royalist historians writing in the years immediately following 1660 was not to eliminate family history from the national histories of the period, but to accommodate it in a way that would render it less destabilizing to their own attempts to create a national vision of the past.

## III

For royalist historians like Clarendon, one means of dealing with the perceived divisiveness of family history was simply to transform that history into a national legacy, drawing on political parallels between inheritance and succession and thereby absorbing divisive loyalties into a unified monarchical vision. One of the scenes of the civil wars most often depicted by royalist historians, repeated with some variations in Clarendon's *History*, is an account of a conversation between Charles I and three of his children, in which the king effectively replaces an alienated succession with the ideological inheritance he transmits to his royal offspring. Always positioned shortly before the execution of the king, the passage depicts Charles I offering specific advice to the children: James, duke of York, the princess Elizabeth, and the seven-year-old Henry, duke of Gloucester. Having warned the duke of York to look for an opportunity to escape his imprisonment at St James, as he later did, and cautioned Elizabeth not to marry without the approbation of her mother and eldest brother, the king speaks earnestly with the duke of Gloucester, and 'after he had given him all the advice he thought convenient in the matter of religion,' commanded his youngest son, 'upon his blessing, never to forget what he said to him upon this occasion, nor to accept, or suffer himself to be made king, whilst either of his elder brothers lived, in what part of the world soever they should be: that he should remember that the prince his brother was to succeed him by the laws of God and man; and, if he should miscarry, that the duke of York was to succeed in the same right; and therefore that he should be sure never to be made use of to interrupt or disturb either of their rights; which would in the end turn to his own destruction.'[35] In this passage the king becomes the epitome of fatherhood, not merely serving as an affective symbol of paternal authority but also actively transmitting to his children his political and religious values and thus recreating through his children the version and person of monarchy that he represents: Protestant, hereditary, and *jure divino* (hence the emphasis on the king's advice to the duke of Gloucester, who, royalists feared, was young enough to be manipulated into breaking the succession by taking the crown and so becoming a puppet for parliament). As such, the king's advice stands in for a succession denied; if Charles cannot pass his crown to the Prince of Wales, he will transmit to his children, in the manner of so many seventeenth-century fathers, the values and beliefs that will ensure the perpetuation of his vision of monarchy into the next generation. And thus, in this case, family history, grounded in continuity and vested in the interests of the family group, is transformed through the use of the royal family into a matter of national interest, creating a common past in which rebellion and the interregnum are altogether elided.

Yet elsewhere in Clarendon's *History*, such attempts to nationalize a private and potentially fractious family inheritance are not always successful, nor necessarily desirable. Not long after recounting his version of the king's conversations with his children, Clarendon includes an aside regarding pressure put on Charles II after the Restoration to recover the body of Charles I, which had been buried at Windsor Castle in an unmarked grave, and to reinter it with ceremony in Westminster Abbey. To have done so would certainly have helped to create a public and national history by rendering private inheritance a matter of political succession and national interest. The recovery of Charles I not only would have permitted his son to render up his own family inheritance, in this case the body of his father, to public view, but would have symbolized the efficacy and public rebirth of the private charms, the locks of hair and the traces of blood, which individuals and families had carefully collected and preserved as mementos of the dead king. Yet such a recovery was not to be. Given the changes that had been rendered upon the chapel at Windsor, those who were sent to recover the body could not find it: 'they could not satisfy themselves in what place or part of the church the royal body was interred: yet, where any concurred upon this or that place, they caused the ground to be opened at a good distance, and, upon such inquiries, found no cause to believe that they were near the place: and, upon their giving this account to the king, the thought of that remove was laid aside; and the reason communicated to very few, for the better discountenancing further inquiry.'[36] Without a body, plans for a public memorial were ultimately laid aside, and one realizes, reading Clarendon, that in the wake of the schisms of the civil wars, reconstructing a sense of national history by making a private inheritance public seemed less simple than it might once have been. At the same time, in Clarendon's account, the failure of royalists who wanted the king disinterred to realize the new conditions under which history was forced to operate after the civil wars ultimately proves disruptive, bringing criticism upon the king from those who felt he did not put enough effort into finding his father's body and thus threatening to undermine the monarchy once again.

As a result, the *History* does not merely appropriate family history to a national vision but rather acknowledges and relies upon the extent to which family loyalties have rendered the processes of writing and reading history fragmentary. Throughout the *History*, Clarendon manipulates perspectives, not just recounting his own version of events, as was common in histories of the period, but depicting history itself as the product of competing constructions of the past by the past, the narrative products of family and party allegiance. The job of the *History* as Clarendon sees it is often to report the accounts of others, as different and as biased as they may be, in order to negotiate between them with a judgment openly based not in some abstract narrative authority but in his own admittedly limited experience of events. When he describes a plot laid by Edmund Waller – a plot coincidentally a family affair concocted by Waller and his brother-in-law, one Mr Tomkins – Clarendon says 'of this plot, there never being such a formed relation made by those who made great use of it, that men can collect what the design was, or that it was laid with any probable circumstances, by which a success might be expected, I shall briefly and faithfully set down all I know, have heard, or can reasonably conjecture to be in it; and it was thought by many, and averred by others who I believe did not think so, 'that I knew as much of it as most men.'[37] Seeking to offer a more thorough explanation of events that will dispel popular rumours about the plot, Clarendon does so only on the authority of what 'was thought by many' and with the admission that the best one can do is to know as much as most men. During an age in which a fragmented history resulting in part from family legacies and material inheritances made it difficult to assemble a common past to be shared among readers, Clarendon seeks instead to expose the limitations of 'known' versions of recent history, assembling a mosaic in which fictions of coherence – the hallmark of polemic – are rendered untenable and both families and individuals are forced to confront narratives – equally limited – set in dialectic opposition to their own. His job as a historian thus becomes not so much to construct the past as to arbitrate among a multitude of competing histories – to become in effect a manager of a public sphere, as David Norbrook and others have defined it – that the historian himself creates, bringing private voices into public view and testing their viability against that of other fragmentary insights.[38]

The perils inherent in such negotiation are duly noted in the preface to the first volume of the first edition of Clarendon's text, which perceives all too well the extent to which national politics and national history are not separate from but rather dictated by family relations, and thus complains that it 'is a difficult Province to write the History of the Civil Wars of a great and powerful Nation, where the King was engaged with one part of his subjects against the other, and both sides were sufficiently inflamed: And the Necessity of speaking the Truth of several Great Men, that were engaged in the Quarrel on either side, who may still have very considerable Relations, descended from them, now alive, makes the Task Invidious, as well as Difficult.'[39] Yet this dangerous game is one in which Clarendon seemingly engages not to present his own version of the truth in any simple way but to spur readers on to actions that will bring about Clarendon's desired vision of a unified and prosperous state. Undoubtedly realizing, as the preface also suggests, that the readers of his *History* would in the end be the wealthy and influential sons and grandsons of its principal actors, Clarendon uses the fragmentation and indeterminacy of family history to remind his powerful readers of the significance of their own legacy. The famous character portraits that run throughout the history, as we have seen, suggest the extent to which the nature and reputation of the individual are inextricably bound in a complex web of family relations, and many read more like the genealogies popular at the time than like personality profiles, involving relatively elaborate descriptions of fathers, brothers, father-in-laws, wives, children, and paths of inheritance. Just as importantly, however, Clarendon also suggests in the *History* the extent to which the reputation of one generation can be affected by the actions of the next, representing the construction of history as an on-going project and thus implicating his readers in the creation of their own family legacy. Speaking of the earl of Bristol, Clarendon says,

> Though he was a man of great parts, and a wise man, yet he had been for

> the most part single, and by himself, in business; which he managed with good sufficiency; and had lived little in consort, so that in council he was passionate, and supercilious, and did not bear contradiction without much passion, and was too voluminous in discourse; so that he was not considered there with much respect; to the lessening whereof no man contributed more than his son, the lord Digby; who shortly came to sit there as secretary of state, and had not that reverence for his father's wisdom, which his great experience deserved, though he failed not in his piety towards him.[40]

Clearly Bristol's alienation from family serves as a major detriment, leaving him isolated and not a little bit difficult, but the effects of lack of family contact upon his reputation pale beside the impact of his own son, whom Clarendon did not like and referred to as a man of 'unhappy temper'; according to Clarendon's account, Digby's actions and reputation have clouded his father's life and history and thus his family legacy. In that sense, Digby himself becomes a warning to readers of the text, appealing to their own fragmented loyalties and dedication to family to lead them into patterns of behaviour that locate the value of the historical inheritance that they themselves will ultimately leave in their proper service to the nation.

In fact, this understanding of the value of one's political legacy to one's more private interests was eventually replicated in the disposition of Clarendon's own manuscripts. In the end, Clarendon chose to make of his massive *History* a family and a private inheritance. Admitting that the text was 'not likely to be published (at least in the age in which it is writ),' Clarendon bequeathed it to his sons in his own last will and testament, leaving all of his papers 'intire to their disposal as they shall be advised either by Suppressing or Publishing by the Advice and Approbation of my Ld. Archbishop of *Canterbury* and the Bishop of *Winchester* whom I do intreat to be Overseers of this my Will, and that they would be both Suitors to his Majesty on my Children's behalf, who have all possible need of his Majestye's Charity, being the Children of a father who never committed fault against his Majestie.'[41] In so doing, Clarendon made of the *History* a political patrimony, designed to instil in his children his own political vision and to connect them more closely to their in-laws the Stuarts by reminding the royal family of the services rendered by the exiled earl. But not, I think, before its very status as family inheritance insured the *History*'s more public role in the age of projects, eventually leading to the heralded publication of a text that set a course for political action in a new age of divided loyalties and challenged subsequent historians to eschew neat fictions of objectivity in favour of a less cohesive respect for the competing voices of the past.

## NOTES

1 Edward Hyde, earl of Clarendon, *The History of the Rebellion and Civil Wars in England*, ed. W. Dunn MacCray, 6 vols. (Oxford: Clarendon Press, 1888), 1:2–3.

2 For readings of Clarendon's text that place his work in a Renaissance historiographical tradition, see Martine Brownley's *Clarendon and the Rhetoric of Historical Form* (Philadelphia: University of Philadelphia Press, 1985), 19 and George Miller, *Edward Hyde, Earl of Clarendon*, (Boston: Twayne Publishers, 1983), 53–4. For more general discussions of the role of history in the seventeenth century see F.J. Levy, *Tudor Historical Thought* (San Marino, CA: Huntington Library Press, 1967), 237 and Joseph M. Levine, *Humanism and History: Origins of Modern English Historiography* (Ithaca, NY: Cornell University Press, 1987), 159.

3 Phillip Hicks, *Neo-classical History and English Culture: From Clarendon to Hume* (New York: St Martins Press, 1996), 12; Levine, *Humanism and History*, 156.

4 Brownley and Miller, for example, each devote an entire chapter to the character sketches: see *Rhetoric of Historical Form*, 145–85 and *Edward Hyde*, 88–103.

5 *History of the Rebellion*, 1:38.

6 Ibid., 1:11.

7 *The Clarendon-family Vindicated, from the gross falshoods and misrepresentations of John Oldmixon ... and George Duckett* (London, 1732).

8 See, for instance, Joan Evans, *A History of Jewelry* (London: Faber and Faber, 1953); Griselda Lewis, *A Collector's History of English Pottery* (New York: Viking Press, 1969); and W.G. Thomson, *Tapestry Weaving in England from the Earliest Times to the End of the VIIIth Century* (New York: Charles Scribners' Sons, [1914?]).

9 Lewis, *History of English Pottery*, 25–6.

10 Evans, *History of Jewelry*, 129; Anna Somers Cocks, *An Introduction to Courtly Jewelry* (London: The Pitman House, 1980), 21.

11 Xanthe Brooke, *Catalogue of Embroideries* (Wolfeboro Falls, NH: Allan Sutton Publishing 1992), 14, 22. The authenticity of *Eikon Basilica* (1649), the supposed spiritual autobiography of Charles I, is still in dispute. It is usually ascribed to John Gauden.

12 'The Will of Thomas Darcy of York,' in *Abstracts of Yorkshire Wills in the Time of*

*the Commonwealth*, ed. John William Clay, Record Series 9 (N.p.: Yorkshire Archaeological and Topographical Association, 1890), 56.

13 Annette Weiner, *Inalienable Possessions: The Paradox of Keeping-while-Giving*, (Berkeley: University of California Press, 1992).

14 *Wills and Inventories from the Registers of the Commissary of Bury St. Edmund's and the Archdeacon of Surrey*, ed. Samuel Tymms, vol. 49 (Camden Society, 1850), 217.

15 Ibid., 219.

16 *History of Rebellion*, 1:29.

17 Martin Holmes and Major General H.D.W. Sitwell, *The English Regalia: Their History, Custody, and Display* (London: Her Majesty's Stationary Office, 1972), 5; Thomas, *Tapestry Weaving in England*, 121.

18 Holmes and Sitwell, *The English Regalia*, 5, 12.

19 *Wills from Doctors' Commons: A Selection from the Wills of Eminent Persons Proved in the Prerogative Court of Canterbury, 1495–1695*, ed. John Gough Nichols and John Bruce, vol. 83 (Camden Society, 1863), 120.

20 Clarendon, *History of the Rebellion*, 4:498.

21 Ibid., 4:498–9.

22 'The Will of Henry Marwood of York,' in *Abstracts of Yorkshire Wills*, 84.

23 George C. Brooke, *English Coins from the Seventh Century to the Present Day* (London: Methuen and Co., 1950), 149; Charles Oman, *The Coinage of England* (Oxford: Clarendon Press, 1931), 294.

24 Christopher Dursten, *The Family in the English Revolution* (Oxford: Basil Blackwell, 1989), 46.

25 William Sanderson, *A Compleat History of the Life and Raigne of King Charles from His Cradle to His Grave* (London, 1658).

26 *The World in a Maize, Or, Olivers Ghost* (London, 1659), 5.

27 *The Case is Altered. Or, Dreadful News from Hell* (London, 1660), 6–7.

28 *Hugh Peters Last Will and Testament* (1660), broadside.

29 Miller, *Edward Hyde*, 92–3; Brownley, *Rhetoric of Historical Form*, 147.

30 Clarendon, *History of the Rebellion*, 2:545–6.

31 Ibid., 1:76.

32 Ibid., 2:534.

33 Ibid., 2:547.

34 Ibid., 1:69–70.

35 Ibid., 4:251–3.

36 Ibid., 4:494.

37 Ibid., 3:38.

38 David Norbrook, *Writing the English Republic: Poetry, Rhetoric, and Politics, 1627–1660* (Cambridge: Cambridge University Press, 1999).

39 Clarendon, *History of the Rebellion*, 1:18.
40 Ibid., 2:532–3.
41 Clarendon, *History of the Rebellion*, 1:3; Clarendon's last will and testament, reprinted in John Burton, *The Genuineness of Ld. Claredon's History of the Rebellion* (Oxford, 1744).

chapter three

# The Interplay of Past and Present in Dryden's 'Palamon and Arcite'

PAUL HAMMOND

The cultural world in which John Dryden lived after the Revolution of 1688–9 was in some important respects an imagined world, a world of his own making. Increasingly in the 1690s Dryden turned to translation, not only to earn a living, but also, imaginatively, to fashion a world that was peopled with congenial companions – Juvenal and Persius, Virgil, and, in the *Fables Ancient and Modern*, published in 1700, just months before his death, Homer, Ovid, Chaucer, and Boccaccio. The textual and temporal world that Dryden created through his translations was eclectic, macaronic: the translation of Virgil was both a foray into the Roman world of the *Aeneid* and a dialogue with it, a partial translation of its key terms, and a space in which Dryden could address the world around him.[1] Translation, for Dryden, was reciprocal, a mutual rendering of past and present, the weaving of a text which looked both ways.

In his poem 'Palamon and Arcite,' the first of the translations in *Fables*, the conceptual space that Dryden creates is particularly complex. The poem is a rendering of Chaucer's 'The Knight's Tale,' which is itself based on Boccaccio's *Teseida*, behind which, in turn, stands Statius's *Thebaid*. It is the story of the two noble kinsmen who both fall in love with the same woman, Emily, and fight a tournament for her hand. In places it is more of a paraphrase than a translation, with some omissions and many expansions and additions. Dryden fashions a poem that moves between three different worlds – the classical world of Athens and Thebes in which the action is nominally set; the medieval chivalric world that Chaucer created; and the world of late Restoration England that Dryden and his readers actually inhabited.[2]

First, he takes some care to add details that evoke a classical world. This is 'a' classical world rather than 'the' classical world, because the Athens and Thebes of Dryden's imagination are hardly historically accurate. Indeed, the gods of the poem are given Roman rather than Greek names (as they are in Chaucer too): Venus, Mars, Saturn, Jupiter. Dryden accentuates this by adding references to Juno (Bk. 1: l. 260; 2:88), Bacchus (3:99), and Actaeon (1:258), and to Rome (2:451), and he uses the Roman greeting to the dead, 'Hail and farewell' (3:994). At the same time, there is a scholarly precision to Dryden's imagination – he was a great frequenter of annotated editions and commentaries – so that we find him deploying the word 'gauntlet' (3:1001) as the native equivalent that he has devised for the Latin *cestus*, the leather thong weighted with iron and lead that was wound around the hand by Roman boxers.[3] He also imports the Roman custom of decorating doorposts with flowers on festivals (3:104).[4] Linguistically, there are many usages that connect the poem to the Latin language and its literature: the word 'labouring' (3:591), which invokes the Latin verb *laborare*, suggesting 'eclipse'; the inter-lingual pun when he says that Venus's month, April, 'opens all the year' (3:134), drawing on the etymological association between 'April' and *aperire*, 'to open'; the use of 'dome' to mean 'temple,' from *domus dei* (2:462). From Virgil's famous designation of woman as *varium et mutabile semper*[5] he takes the term 'various' to describe Fortune (1:408). The idea that each of the champions 'an army seemed alone' comes from Virgil,[6] as does the phrase 'barb'rous gold' to describe the trappings on a horse.[7] But 'barb'rous gold' is not just a rendering of *barbarico auro*: it also recalls Milton's 'barbaric pearl and gold.'[8] It is characteristic of Dryden's working methods that he associated passages and turns of phrase from diverse poets,[9] and here Dryden's imagination is fusing Chaucer, Virgil, and Milton to create an heroic idiom.[10]

The texture of medieval romance is similarly evoked by a careful use of technical vocabulary. Some of these words are taken or adapted from Chaucer, for Dryden at times re-introduces to the language words for which the *OED* found no evidence in the period since Chaucer: for example, no one seems to have used 'knares' (2:536) for the knots in wood until Dryden picked it up from Chaucer.[11] But various chivalric terms are added by Dryden himself: Theseus's banner has a field of 'argent' (1:109), where Chaucer just says 'white' (976); the knights' armour includes 'surcoats' (1:148), a 'jupon' (3:28), and 'jambeaux' (3:35), the latter being a term that Dryden probably found in Spenser;[12] Arcite imagines his defeated opponents with 'arms reversed' and their 'achievements' disgraced (3:344); the rivals wear 'hawberks' (3:603) and wield 'falchions' (2:244, 599; 3:602). The field is 'listed' (2:258), that is, prepared with lists for a tournament, a usage previously recorded only in

Milton.[13] In such ways, Dryden accentuates the poem's medievalism, a medievalism re-imagined via Spenser and Milton.

As for those allusions that reach out to the world of Dryden's contemporary readers, the ones that are immediately striking are the political references, notably those that allude to the deposition of James II and the usurpation (as Dryden saw it) of William III. A recurring element of contemporary commentary is permitted by Dryden's decision to call Theseus 'Prince' or 'King' rather than 'Duke,' as he is in Chaucer, so that the poem's exploration of the virtues of the ideal king (virtues that Theseus sometimes has to struggle to display, as when he is almost overcome by anger) functions as a running contrast with the kingship of William III. Sometimes the connections are overt allusions, as in the lines from Saturn's speech:

> When churls rebel against their native prince,
> I arm their hands, and furnish the pretence ...
> Bought senates and deserting troops are mine. (3:408–9, 411)

Here Chaucer's single line – 'The murmure, and the churles rebelling.' (2459) – only partly authorizes Dryden's version. Creon 'usurps the land' (1:82), and exercises 'tyranny' (1:86). The two concepts are linked, for 'tyranny' not only means oppressive or severe rule (*OED*, tyranny 2, 3a), but also rule by one who seizes the sovereign power without legal right, that is, a usurper (*OED*, tyrant 1). The word 'tyrant' and its cognates echo through the poem (1:86, 270; 2:168; 3:228, 671), and at one point lines on Jupiter's seizure of power clearly allude to William. Mars, says Dryden, laughed

> ... when the rightful Titan failed,
> And Jove's usurping arms in heaven prevailed;
> Laughed all the powers who favour tyranny,
> And all the standing army of the sky. (3:669–72)

The reference to a standing army recalls opposition to William's desire to maintain an army at home after the European wars had been ended by the Treaty of Ryswick in 1697. Another passage reflects on the failure of enlightened kingly patronage in the reign of William:

> Theseus beheld the fanes of every god,
> And thought his mighty cost was well bestowed:

So princes now their poets should regard;
But few can write, and fewer can reward. (2:659–62)

William, it seems clear, is not going to show the Aristotelian virtue of 'vast magnificence' (2:664) that Theseus displays.[14]

Other comments are less direct, their aphoristic phrasing permitting but not insisting upon a contemporary application: 'Kings fight for kingdoms, madmen for applause' (2:322) associates the two categories rather uncomfortably, while 'All own the chief, when Fortune owns the cause' (3:666) reflects on the readiness of the people to follow William once events had moved in his favour. The reference to 'successless wars' (2:587) may bring to mind William's long-drawn-out military campaigns on the continent.[15] In a couplet from Egeus's speech,

Ev'n kings but play, and when their part is done,
Some other, worse or better, mount the throne. (3:889–90)

we hear the voice of those who argue that there is nothing sacred about kingship, that kingship is merely a role that is well or badly performed. But such a Whiggish view of kingship is immediately presented as superficial, as satisfying only a superficial audience:

With words like these the crowd was satisfied,
And so they would have been had Theseus died. (3:891–2)

Among those whom Dryden sees as complicit in William's seizure of power are the clergy, and the allegorical figures depicted on the walls of Mars's temple include clerical assassins:

... Hypocrisy, with holy leer,
Soft, smiling, and demurely looking down,
But hid the dagger underneath the gown:
...
Contest, with sharpened knives in cloisters drawn,
And all with blood bespread the holy lawn. (2.564–6, 571–2)[17]

So much for the clergy.

But a contempt for the Restoration clergy by no means entails a contempt for religion. Quite the opposite, for 'Palamon and Arcite' is a thoughtfully religious poem, albeit not in a conventionally religious idiom. For running through 'Palamon and Arcite' is another form of contemporary vocabulary that Dryden has drawn from *Paradise Lost.*[17] Though Dryden establishes a verbal connection between the gloomy temple of Mars and the gloomy abode of hell (3:359; cf. *PL* 2:877–82), he more often turns to Milton for phrases that help him to depict an ideal landscape, phrases that open the poem out towards a version of Eden, a world where man seems still in tune with divinity. Arcite says of May, 'For thee the Graces lead the dancing hours' (2:55; cf. *PL* 4:266–8). When Emily goes to invoke Diana, 'Now morn with rosy light had streaked the sky' (3:189; cf. *PL* 5:1, 4:623), and she worships the goddess with 'Incense, and od'rous gums, and covered fire' (3:194; cf. *PL* 4:248). In the temple of Venus, 'Her turtles fanned the buxom air above' (2:519; cf. *PL* 5:269–70). In the first three instances, Dryden is echoing phrases from Milton's description of Eden before the fall; in the fourth, he draws upon Milton's account of Raphael descending to earth to speak to Adam. What lodged in Dryden's memory, and what lodges now in his text, is a way of imagining a harmonious relationship between man and God. These are more than useful turns of phrase, but less than overt allusions; they contribute a Miltonic lexis and inflection to Dryden's poem for anyone with an ear for the distinctive idiom of *Paradise Lost.* They signal a world elsewhere, a paradise temporarily regained in the text.

All these examples show that Dryden was expecting his readers to enter into the intricately textured space of 'Palamon and Arcite,' and to be alert to the ways in which literary materials and their associated values, drawn from different times and cultures, may speak to one another. Indeed, one of the most remarkable elements of 'Palamon and Arcite' is the way in which its language of theology and philosophy functions, drawn, as it is, from varied and often incompatible sources. Perhaps one of the facets of Chaucer's poem that originally attracted Dryden was its philosophy, in particular the way in which it dramatizes the problem of man living in a world beyond his control, where prayers to the gods may be answered in uncertain, or unexpected, or unwelcome ways, and where the gods themselves quarrel over who is to rule.[18]

Who are the supernatural forces in Dryden's poem? In part they are the Roman deities Mars, Venus, and Diana, to whom the three protagonists pray before the tournament, and Jupiter and Saturn, who have to resolve the problem created when Mars and Venus both grant the incompatible prayers of their votaries. But Dryden also refers to 'gods' (ten times)[19] and 'deities' (1:470)[20] – plural and unnamed – and to 'God' in the singular three times (1:422, 2:210, 3:1042) in the sense of a single supreme deity, otherwise unnamed.[22] Then there are other named forces, including Chance, Destiny, Fate, and Fortune.

Dryden has clearly rethought the domain of the principal Roman deities. He does this partly by highlighting their astrological functions (Dryden was himself an amateur astrologer),[23] accentuating what was already there in Chaucer, and so providing one model of how human lives are influenced by supernatural powers. He also rethinks the gods by imagining how the forces of growth and destruction in the natural and human worlds might be thought of as aspects of divinity. Mars is more than the god of war: he is a widely destructive force. Some details in the description of Mars show that both Chaucer and Dryden were drawing on the characteristics that astrologers attributed to the planet Mars and its influence. William Lily in his widely read handbook *Christian Astrology* (1647) not only associates Mars with war but calls the man influenced by Mars 'a lover of Slaughter and Quarrels, Murder, Theevery, a promoter of Sedition, Frayes and Commotions ... a Traytor, of turbulent Spirit, Perjured ... neither fearing God or caring for man, Unthankful, Trecherous ... Furious, Violent.' Mars is associated not only with soldiers, but also with 'Cutlers of Swords and Knives' and cooks; his stone is adamant, his metal is iron and he produces 'all hurts by Iron'; he promotes anger; and 'he delighteth in Red colour'. Trees associated with Mars are 'prickly,' and he causes 'Thunder, Lightning ... pestilent Aires.'[23] All these details of the widespread influence of Mars find an echo in Dryden's passage. When he describes the temple of Mars, he introduces the astrological term 'mansion,'[24] and also adds details of the sinister, uncanny landscape of Thrace in which Mars dwells: in the following quotation Dryden's more substantial additions to Chaucer are emphasized:

> For that cold region was the loved abode
> And sovereign mansion of the warrior god.
> The landscape was a forest wide and bare,
> Where neither beast nor human kind repair;
> *The fowl that scent afar the borders fly,*
> *And shun the bitter blast, and wheel about the sky.*
> *A cake of scurf lies baking on the ground,*
> And prickly stubs, instead of trees, are found,
> Or woods with knots and knares deformed and old;
> Headless the most, and hideous to behold.
> A rattling tempest through the branches went,
> That stripped 'em bare, and one sole way they bent.

*Heaven froze above, severe, the clouds congeal*
*And through the crystal vault appeared the standing hail.*
*Such was the face without, a mountain stood*
*Threatening from high, and overlooked the wood.* (2:528–43)

Mars, then, is the force that turns nature into a landscape of deformity and lifelessness, of grotesque transformations.

By contrast, in the account of Venus's temple, there is a substantial addition to Chaucer when Palamon invokes the goddess. This entire passage is new:

Creator Venus, genial power of love,
The bliss of men below, and gods above,
Beneath the sliding sun thou runn'st thy race,
Dost fairest shine, and best become thy place.
For thee the winds their eastern blasts forbear,
Thy month reveals the spring, and opens all the year.
Thee, goddess, thee, the storms of winter fly,
Earth smiles with flowers renewing, laughs the sky,
And birds to lays of love their tuneful notes apply.
For thee the lion loathes the taste of blood,
And roaring hunts his female through the wood:
For thee the bulls rebellow through the groves,
And tempt the stream, and snuff their absent loves.
'Tis thine, whate'er is pleasant, good, or fair:
All nature is thy province, life thy care;
Thou mad'st the world, and dost the world repair. (3:129–44)

Here too Dryden is writing of the divinity as a planet with astrological significance (3:131–2), and this is linked to an evocation of the procreative ('genial': 3:129) power of Venus in the natural world. But the passage also has a complex literary texture, for it echoes Dryden's own earlier praise of love in his translation of 'Lucretius: The Beginning of the First Book,' and preserves some of that earlier poem's rhetorical borrowings from Spenser;[25] while the interest in the details of animal behaviour also betrays the fact that Dryden had recently been translating the *Georgics*.[26] So in such passages (and in the account of the activity of Saturn) Dryden reinvigorates the idea of these classical deities by showing their power over the human and natural worlds, and does so through a series of literary echoes – a macaronic literary idiom.

As we have already seen, religious ritual is important to the imagined world of 'Palamon and Arcite.' Perhaps this specially concerned Dryden at a time when his own access to the rituals of the Roman Catholic church – to the sacraments that for him were the means of grace – was difficult or impossible. There is in this poem a special awareness of the importance of correct ritual in man's approach to the gods. When Chaucer says that Creon

Hath all the bodies on a heap ydrawe;
And will nat suffer hem by none assent
Neither to be buried, ne to be brent; (944–6)

Dryden provides a different perspective:

But Creon, old and impious, who commands
The Theban city, and usurps the lands,
Denies the rites of funeral fires to those
Whose breathless bodies yet he calls his foes.
Unburned, unburied, on a heap they lie;
Such is their fate, and such his tyranny. (1.81–6)

The word 'rites' quietly draws our attention to proper religious ceremony, and other additions by Dryden portray Creon as 'impious,' a usurper, and a tyrant: to deny the use of religious rites is an example of impious and tyrannical rule. When Theseus rectifies Creon's impious neglect of ritual, Dryden mentions 'with what ancient rites they were interred' (1:129).[27] The phrase 'ancient rites' evokes immemorial ritual, allowing, though not insisting upon, a recollection of the old Catholic ceremonies of England. So too when describing the worship of Diana, Dryden adds some details that connect with Roman Catholic liturgical practice. Emily goes to Diana's temple

In state attended by her maiden train,
Who bore the vests that holy rites require,
Incense, and od'rous gums, and covered fire.
The plenteous horns with pleasant mead they crown,
Nor wanted ought besides in honour of the moon.
Now while the temple smoked with hallowed steam,
They wash the virgin in a living stream;
The secret ceremonies I conceal,
Uncouth, perhaps unlawful to reveal.

But such they were as pagan use required,
Performed by women when the men retired,
Whose eyes profane their chaste, mysterious rites
Might turn to scandal, or obscene delights.
Well-meaners think no harm, but for the rest
Things sacred they pervert, and silence is the best. (3:192–206)[28]

As we have seen, some of these details are drawn from Milton's unfallen world, but Dryden also adds several terms that show that he is thinking quite precisely about the elements of religious ceremony: 'vests,' in the sense of 'ecclesiastical vestments' (*OED*, 2);[29] 'crown,' meaning 'fill to the brim,' which is Dryden's coinage;[30] 'living,' meaning 'constantly flowing' (*OED*, 2d), but also drawing on the biblical sense of the 'living water' that is the gift of the Holy Spirit (John 4:10). The fire is 'covered' as it would be in a thurible in the mass. The cult of Diana is described as a 'use,' a term for a religious ceremony which the citations in the *OED* suggest was reserved for the Catholic liturgy.[31] These chaste and mysterious rites are vulnerable to obscene detraction by scoffers, as happened so routinely through the Restoration period to Catholic worship. Subtly Dryden opens out a field of Catholic memory (perhaps a yearning for unavailable sacraments) within the field of the poem. But this is neither an obtrusive allusion to Catholic practice nor a direct translation of the classical-Chaucerian ritual into contemporary terms: rather, it is the creation of a textual field that allows the sympathetic reader to bring his own experience into contact with the world of the poem.

Besides the gods, there are other forces that act upon the lives of the characters. We know that Fortune played an important part in Dryden's thinking throughout his career.[32] She is the capricious power that arbitrarily gives man good things, and just as arbitrarily snatches them away. She features thus in Dryden's translation of Horace's *Odes*, 3:29, and in several of the plays. In 'Palamon and Arcite' Dryden introduces Fortune twenty-one times. In part this is a response to an element that was already present in Chaucer's poem, as when he says that the queens who supplicate Theseus owe their wretched condition to 'fortune, and her false whele' (925). Elsewhere, Dryden turns Chaucer's 'mischaunce' into 'misfortune' (2:580), the combatants try 'the fortune of the field' (3:579, 655) and Fortune awards the bride to the victor (3:1). Arcite complains:

To die when heaven had put you in my power,
Fate could not choose a more malicious hour!

What greater curse could envious Fortune give,
Than just to die when I began to live! (3:790–3)

This is Dryden's addition.[33] Fortune is brought into the poem's political thinking, too, when the narrator remarks that 'All own the chief, when Fortune owns the cause' (3:666): so when force of arms prevails, and the rewards are assigned by Fortune (rather than by divine Providence, or according to merit), the onlookers supinely – or opportunistically – acquiesce. The comment in part recalls how the people of England went along with William's usurpation because the fortune of the field has awarded him victory. One is reminded of Achitophel's preference for Fortune's ice over virtue's land,[34] but this may also be a riposte to William Sherlock's much-derided argument for submission to the new regime, for Sherlock had claimed (with at least one eye on his ecclesiastical career) that such de facto powers are attributable to the work of divine Providence, and should therefore be obeyed.

> But now God governs the rest of the world, removeth Kings, and setteth up Kings, only by his Providence; that is, then God sets up a King, when by his Providence he advances him to the Throne, and puts the Sovereign Authority into his hands; then he removeth a King, when by his Providence he thrusts him from his Throne, and takes the Government out of his hands: for Providence is God's Government of the world by an invisible influence and power, whereby he directs, determines, over-rules all Events to the accomplishment of his own Will and Counsels.[35]

Such de facto powers, Dryden seems to be suggesting, owe their success to Fortune, not to Providence, to arbitrary occurrences, not to right. It is, finally, part of Theseus's task in settling Emily upon Palamon in marriage to repair the damage caused by Fortune:

Long love to her has borne the faithful knight,
And well deserved, had Fortune done him right:
'Tis time to mend her fault. (3:1123–5)

Such mending of the damage wrought by Fortune is the work of the thoughtful and pious king.

The fact that Dryden at one point changes Chaucer's 'fortune' into 'giddy chance' (1:67) might suggest that he regarded the two ideas as synonymous, but it seems not. For him, Fortune is always hostile, whereas he can write of 'wondrous chance' (2:216) or 'happy chance' (2:411). Palamon says that he has been freed from prison 'by chance' (2:137).[37] Yet there are several points where the poem offers several alternative explanations for the forces that govern the outcome of its narrative: 'Fate or happy Chance' (2:411, spoken by the herald); 'Fate, or partial Chance' (3:162, spoken by Palamon); 'Were it by chance, or forceful destiny' (2:11, spoken by the narrator). Here the poem refuses to be a philosophically closed text, and suspends the impression of narratorial omniscience, opening out a field of multiple explanations for the reader to consider.

Besides Fortune and Chance, another of the poem's principal concepts is Fate. Several of Dryden's additions strengthen the impression that the events are fated: Capaneus falls at Thebes on 'the fatal day' (2:77), where 'fatal' might just mean 'resulting in death' (*OED*, 6), but also carries the sense 'allotted or decreed by fate' (*OED*, 1). It is open to us to decide whether we hear both meanings, or only one. The narrator comments that

> At length, as Fate foredoomed, and all things tend
> By course of time to their appointed end. (3:636–7)

Fate may produce happiness: Arcite exclaims to Palamon that 'much more happy fates thy love attend!' (1:398). Yet in Arcite's case it is a 'fatal blessing' (3:440) that simultaneously released him from prison and robbed him of the sight of Emily. In Arcite's dream, Hermes tells him to go to Athens, where 'Fate appoints an end of all thy woe' (1:554). Perhaps Arcite lives (and dies) rather more in a world ruled by Fate than do the other characters. He prays:

> So may my arms with victory be blessed,
> I ask no more; let Fate dispose the rest. (3:355–6)

But Arcite also senses that he and Palamon are subject to forces that cannot easily be named:

> So thou, if *Fortune* will thy suit advance,
> Love on; nor envy me my equal *chance*:
> For I must love, and am resolved to try
> My *fate*, or failing in th' *adventure* die. (1:348–51, emphasis added)

Arcite uses three linked but not synonymous words, 'Fortune,' 'chance,' and 'fate,' together with a fourth, 'adventure,' which here primarily means not just 'exploit', 'hazardous enterprise' (*OED*, 5) but 'trial of one's chance, hazard, venture' (*OED*, 3), recalling the fundamental meaning (common in Chaucer) of 'chance' or 'fortune' (*OED*, 1).[37] Similarly, Arcite cries in despair:

Fire, water, air, and earth, and force of fates
That governs all, and heaven that all creates,
Nor art, nor Nature's hand can ease my grief,
Nothing but death, the wretch's last relief:
Then farewell youth, and all the joys that dwell
With youth and life, and life itself farewell.
But why, alas! do mortal men in vain
Of Fortune, Fate, or Providence complain?
God gives us what he knows our wants require,
And better things than those which we desire. (1:414–23)

The first three lines have no equivalent in Chaucer. Here Dryden's Arcite imagines (only to dismiss) various sources of comfort: the four elements, the fates (the plural suggesting the three Greek Moirai), heaven, art, and Nature. In this eclectic philosophy, heaven creates, but fate or the fates govern. He then speaks of men railing against Fortune, Fate (now in the singular) or Providence, thus mixing classical and Christian vocabulary, before using the singular 'God' for his positive, reverential submission to divine dispensations, a usage that suggests a vision that is at least compatible with Christianity.[38]

It is Palamon, in his prayer to Venus, who disregards Fate, and places himself emphatically in a world where the ruling power is Venus:

In my divine Emilia make me blest,
Let Fate, or partial Chance, dispose the rest;
...
With smiling aspect you serenely move
In your fifth orb, and rule the realm of love.
The Fates but only spin the coarser clew,
The finest of the wool is left for you.
Spare me but one small portion of the twine,
And let the sisters cut below your line:
The rest among the rubbish may they sweep,
Or add it to the yarn of some old miser's heap. (3:161–74)

For Palamon, Venus has the power to bestow possession of Emily, and

Fate or Chance, or the Fates (plural, and expicitly the Greek Moirai) are subordinate to her. Like Arcite, Palamon has an eclectic – or perhaps just confused – philosophical vocabulary, suspending any judgement about which of these classical forces do actually have some influence over his life.

There are, however, some passages where Dryden – writing now in the person of his narrator rather than one of his characters – opens out the complexity of the philosophical vision that animates the poem. One such comes in book 2, when he describes the fight between the two kinsmen:

> So fought the knights, and fighting must abide,
> Till Fate an umpire sends their diff'rence to decide.
> The power that ministers to God's decrees,
> And executes on earth what heaven foresees,
> Called Providence, or Chance, or Fatal Sway,
> Comes with resistless force, and finds or makes her way.
> Nor kings, nor nations, nor united power
> One moment can retard th' appointed hour.
> And some one day some wondrous chance appears,
> Which happened not in centuries of years:
> For sure, whate'er we mortals hate or love,
> Or hope, or fear, depends on powers above:
> They move our appetites to good or ill,
> And by foresight necessitate the will. (2:208–21)

Here there is again an insistence on the inevitability of 'th' appointed hour' (2:215; cf. 'appointed end,' 3:637), but now Dryden carefully indicates that what God decrees (and what heaven foresees) is carried out on earth by a lesser power that men variously label providence, chance, or fate.[39]

As this passage suggests, the question of free will is one of the recurring motifs in Dryden's poem, and it was surely one of the motifs that drew him to Chaucer, for it is also prominent in 'The Cock and the Fox.' In the world of 'Palamon and Arcite,' freedom is a much-used but uncertain concept. Arcite laments the 'unwelcome freedom' (1:388) that he enjoys on his release; once he thought that 'our utmost good / Was in one word of freedom understood' (1:438–9), but now he knows better, as his release from prison has deprived him of the sight of Emily. Palamon too addresses the subject of freedom and compulsion when he tells Theseus:

Our life's a load; encumbered with the charge
We long to set th' imprisoned soul at large.
Now as thou art a sovereign judge, decree
The rightful doom of death to him and me:
Let neither find thy grace, for grace is cruelty.
...
Arcite of Thebes is he, thy mortal foe,
On whom thy grace did liberty bestow,
...
Know me for what I am: I broke thy chain,
Nor promised I thy prisoner to remain:
The love of liberty with life is given,
And life itself th' inferior gift of heaven.
Thus without crime I fled. (2:265–93)

Theseus initially reacts angrily to the revelation that these two combatants are his sometime prisoners, but then reflects that the law of Nature commands man to seek liberty: 'The pris'ner freed himself by nature's laws: / Born free, he sought his right' (2:337–8). Liberty is part of the law of nature, and the gift of heaven: the concept is, Dryden suggests, not simply the property of the Whigs.

Theseus, who struggles against his baser instincts to bring justice into a world fraught with pain and disappointment, is in many respects the ideal king. And in one particular respect Dryden associates him with an important theological concept, drawing out the analogy between human and divine kingship. This concept is grace.[40] For Dryden, brought up in a Puritan family, the Calvinist notion of grace would have been early and deeply imprinted on his soul. But it was an imprint that faded with the years, or from which he actively recoiled, as his thinking grew more latitudinarian and sceptical. By the time he wrote *Religio Laici* in 1682, the idea that a man might be damned for an accident of history – living before the birth of Christ – was clearly abhorrent to him, and he made his disgust at the Athanasian Creed clear in the poem's Preface.[41] There are many indications in 'Palamon and Arcite' that, at the end of his life, Dryden is rethinking the notion of grace, and attempting to prise it away from the grip of Protestant dogmatism.[42] Theseus is often the focus of this line of thought. The widowed queens approach Theseus as if he were a benevolent deity:

Tis thine, O King, th' afflicted to redress,

And fame has filled the world with thy success:
We wretched women sue for that alone
Which of thy goodness is refused to none:
Let fall some drops of pity on our grief,
If what we beg be just, and we deserve relief:
For none of us, who now thy grace implore,
But held the rank of sovereign Queen before;
Till, thanks to giddy Chance, which never bears
That mortal bliss should last for length of years,
She cast us headlong from our high estate,
And here in hope of thy return we wait:
And long have waited in the temple nigh,
Built to the gracious goddess Clemency.
But rev'rence thou the power whose name it bears,
Relieve th' oppressed, and wipe the widow's tears. (1:59–74)

The queens regard Theseus not simply like a deity, but like a deity who pities the afflicted and freely bestows grace on mankind: the queens implore his 'grace,' invoke 'the gracious goddess Clemency,' and remind him that his grace 'is refused to none'. Moreover, they claim to 'deserve' it, which no Protestant would ever say. This is a long way from the theology of grace prevalent in Dryden's childhood, and it suggests that Theseus in his compassion is a godlike king who has the capacity to redress the wrongs brought about by Fortune and Chance.[43] Chaucer in the equivalent passage twice uses 'mercy': 'grace' is Dryden's word.[44] 'Grace' is being used in such contexts primarily with the meaning 'mercy, forgiveness' (*OED*, 15), but it also potentially carries the theological meaning 'divine favour and influence' (*OED*, 11). And when Palamon explains to Theseus that his rival is Arcite, 'On whom thy grace did liberty bestow' (2:275), he is both referring to Theseus's mercy in releasing Arcite and using a royal form of address (*OED*, 16b). Arcite himself seems to regard his predicament as an inherited curse, a version of the idea of original sin:

... all who come
From Cadmus are involved in Cadmus' doom.
I suffer for my blood: unjust decree
That punishes another's crime on me! (2:94–7)

The term 'decree' (here *OED*, 3) suggests the Augustinian and Calvinist doctrine of predestination. It is over against such a world view of a punitive power acting unjustly that Theseus has the opportunity to bring grace: grace in the sense of mercy, instead of grace in the sense of divine approbation that is restricted arbitrarily to the elect regardless of their merits.[45] When Theseus has made his decision, it is reported by the herald in terms with religious connotations: 'of his grace and inborn clemency / He modifies his first severe decree' (3:500–1). Theseus has been swayed by compassion to abandon a punitive stance and instead exercise mercy. And when he rises to speak the final, profoundly philosophical speech of the poem, Theseus displays 'awful grace' (3:1020); another of Dryden's additions.

The speech that Theseus then makes articulates a complex vision of universal harmony that weaves together material from several different sources, and constitutes a substantial expansion by Dryden of the speech he found in Chaucer, which is itself Boethian. Theseus refers to 'what the gods decreed' (3:1104) and thanks 'the gracious gods for what they give' (3:1113),[46] yet in some important respects his is a monotheistic vision, referring to 'the cause and spring of motion' and 'the same first mover' (3:1024, 1032). This modulates into a form of panentheism in the lines

> Parts of the whole are we, but God the whole,
> Who gives us life, and animating soul.[47] (3:1042–3)

While man and God may be asssociated in this way, as part and whole, they are also clearly distinct, for man is an imperfect part of an ordered universe:

> He perfect, stable; but imperfect we,
> Subject to change, and different in degree:
> Plants, beasts, and man; and as our organs are
> We more or less of his perfection share. (3:1046–9)

It is Dryden rather than Chaucer who writes of the triple soul, vegetal, animal, and intellectual (cf. 3:1076–7). The vision is hierarchical, but also imbued with a profound sense of an eternal rhythm, in which the transitoriness of particular human forms is set against the permanence of eternal matter (3:1030–3).[48] This is a mode of succession or inheritance, for God has decreed that 'every kind should by succession live' (3:1055). If this carries a political implication – that true succession is part of the natural and God-given order of things – it is not the only point in the speech where Dryden is thinking once again about the nature of kingship. The oak is called 'monarch' and 'patriarch' (3:1058), while the egoistic individual who begrudges his own parents their life is a usurper,

> Retchless of laws, affects to rule alone,
> Anxious to reign, and restless on the throne. (3:1074–5)

True kingship, it seems, is part of a God-given natural order, against which may be contrasted the greedy impatience of usurpers. But this speech reaches out beyond the local and the contemporary to embrace a larger vision of growth and decay into which Dryden has drawn elements from various classical and contemporary writers. What Chaucer called a 'faire chain' joining the constituent parts of the universe is for Dryden a 'golden chain,' which he has taken over from Homer and Milton.[49] Milton also provided the vocabulary for the idea that love 'the new creation crowned' (3:1029).[50] The passage that likens the growth of the human foetus to that of a chick inside an egg –

> Secret he feeds, unknowing in the cell,
> At length, for hatching ripe, he breaks the shell (3:1068–9)

– comes from a poet who has closely pondered Ovid. The first mover sends 'peace among the jarring seeds' (3:1027), and 'seeds' for atoms (not recorded in this sense in the *OED*) is a Lucretian usage.[51] Lucretian, too, is the notion that after death the matter that made up the body is recycled so that 'the same matter makes another mass' (3:1053),[52] and the argument in favour of welcoming a timely death (3.1084–1101).[53] And Theseus tells his hearers that by adopting the right attitude towards our mortality we 'leave no more for Fortune to dispose' (3:1093). Yet while Theseus's long *consolatio* may draw upon Lucretius for some of its arguments, it is also biblical in its evocative phrasing: 'few arrive to run the latter stage' (3:1079) uses the image of life as a race from Hebrews 12:1,[54] and his injunction to 'Possess our souls' (3:1114) draws upon Luke 21:19: 'In your patience possess ye your souls.'

The note on which Theseus ends – 'As jarring notes in harmony conclude' (3:1118) – is an appropriate designation for Dryden's own method. Dryden's poem has been open to the jarring elements that destroyed the life of Arcite, and brought grief to his friends. It has canvassed various powerful but mutually incompatible readings of the superhuman forces that have contributed to the development of the story, and has shown us human beings trying to make sense of the painful world in which they live, fumbling with their philosophical vocabularies, but reverently worshipping their gods in the ancient rites which they have inherited. Recovering the past is, for Dryden, not simply a matter of translating a Chaucerian poem into a Restoration idiom; rather, he has woven a text that has strong classical, biblical, Spenserian, and Miltonic echoes, a macaronic fabric into which he has translated classical, medieval, and Restoration worlds, an intricate commentary on past and present. In so doing, he wittily and reverently fashioned for himself a more congenial home than the actual England that met his gaze when he closed his books and laid down his pen.

## NOTES

This essay has appeared in the journal *The Seventeenth Century* (2008).

1 See Paul Hammond, *Dryden and the Traces of Classical Rome* (Oxford: Clarendon Press, 1999), 218–82.

2 Dryden is quoted from *The Poems of John Dryden*, ed. Paul Hammond and David Hopkins, Longman Annotated English Poets, 5 vols. (London: Longman, 1995–2005), hereafter cited as *Poems*. In referring to Dryden's 'Palamon and Arcite: or The Knight's Tale,' I have put the book and line references within parenthesis in my text. Chaucer is quoted from the text reprinted from Speght's edition in Dryden's *Fables Ancient and Modern* (London, 1700), with line numbers added from *The Works of Geoffrey Chaucer*, ed. F.N. Robinson, 2nd ed. (London: Oxford University Press, 1966).

3 He had first used the term in 'The Fifth Book of the *Aeneis*,' l. 88.

4 As Kinsley notes in *The Poems of John Dryden*, ed. James Kinsley, 4 vols. (Oxford: Clarendon Press, 1958), 2068.

5 *Aeneid* 4.569.

6 'Palamon and Arcite' book 3: l. 8 echoes *Aeneid* 7.707, as W.D. Christie noted in *The Poetical Works of John Dryden* (London: Macmillan, 1870), 539.

7 'Palamon and Arcite' 3:65 echoes *Aeneid* 2:504, as Kinsley noted.

8 *Paradise Lost*, ed. Alastair Fowler, 2nd ed. (London: Longmans, 1998), 110 (2:4), as Kinsley noted.

9 Hammond, *Dryden and the Traces of Classical Rome*, 173–4.

10 Similarly, the lines listing the trees that are cut down to make Arcite's funeral pyre draw on a comparable passage in Spenser, with a phrase added from Virgil to characterize alders ('Palamon and Arcite' 3:959–64; *The Faerie Queene* bk. 1, canto 1, st. 8–9; *Georgics* 1:136; noted by Kinsley).

11 Athens has a 'master street' (3:940), a phrase no one seems to have used between Chaucer and Dryden (*OED*).

12 *The Faerie Queene* 2, canto 6, st. 29.

13 Milton, *Samson Agonistes* l. 1087.

14 'magnificence' is the Aristotelian virtue of liberality of expenditure combined with good taste (*OED*, 1); sovereign bounty or munificence (*OED*, 2): Aristotle, *Nicomachean Ethics*, ed. H. Rockham, Loeb Classical Library (Cambridge, MA: Harvard University Press, 1962), 204–13 (4:2).

15 Compare Momus's comment to Mars in 'The Secular Masque' (1700) that 'Thy wars brought nothing about' (l. 88).

16 Compare 'The Cock and the Fox' ll. 480–5, 492 for similar comments on Reynard as a religious hypocrite, and 'The Character of a Good Parson' for a contrasting depiction of the honest and dutiful clergyman.

17 I am indebted here to J.R. Mason, 'To Milton through Dryden and Pope,' unpublished PhD dissertation, University of Cambridge, 1987, and gratefully acknowledge Dr Mason's kind permission to use his thesis in the annotation to the Longman edition.

18 For a discussion of the philosophical thinking of Chaucer's poem see Douglas Brooks and Alastair Fowler, 'The Meaning of Chaucer's *Knight's Tale*,' *Medium Ævum* 39 (1971): 123–46. For the supernatural forces in Dryden's version, see Earl Miner, 'Chaucer in Dryden's Fables,' in *Studies in Criticism and Aesthetics, 1660–1800*, ed. Howard Anderson and John S. Shea (Minneapolis: University of Minnesota Press, 1967), 58–72, esp. 69–71; Cedric D. Reverand II, *Dryden's Final Poetic Mode: The 'Fables'* (Philadelphia: University of Pennsylvania Press, 1988); and Tom Mason, 'A Noble Poem of the Epique Kind? *Palamon and Arcite*: Neoclassic Theory and Poetical Experience,' in *Dryden and the World of Neoclassicism*, ed. Wolfgang Görtschacher and Holger Klein (Tübingen: Stauffenburg, 2001), 181–91.

19 See Guy Montgomery, *Concordance to the Poetical Works of John Dryden* (New York: Russell and Russell, 1957), 260.

20 The plural is used bitterly by Palamon when he rails against the 'eternal deities / Who rule the world with absolute decrees' (1:470–1), though not so bitterly as Chaucer's Palamon, who calls them 'cruell goddes' (1303). The modification of Palamon's bitterness against the gods continues when Dryden has him ask: 'Or does your justice, power, or prescience fail, / When the good suffer, and the bad prevail?' (1:480–2), making this a question, whereas Chaucer's Palamon makes a bitter statement of fact, that the gods do actually torment the innocent: 'What gouernance is in this prescience, / That giltlesse turmenteth innocence' (1313–14). Dryden then adds the idea (1:484) that things could not be worse for mankind if fate or fortune governed all, rather than the gods; but of course this is Palamon speaking, not the narrator.

21 When Chaucer's Arcite swears by 'God, that sitteth aboue' (1599), Dryden changes this to 'the gods who govern heaven above' (2:143), probably wanting to preserve the dignity of the single God.
22 *Poems* 1:76.
23 William Lily, *Christian Astrology* (London, 1647), 66–8; cf. 'Palamon and Arcite' 2:524–97.
24 *OED*, mansion 5a; cf. *OED*, house 8, part of the zodiac where a planet lodges.
25 *Faerie Queene* 4, canto 10, st. 44–6; the links with Lucretius and Spenser are noted by Kinsley.
26 For example, 'snuff' meaning 'detect or anticipate by inhaling the odour of' (*OED*, 4) is first recorded in 1697 in Dryden's 'The First Book of the *Georgics*' l. 519.
27 Chaucer says, 'To done obsequies, as tho was the gise' (993).
28 The last eight lines of the quotation expand significantly on Chaucer's equivalent:

> But how she did, right I dare not tell;
> But it be any thing in generall,
> And yet it were a game to here it all:
> To him that meaneth wel it were no charge,
> But it is good a man be at his large. (2284–8)

Again Dryden takes phrasing from Milton: cf. 'The secrets ... / Not lawful to reveal' (*Paradise Lost* 5:569–70); noted by J.R. Mason.
29 Also cited in *OED*, 1b as the first example of the word used for a woman's robe.
30 *OED*, 8; first example 1697 from Dryden's 'The Fifth Pastoral of Virgil' l. 108.
31 *OED*, 12. It cites no examples between 1643 and 1849, suggesting that this term was uncommon at the time Dryden was writing. The latter date is after the Tractarian revival of Catholic liturgy in the Church of England. The *OED*'s examples from 1636 and 1643 both refer historically to medieval Catholicism.
32 Paul Hammond, 'Dryden's Philosophy of Fortune,' *Modern Language Review* 80 (1985) 769–85; for a fuller discussion of Fortune in 'Palamon and Arcite' see 781–4.
33 It is Dryden who stresses that Fortune is 'envious,' and elsewhere he adds the epithet 'treacherous' (2:36).
34 *Absalom and Achitophel* ll. 199, 252–65.
35 William Sherlock, *The Case of the Allegiance due to Soveraign Powers* (London, 1691), 12. For Dryden's engagement with Sherlock elsewhere, see 'The Character of a Good Parson,' and the notes in the Longman edition.

36 For 'by grace' in Chaucer. Dryden may be reserving the concept of grace for more serious treatment: see below.

37 The sense *OED*, 1 is not recorded after 1594 except in Dryden's translation of the apocryphal Chaucerian poem 'The Flower and the Leaf' l. 605.

38 In annotating this passage, Kinsley suggests that Dryden recalls Boethius's distinction between *providentia* and *fatum*.

39 The narrator explicitly says that our appetites are determined by supernatural forces, and that by foreknowledge they determine our actions. But many theologians had insisted that divine foreknowledge does not determine men's actions (cf. *Paradise Lost* 3:117–19), and in 'The Cock and the Fox' a long interpolation explicitly refutes the idea that God's foresight constrains human will (ll. 525–51).

40 Other words that Dryden is redefining include 'redeem': Theseus loved Pirithous so much that he 'to redeem him went to hell' (1:364).

41 *Poems* 2:88–93 and nn. In 'Palamon and Arcite' 3:844–53 the narrator professes agnosticism about the fate of virtuous pagans, but clearly implies that good works are more important to salvation than faith, a position that at least approximates to the Roman Catholic position.

42 Or, indeed, of Jansenist dogmatism, if he was familiar with the recent debates in France. The Augustinian element in French Catholicism would certainly not have appealed to him.

43 For mercy as a characteristic of Charles II and James II, in Dryden's eyes, see *Poems* 2:396.

44 Similarly, when Dryden's Palamon says to Theseus, 'Let neither find thy grace, for grace is cruelty' (2:269), Chaucer's character says: 'Ne yeue us neither mercie ne refuge' (1720).

45 Venus 'is the gracious goddess' (3:183), while the malevolent Saturn is 'sparing of his grace' (3:383). Both are Dryden's additions.

46 Chaucer's equivalent line refers to Jupiter.

47 'animating soul' is an interlingual tautology.

48 This is the classic distinction between matter and form familiar in scholastic philosophy (see *OED*, form *n.* 4a), but the particular emphasis on matter also recalls Lucretius's vision of the material universe.

49 Zeus (*Iliad* 8:18–27; 338–41 in Loeb Library ed., Cambridge: Harvard University Press, 1946) claimed that he could hang the other gods, the earth, and the sea from Olympus by a golden chain, and Milton imagines the earth hanging by a golden chain in *Paradise Lost* 2:1051.

50 *Paradise Lost* 3:661, 7:386.

51 Dryden used 'seeds' for Lucretius's *semina* in 'Lucretius: Against the Fear of Death' l. 35; see *Poems* 2:318.

52 Cf. 'Lucretius: Against the Fear of Death' ll. 9–48.
53 With 'Throw off the burden' (3:1039), cf. 'Lay down thy burden' in 'Lucretius: Against the Fear of Death' l. 137.
54 Cf. also 'To the Memory of Mr Oldham' ll. 7–10*n* in *Poems* 2:230. The phrase 'early won' (3:1099) may recall 'early ripe' in 'To the Memory of Mr Oldham' l. 11.

chapter four

# Trojan Originalism: Dryden's *Troilus and Cressida*

ELLIOTT VISCONSI

'He laid aside his Epick Poem, perhaps without much loss to mankind; for his hero was Brutus the Trojan, who, according to a ridiculous fiction, established a colony in Britain. The subject, therefore, was of the fabulous age; the actors were a race upon whom imagination has been exhausted and attention wearied, and to whom the mind will not easily be recalled when it is invited in blank verse.' So writes Samuel Johnson in his *Life of Pope*, speaking of Pope's flirtation with the English national myth of its own Trojan origins.[1] Pope never completed his much-planned epic poem describing the peregrinations of the dispersed Trojan Felix Brutus, the supposed founder of ancient Britain. He came to this myth belatedly and with full knowledge of its historically spurious nature. This 'ridiculous fiction' of *translatio imperii* had hit its apex in late Tudor England, when the project of proto-nationalist cultural legitimation was in full swing, as Richard Helgerson has pointed out in *Forms of Nationhood*.[2]

In *The Faerie Queene*, for instance, Edmund Spenser supplies the Tudor dynasty with a Trojan lineage through the figure of Britomart, the allegorical knight of chastity. She is destined to be the mother of 'A famous progenie ... out of the ancient Trojan blood, / which shall revive the sleeping memory / of those antique peers, the heaven's brood, / which Greek and Asian rivers stained with their blood.'[3] Here the 'sleeping memory' of England's Trojan progenitors commands both bravura and melancholy – Britomart's progeny, figuratively heaven-born like their Trojan forebears, will be 'renowned kings and sacred emperours' who create a rival to both Troy and imperial Rome: 'A third kingdom yet is to arise, / Out of the Trojans scattered offspring / That in all glory and great enterprise, / Both first and second Troy shall dare to equalise' (3:9, st.

44). But like those ill-fated 'antique peers' whose blood stains the rivers, Britomart's descendents will suffer misery and slaughter as a consequence of their excellence. Even within the frame of Spenser's Tudor boosterism, the idea of Trojan origins has as its affective core an unusually vivid combination of pathos and *aristeia*, melancholy and honour. The English diasporic myth of Trojan origins, like the notion of racial diaspora itself, relies upon a mixture of trauma and typology: as Svetlana Boym puts it, 'The *nostos* of a nation is not merely a lost Eden but a place of sacrifice and glory, of past suffering ... Defeats in the past figure as prominently as victories in uniting the nation. The nation-state at best is based on the social contract that is also an emotional contract, stamped by the charisma of the past.'[4] The Trojan diaspora is more than a charismatic nationalist fiction, however; even while it bathes the mythic past in sacrificial glory, it also promises a forward-looking racial destiny imbued with historical providentialism that David Quint has described as the formal core of epic.[5]

The Trojan elements of *The Faerie Queene* are one small aspect of a much broader Tudor and early Stuart controversy over the Brutus legend, the myth of English national origins invented by Geoffrey of Monmouth in his twelfth-century *Historia Regum Britannia*, versified in the form of Layamon's *Brut*, and subsequently bandied about in a historiographical and political argument until the seventeenth century, when the story seems finally emptied of plausibility. Here is a caricature of this story: Brute, Brutus, or sometimes Felix Brutus, the grandson of Aeneas, gathers to him some of the dispersed remnants of the Trojan race, receives an oracular vision of Albion, and after a number of prescribed epic adventures sails with his band out of the Mediterranean. Eventually he lands in England, where he crushes the indigenous giants and creates a new Troy ('Troynovant,' later London) on the banks of the Thames. Geoffrey of Monmouth's *Historia* had met with scepticism in its contemporary moment, and in Tudor England the legend's historically dubious quality instigated a number of passionate refutations by, for example, Polydore Vergil, Ben Jonson, and William Camden. As Heather James has written of this controversy, the Tudors 'were willing to host the legend, as long as they were not required to give it credence.'[6]

Some did, of course, try to cultivate favour with Elizabeth or James I by providing them with a Trojan heritage. George Owen Harry, for example, composed a genealogical chart linking the new King James I with the legendary Brute, just as an engraving in Richard Grafton's 1569 *Chronicle* would portray Elizabeth as a direct descendant of the mythic founder. While such gestures are mostly legible as spurious flattery, and there is little evidence that many peoples took these legends seriously as historiography, the Brutus myth persists as a useful and labile fiction of national origins. The story survives through the seventeenth century and long enough into the eighteenth to command Pope's attention towards the end of his life, even if the project ultimately frustrated him. John Milton flirted with the idea of a Brutus-poem in works such as 'Epitaphis Damonis' and 'Mansus.' Milton's critical evaluation, in the opening pages of his 1670 *History of Britain*, is emblematic of the legend's heuristic utility in the seventeenth century. The fiction of Trojan origins is most relevant to 'our English Poets and Rhetoricians ... who by their art will know how to use [it] judiciously.'[7] Milton's *History* is at least in part a work of analytical ethnography, and he sees such dubious legends as usefully descriptive markers of the English national character, but it is his argument that such legends merit 'judicious use' by poets that explains the persistence of the Brutus plot.

## I

After decades of relative dormancy, the myth of Trojan origins re-emerges in 1678 and 1679, those explosive, paranoid years dominated by the Popish Plot and by the beginnings of the Exclusion Crisis. With substantial hesitation, I am tempted to identify a 'Trojan moment' on the London stage in these years. While in 1675 John Crowne had broached the subject in his Racininan tragedy *Andromache*, three Trojan plays are more or less coeval with the leading edge of the Exclusion Crisis. Nahum Tate's 1678 *Brutus of Alba* is a melodramatic rehearsal for his later opera *Dido and Aeneas*, while John Banks's 1679 *The Destruction of Troy* is an overt 'parallel' or topical commentary on London factionalism and city politics that might have been more accurately titled 'The Destruction of Troynovant.'[8] The third play, easily the richest and most judicious use of the Trojan theme, is John Dryden's 1679 *Troilus and Cressida*, a major overhaul and reinvention of Shakespeare's anti-aristocratic play.[9]

Here, as throughout his career as a dramatist, Dryden is uninterested in the crude 'parallel play' as a method of philosophical or political argument – and I want to avoid the assumption or even tacit implication that his *Troilus and Cressida* is such a parallel along the lines of its near-contemporary *Absalom and Achitophel*.[10] The Shakespearean precedent is notoriously difficult and corrosive: in his preface to the play, Dryden argues, somewhat confidently, that 'because the play [*Troilus and Cressida*] was

Shakespeare's, and that there appeared in some places of it, the admirable Genius of the author, I undertook to remove that heap of rubbish under which many excellent thoughts lay wholly buried.'[11] But what are those many excellent thoughts that Dryden tries to bring forward into clarity? Unlike the Shakespearean precedent, Dryden's play is deeply concerned with posterity, the transmission of the soul across generations, and the exemplarity of a Trojan house divided against itself. It is through such figurative additions that Dryden hopes to foreground his play's commentary upon the fiction of Trojan origins without violating the rules of neoclassical fidelity, and that commentary becomes an occasion for Dryden to unfold a theory of fiction that casts the poet as the privileged interpreter of not just the historical past and the literary tradition, but also of the idea of national origins itself.

Dryden's *Troilus and Cressida* is thus a case of 'Trojan originalism' in two discrete senses. First, it is a play in which Dryden adapts the pliable fiction of national origins into a pseudo-ethnographical political critique; avoiding an overt discussion of the English national character, Dryden describes instead the often inconsistent and finally self-consuming Trojan *ethos* as an obvious surrogate, a stand-in made plausible by the enduring presence of the Brutus myth. Dryden finds much to despise and much to celebrate in this national character, which is composed of the love of honour and of violence, of painful credulity and wilful independence. In this sense, Dryden's 'Trojan originalism' is thematic – he turns the plot of England's Trojan origins from a script of proto-nationalist chest-thumping into a moment of analytical reflection on an ambivalent choice of inheritance.

In addition, however, Dryden sees the chance to speculate on national myths and foundational fictions as an unusually rich opportunity to unfold, in the figural and narrative construction of his play (as well as in the prefatory essay *The Grounds of Criticism in Tragedy*), his well-developed theory of dramatic poetry as civic enlightenment. *Troilus and Cressida* is rich with meta-theatrical and self-reflexive discussions of the fictional method, and in the play Dryden casts the dramatic poet as the intellectual agent best suited to interpret the origins of laws, nations, and peoples. In this sense, the play resembles what has been called 'originalism' in contemporary constitutional scholarship – it works as an account of first principles or original qualities that has normative consequences for the present, and for Dryden, the poet is the individual best able to translate those norms into accessible and emotionally commanding fictions.

## II

*Troilus and Cressida* is a 'performance of cultural memory,' to borrow Joseph Roach's phrase, which emphasizes the persistent consequences of a foundation myth built upon catastrophe and diaspora.[12] Dryden traces English cultural memory back as far as it will go in order to explain the origins of a national character both meritorious and regrettable. The diasporic myth provides the English with a charismatic moment of inception, but that origin is an ignoble disaster. The Trojan *oikos*, fractured fatally by the love of Troilus and Cressida, reflects this deep structural opposition in the familiar idiom of Restoration heroic drama. Troilus cannot be both a faithful lover and a dutiful Trojan, and his formulaic struggle between passion and duty is an emblem of the self-consuming inconsistency of a plot based on simultaneous catastrophe and inception.

Dryden's play advertises its relationship to the diasporic theory of national origins in its prefatory matter. Consider, for example, the commendatory verses by Randolph Duke printed with the play in 1679: '*Hector* and *Troylus* darlings of our Age, / Shall hand in hand with *Brutus* tread the stage. / *Shakespeare* 'tis true this tale of *Troy* first told, / But, as with *Ennius Virgil* did of old, / You found it dirt but have made it gold.'[13] Duke (not so subtly) praises Dryden for removing the Shakespearean 'heap of rubbish,' and in so doing reminds us of the play's implications: Hector, Troilus, and Brutus are together the 'darlings of our Age ... hand in hand,' one coherent vision of the Trojan *ethos.* I think, however, Duke misses the subtlety of Dryden's Troy, and the manner in which it presents an ambiguous portrait of a self-consuming Trojan national character. The famous prologue 'spoken by Mr. Betterton, representing the ghost of Shakespeare,' ratifies Dryden's thesis ironically. Concluding the prologue, the ghost asks his audience to

> *Sit silent then, that my pleas'd Soul may see*
> *A Judging Audience once, and worthy me:*
> *My faithfull Scene from true Records shall tell*
> *How* Trojan *valour did the* Greek *excell;*
> *Your great forefathers shall their fame regain,*
> *And* Homers *angry Ghost repines in vain.*[14]

It is essential to remember that Dryden has publicly advertised his desire to supplement Shakespeare's crudeness in *this particular play,* even though he venerated his great predecessor as a 'divine poet' equipped with a 'universal mind, which comprehended all characters and passions.'[15] Here Shakespeare's ghost voices a simple idealism consistent with the myth of Trojan origins – 'how Trojan valour did the Greek excell' – makes a claim of historical fidelity – 'my faithfull scene from true records shall tell' – and links the Trojan past with the English present – 'Your great forefathers shall their fame regain.'

These lines ending the ghost's prologue are to double business bound. On the one hand, they make a clear link between the play and the fiction of Trojan origins, and this link is asserted by the authoritative voice of Shakespeare, who here appears in the guise of a subtle advocate for the Brutus legend. But it is nearly impossible to read Dryden's *Troilus and Cressida* as in any way supporting the crude thesis that 'Trojan valour did the Greek excell,' for the Trojan *oikos* fatally divides against itself. In fact, the wooden moral with which Ulysses closes the play is the opposite of the ghost's claim, for it celebrates the Greeks for their ability to triumph over aristocratic infighting: 'Now peacefull order has resumed the reins, / old time looks young, and Nature seems renewed; / Then, since from homebred factions ruin springs, Let subjects learn obedience to their kings' (5.2). The play does not bear out the ghost's thesis, which looks simple and idealistic in comparison with the moral ambiguity that follows. The most important work achieved by the ghost's prologue is to cast the very idea of Trojan valour and Trojan diaspora, like Shakespeare's play itself, as a primitive vision that Dryden's *Troilus and Cressida* will remedy and purify.

Heather James has argued that Shakespeare's play contaminates the Troy legend to such a degree that 'the late Elizabethan audience should mortally fear that England has indeed inherited its national identity' from the Trojans.[16] Dryden's version of the plot is less sexualized, and also much more ambiguous in its description of the implied national character. Dryden's audience, I think, would not have 'mortally feared' his description of their national character because the tone and objective of the play is measured in contrast with Shakespeare's broad deictic calumny. And consider the epilogue, spoken by Thersites, the vulgar fool: 'You British Fools, of the old Trojan stock / That stand so thick one cannot miss the flock, / Poets have cause to dread a keeping Pit, / When Women's cullyes come to judge of Wit.' The high seriousness of the play's ending, with Troy in ruins and Ulysses moralizing about the obedience of subjects, turns over in these lines to satire when Thersites mocks the 'old Trojan stock' in the pit as a clot of 'women's cullyes.' Of course, part of this tone is *pro forma* – epilogues in Restoration tragedy tend to lighten the mood with wit and irony – but it is also the case that Dryden's play tolerates such satire in a manner that Shakespeare's does not.

*Troilus and Cressida* relies upon the diasporic genealogy to realize an ambiguous historiography of the English race at its moment of inception. In his treatment of the doomed love plot of the eponymous lovers and in the equally crucial and vexed relationship between Hector and Troilus, Dryden describes the essential character of Trojan culture as an inevitably self-consuming blend of liberty and hedonism, of sacrifice and ingratitude, of generosity and indolence. Hector and Troilus are each princes in whom *psychomachia* is visible; instead of a too easy opposition in which Hector stands for martial heroism and Troilus stands for hedonistic private passion, Dryden offers us the wild oscillations of heroic personality one comes to expect from his serious plays. Troilus, whom we first meet rejecting the desire to fight – 'why should I fight without the Trojan walls, / who, without fighting, am overthrown within: / The Trojan who is master of a Soul, / Let him to battle, Troilus has none' – is at turns martial, raging, uxorious, honourable, and equipped with bad judgment. He rejects his ascribed role as the man 'made a public sacrifice for Troy' when his beloved Cressida is exchanged, and pursues his revenge upon Diomedes to such a degree that he carries Hector and all of Troy with him into catastrophe. As he seeks revenge ungratefully, his complex blend of excellence and solipsism increasingly becomes evident.

Hector likewise demonstrates a combination of characteristics long imputed in classical and early modern ethnography to the gothic northerner – a warlike love of liberty and independence coupled with strikingly bad judgment. Early in the play, the Trojan princes discuss returning Helen to the Greeks, and Hector argues in favour of this remedy as consistent with 'the moral law of nature and of nations' (2.1). But he seems unable to tolerate a long-term embrace of the rule of law, and comes quickly to concur with Troilus, hoping to keep Helen as 'a cause on which our Trojan honour / And common reputation will depend.' In the final act of the play, Hector is easily gulled into seeking the field of honour and revenge with his brother, despite Priam's clear injunction that 'the general safety upon [his] life depends; / and should you perish in this rash attempt, / Troy with a groan, would feel her Soul go out' (2.1). These words are prophetic, and Hector's manly excellence is not enough to defend him from vanity and bad judgment – he pursues and kills an anonymous Greek for his beautiful armour, and when he strips to put on this shiny new armour, the Myrmidons surprise and kill him.

Taunting Troilus on the final battlefield, Diomedes says, 'I triumph in thy vain credulity, / Which levels thy state to mine' (5.2). His analysis is correct – Troilus is both vain and credulous – but these lines are also an analysis of the collective character of the Trojan aristocracy – it is indeed heroic vanity and bad judgment that has reduced the glory that was Troy to the condition of the petty, factious Greeks. Dryden provides Hector and Troilus with a conventional heroic *psychomachia*, but draws the terms of that inner struggle from ethnographic portraits of the English national character – warlike, vain, independent, lawless, and credulous. Early in 1679 such characteristics were for Dryden both laudable and disturbing; at this point, in his view, the national character was still capable of displaying both heroism and hedonism.[17]

*Troilus and Cressida* is thus legible as part of a pro-Stuart initiative mounted during and after the Exclusion Crisis by Dryden, Roger L'Estrange, and Aphra Behn among others, designed to link political subversion to the inconsistency of the English people's racial character, and in so doing promote the thesis that only a disinterested *jure divino* sovereign could discipline such a group. This mode of argument is present clearly in later works such as Dryden's 1685 opera *Albion and Albanius* and Behn's 1688 novella *Oroonoko*.[18] From 1679 to 1685, when James II takes the throne and his poet laureate converts to Roman Catholicism, political crises and seditious plots convince Dryden of the essentially lawless and primitive character of the English race. The qualified approval, or at least toleration, in *Troilus and Cressida*, of the English people's inherited blend of martial honour and self-interest, of pusillanimity and self-sacrifice, vanishes before the amassing evidence of simple barbarism. As Dryden writes not much later in *Absalom and Achitophel*, the English are a 'Headstrong, Moody, Murmuring race, / ... [and] God's pamper'd people, whom debauch'd with ease, / No King could govern, nor no God could please / ... [who] thought that all but Savages were Slaves.'[19]

As the philosopher Giorgio Agamben has pointed out in *Homo Sacer*, 1679 also saw the passing of the Act of Habeas Corpus in parliament.[20] Agamben reads this law, hyperbolically I think, as a democratic breakthrough, a transformation in the normative status of human life in which the *corpus*, the body that signifies *zoe* or bare life, becomes the legal index of personhood in place of an older ideal of citizenship and ethical pursuit of the good. But he is right to notice the law, and while habeas corpus was enacted to restrict the crown's ability to do away with its enemies, Agamben sees the law as evidence of a mounting concern with the essential – with the simplest, most pure, and qualitative description of the human person replacing the outward ornaments of citizenship and public duty.[21] For Dryden, racial character is an essential and descriptive physiological fact, legible in Agamben's terms as *zoe* writ large in the body of the nation. *Troilus and Cressida* is obsessed with essential qualities and irreducible phenomena, and in this play Dryden uses the figure of soul to describe a national ethos that links past and present in a mythic, if spurious, genealogy.

'Soul' abounds in Dryden's play, and the princely Trojans seem most determined to cultivate their souls in as many permutations as possible. Hector describes his unwillingness to leave Andromache before battle, for example, as 'struggling in my manly Soul.' This is a conventional and unremarkable reference; of course he would describe a psychological struggle as a torment of the soul. Andromache rebuts his claim in similar terms, suggesting that his 'Soul is proof to all things but to kindness.' Hector's soul, however, quickly turns over into something else entirely, into a startling claim of collective *megalopsychia* as he leads the Trojans into their last battle:

> To Arms, to arms, the vanguards are engaged:
> Let us not leave one man to guard the walls,
> Both old and young, the coward and the brave,
> Be summoned all, our utmost fate to try;
> As one body move whose soul am I. (5.1)

Here Hector understands himself as the soul of the Trojan state – not an inaccurate thesis, according to Priam – and presents the combat as a judgment of that collective soul, the trial of '*our* utmost fate.' But it is no accident that Hector's heroic call to arms ends with the pronominal expression of his real concern –'whose soul am *I*.' Hector's soul becomes the soul of the Trojan polity; his manly excellence and soft-headed credulity radiate outward to become the character of the whole nation.

This speech is one of the many such emblematic moments in the play where the soul is a marker of transmitted essential quality, of radiant personality that echoes down through generations. For example, after they argue over the exchange of Cressida, Hector insults his brother, after having branded him as a traitor, by swearing an oath on 'our father's soul, of which no part / did ere descend to thee,' to which Troilus replies, 'if urged beyond my temper, [I'll] prove my daring, / And see which of us has the larger share / of our great father's Soul.' At this point, Troilus has refused to part with Cressida, whom he calls 'my life, my being, and my soul,' despite Hector's argument that the public good demands her exchange, and he defends his position with a withering assault on the idea of the public:

> The public is the Lees of vulgar slaves:
> Slaves, with the minds of slaves: so born, so bred:
> Yet such as these united in a herd,
> Are called the public; millions of such cyphers
> Make up the public sum: an Eagle's life
> Is worth a world of crows: are princes made
> For such as these, who, were one Soul extracted
> From all their beings, could not raise a Man. (3:2)

Troilus's most potent damnation of this vulgar, slavish public comes in the last two lines, where the mob of millions is found wanting soul enough to 'raise a man.' Later Troilus uses the same figure in vulgar fashion to explain Cressida's apparent infidelity with reference to her apostate father Calchas: 'with her mother's milk / She sucked the infusion of her father's soul. / She only wants an opportunity, / Her soul's a whore already' (4.2). Troilus and Hector, in antithetical terms, each describe an exclusively descending relationship between the exemplary man and his posterity – the actual or collective progeny his transmitted soul comes to inhabit and constitute. Through the figure of radiant soul, Dryden is able to describe the transmission of *essential* qualities (racial character) across time, and in so doing reinforce his play's status as a fiction of national origins.

Dryden's play is a complex portrait of the English national character that makes judicious use of the diasporic legend while acknowledging its fictionality. It is also an occasion for Dryden to outline self-reflexively his fictional method, and assert the dramatic poet's privileged ability to interpret the inscrutable normative origins of law, race, and political society. Attached to the preface of the play is 'The Grounds of Criticism in Tragedy,' which along with the *Essay of Dramatick Poesy* is Dryden's most important piece of literary criticism. In this short essay, Dryden outlines a neo-Aristotelian theory of tragedy in which fear and pity are the affective vehicles of systematic public enlightenment.[22] Tragedy cannot be one long hyperbolic rant, Dryden argues, but must be measured: 'he who would raise the passion of a judicious audience ... must be sure to take his hearers along with him; ... he must move them by degrees, and kindle with 'em; otherwise he will be in danger of setting his own heap of stubble on fire, and of burning out by himself, without warming the company that stand about him' (*Grounds*, 23). It may seem obvious, but here Dryden is defending the use of dramatic irony, suspense, and the emotional crescendo – all strategies that he had put to masterful use in his earlier heroic plays. He puts the matter quite bluntly – ''tis necessary therefore for a poet, who could concern an audience by describing of a passion, first to prepare it, and not to rush upon it all at once' (23–4).

The most wrenching and passionate scene in *Troilus and Cressida* (3.2) is a set piece of this dramatic theory. Hector prepares Troilus for the shattering news of Cressida's exchange *en crescendo*, moving his brother by 'degrees and glimpses' ostensibly to prevent his intemperate and unmanly rage:

*Hector* And I will tell my news, in terms so mild,
So tender, and so fearful to offend
As Mothers use to sooth their froward Babes;
Nay I will swear as you have sworn to me,
That if some gust of passion swell your soul
To words intemperate, I will bear with you.
*Troilus* What would this pomp of preparation mean?
Come you to bring me news of *Priam*'s death,
Or *Hecuba*'s?
*Hector* The Gods forbid I should;
But what I bring is nearer, more close,
An ill more yours.
*Troilus* There is but one that can be.
*Hector* Perhaps 'tis that.
*Troilus* I'll not suspect my fate
So far, I know I stand possesst of that.
*Hector* 'Tis well: consider at whose house I find you.
*Troilus* Ha!
*Hector* Does it start you? I must wake you more:
Anthenor is exchanged.
*Troilus* For whom?
*Hector* Imagine.
*Troilus* It comes like thunder grumbling in a cloud,
Before the dreadful break; if here it fall,
The subtle flame will lick up all my blood,
And in a moment turn my heart to ashes.

*Hector* That *Cressida* for *Anthenor* is exchang'd
Because I knew 'twas harsh, I would not tell
Not all at once, but by degrees and glimpses
I let it in, lest it might rush upon you
And quite overpower your Soul.[23]

Here Dryden casts Hector as poet and Troilus as audience, using the theoretical language of his *Grounds of Criticism* to announce the play's status as a meditation on the fictional method and its affective tactics. At the fatal moment, Hector refuses to even say Cressida's name, substituting a cruel 'imagine.' This one word, which for Troilus connotes disaster but denotes endless possibility, refers to the affective core of Dryden's tragic theory – it is only through the imaginative experience of disaster that fear and pity arise for the audience.

This exchange is the beginning of a long quarrel and reconciliation scene between Troilus and Hector, a scene that attracted substantial critical notice. In the preface Dryden admits that 'the occasion of raising [the scene] was hinted to me by Mr. Betterton: the contrivance and working of it was my own. They who think to do me an injury by saying that it is an imitation of the scene betwixt Brutus and Cassius, do me an honour, by supposing that I could imitate the incomparable Shakespeare' (Preface, 12–13).[24] In short, this scene of almost unimaginable pathos turns out to be the play's most meta-theatrical moment, a scene in which Shakespeare's genius (by way of his *Julius Caesar*) and Dryden's affective method are equally on display.

It is also worth recalling at this point that in the prologue, Shakespeare's ghost asks the audience to 'sit silent then, that my pleased Soul may see / a judging audience once, and worthy me.' In these lines Dryden echoes his own dramatic thesis in the *Grounds of Criticism*, advertising his courtship of a 'judicious audience' and promising to 'take his hearers along with 'em, move them by degrees, and kindle with 'em.' And it is no coincidence that Shakespeare delivers these lines. For Dryden, Shakespeare 'had a universal mind, which comprehended all characters and passions.' He was an unsurpassed genius writing in a barbarous age whose 'magic could not copied be' and whose 'power is as sacred as a king's,' to quote from the prologue to Davenant and Dryden's 1667 *Tempest*. For while individual plays such as *Troilus and Cressida* might be buried beneath rubbish, Shakespeare is Dryden's ideal poet, described admiringly as a 'masculine soul' able to conjure into drama 'just and lively images of human nature,' Dryden's definition of poetical excellence.

In fact, the transmission of soul is Dryden's preferred characterization of his relationship to Shakespeare, as in these lines from his poem 'To Sir Godfrey Kneller':

> Shakespeare, thy gift, I place before my sight;
> With awe I ask his blessing ere I write;
> With reverence look on his majestic face
> Proud to be less, but of his godlike race.
> His soul inspires me, while I thy praises write. (73–7)

Describing a drawing of Shakespeare that Kneller had presented to him, Dryden casts himself in a posture of abjection strikingly at odds with the confidence of his screeds against Shakespeare's *Troilus and Cressida.* Dryden is 'of his godlike race,' the last term a fairly conventional claim of poetic filiation akin to the 'tribe of Ben,' and he is inspired by Shakespeare's comprehensive soul. The conjunction of race and soul in this later poem echoes Dryden's nearly ubiquitous use of this figure in *Troilus and Cressida,* where the exemplary soul resonates outward to give coherence and identity to a whole nation.

I have argued above that Dryden's *Troilus and Cressida* is an example of the 'judicious use' of the Brutus legend in which we are provided with a portrait of the English national character at its moment of inception and a self-reflexive commentary on the fictional method. I would like to add to these claims a third – that the play is symptomatic of Dryden's broadest argument for the socio-political and ethical function of the poet, a view shared by contemporaries such as John Milton. This is the idea that the serious poet (the writer of epic or tragedy) is unusually well qualified to apprehend the soul of the great man, interpret the founding intentions of law or polity, and grasp the universally valid thesis behind the facts and circumstances of a historical narrative of origins. Dryden naturalizes this role: he repeatedly insists that 'genius must be born and never can be taught,' even though to natural perceptiveness one must add experience and method. He asserts that poets such as Shakespeare and himself are members of a 'godlike race,' and links the fictional method persistently to an ethical mandate: the poet may 'add, alter or diminish' historical facts so that his work will be a 'just and lively image of human nature.' And serious poetry should 'resemble natural Truth but it must *be* Ethical.'[25] The broader goal of such activity is, as Dryden's mentor Sir William Davenant had written in his *Preface to Gondibert,* is to 'persuade an amity in divided nations.'[26]

Dryden devoted a good deal of critical attention to the theory of translation late in his career, when he supported himself with editions of Virgil and Ovid, most notably his 1697 translation of the *Aeneid*. In the Dedication to his *Aeneis*, Dryden suggests that the anachronistic meeting of Dido and Aeneas is wholly appropriate, since Virgil may 'supersed[e] the mechanic Rules of Poetry, for the same Reason, that a Monarch may dispense with, or suspend his own Laws, when he finds it necessary to do so; especially if those Laws are not altogether fundamental' (*California Dryden*, 5:300–1). A bit later Dryden puts himself in the company of both Virgil and the absolute monarch who may dispense with his laws, when he describes his approach to translation: 'taking all the materials of this divine Author, I have endeavoured to make *Virgil* speak such *English*, as he would himself have spoken, if he had been born in *England*, and in this present Age' (ibid., 5:330–1). I think these lines are representative of not just Dryden's theory of translation, but of his fictional method itself – he understands *poesis* as a school of equity, that habit of judgment practised by sovereigns who look past the letter of the law to the spirit of its original intent.[27] Dryden's comparison of Virgil to a monarch who may 'dispense with or suspend his laws when he finds it necessary to do so,' a phrase literally drawn from the legal description of royal prerogative, makes this nexus apparent – a wide-ranging ability to conjure original intentions and universal norms from a murky, inscrutable historical past is the sign of naturalized poetic genius and the highest ascriptive goal of poetry.

Above I suggest that Shakespeare's ghost establishes a crude thesis of Trojan glory that Dryden's play is intended to supplement, and also that Shakespeare was a venerated deity in Dryden's poetic pantheon. These two apparently opposed sentiments of adoration and correction reveal the ghost's prologue as an emblem of what I call Dryden's imaginative originalism. Betterton is playing a ghost, after all, an effigy of uncertain provenance lurking at the opening of the play, casting a long shadow over the subsequent drama. Shakespeare's ghost claims his authority in clear verse, but that plain intent is obscured by the spectral source of its utterance. The past – whether in the form of predecessor poets, history, prescriptive regimes, or mythic diasporas – exerts a normative claim upon the present, and it cannot be dispelled once conjured. It falls to the godlike, sovereign poet, in Dryden's view, to translate murky apparitions into ethical clarity, honouring the past while purifying it for the use of the present. The poet's ultimate goal, for Dryden, is systematic public enlightenment, and he relies upon the affective method of tragedy to achieve the 'unacknowledged legislation' of the English soul.

## III. Epilogue: An Idea for Universal History

The Patient Chief, who lab'ring long, arrived
On Britain's Shore and brought with fav'ring Gods
Arts Arms and Honour to her Ancient Sons:
Daughter of Memory! from elder time
Recall; and me, with Britain Glory fired,
Me, far from meaner Care or meaner song,
Snatch to thy Holy Hill of Spotless Bay,
My Country's Poet, to record her Fame.

Alexander Pope, Fragment of *Brutus, An Epic*[28]

This eight-line invocation is all that exists in verse of Alexander Pope's perpetually deferred and ultimately abandoned epic poem *Brutus.* In these blank verse lines he seeks a mantle that would lay uneasily upon his shoulders later in life – the role of his 'Country's Poet.' But the poet's self-regard, whatever his ambition, is only an epiphenomenon of the epic occasion itself, and in these lines the young Pope demonstrates his awareness of the genre's teleological and single-minded view of history and nationhood. Towards the end of his life, the national poet claimed to be perpetually at work on the project and nearly ready to send the manuscript off to his publisher.[29] But this was not the case – the idea never seems to have been versified beyond the fragment above, although the poet's notes are extant in two forms – an extensive outline redacted and included in Owen Ruffhead's 1769 *Life of Alexander Pope* and a four-page manuscript of notes and jottings held by the British Library.[30]

Ruffhead's sketch gives a bit of the flavour of Pope's thesis, as, for example, when Brutus rejects settling in the Canary Islands: such a new Troy would be a 'narrow and selfish proposition, as incompatible with his generous plan of extending benevolence, by instructing and polishing uncultivated minds. He despises the mean thought of providing for the happiness of themselves alone.'[31] Here, Pope's Brutus echoes the philosophical thesis of the *Essay on Man,* the predecessor work in the poet's planned magnum opus of ethical poetry. Brutus rejects insular hedonism in favour of generosity and benevolence that will radiate outward, supplying the national epic with a universally normative dimension. Here, for Pope, nationalist and cosmopolitan sentiments are integrated without contradiction, just as they are towards the end of the *Essay on Man,* in the well-known description of radiant humanitarian love:

God loves from Whole to Parts: but human soul
Must rise from Individual to the Whole.
Self-love but serves the virtuous mind to wake,
As the small pebble stirs the peaceful lake;
The centre mov'd, a circle strait succeeds,
Another still, and still another spreads;
Friend, parent, neighbour, first it will embrace,
His country next, and next all human race. (4:361–8)[32]

At its best, the sketch of Pope's epic describes a belated treatment of the diasporic myth of national origins. In the radiant benevolence and humane generosity of its eponymous subject, the unrealized poem *Brutus* seems devoted to integrating the ethnic providentialism at the core of epic into a much broader philosophical idea of world history from a cosmopolitan point of view. Pope's *Brutus* is in many ways the epic most descriptive of the 'Age of Projects' – optimistic but unfulfilled, forward-looking but bound to the past, benevolent but finally inchoate.

## NOTES

I would like to acknowledge debts of gratitude to Max Novak for the chance to present this material at the William Andrews Clark Library; to the audience of The Age of Projects Conference for comments and queries, especially Felicity Nussbaum, Kirstie McClure, and Jayne Lewis; and to Elizabeth Dillon, Amy Hungerford, and Pericles Lewis for their comments on earlier versions of this essay. A version of this essay appears in my book *Lives of Equity: Literature and the Origins of Law in Later Stuart England* (Ithaca: Cornell University Press, 2008). This essay appears with permission of that Press.

1 Samuel Johnson, 'Life of Pope,' in *Samuel Johnson*, ed. John Butt (New York: Oxford University Press, 1988), 433.
2 Richard Helgerson, *Forms of Nationhood: The Elizabethan Writing of England* (Chicago: University of Chicago Press, 1992).
3 Edmund Spenser, *The Faerie Queene* (London, 1590), book 3, canto 3, stanza 22.
4 Svetlana Boym, *The Future of Nostalgia* (New York: Basic Books, 2001), 15.
5 David Quint, *Epic and Empire: Politics and Form from Virgil to Milton* (Princeton: Princeton University Press, 1993).
6 Heather James, *Shakespeare's Troy* (New York: Cambridge University Press. 1997), 87.

7 John Milton, *The History of Britain* (1670), ed. Graham Parry (New York: AMS Press 1996), 1, 2. See also Nicholas von Maltzahn, *Milton's History of Britain: Republican Historiography in the English Revolution* (New York: Oxford Univ. Press 1991), 104.

8 John Wallace, '"Examples are Best Precepts": Readers and Meanings in Seventeenth-Century Poetry.' *Critical Inquiry* 1 (1973).

9 Michael Dobson, *The Making of the National Poet* (New York: Oxford University Press, 1992); see also Philip Harth, *Pen for a Party: Dryden's Tory Propaganda in its Political Context* (Princeton; Princeton University Press, 1993).

10 For discussion of the political 'parallel,' see, for example, Susan Staves, *Players' Scepters: Fictions of Authority in the Restoration* (Lincoln: University of Nebraska Press, 1979).

11 John Dryden, 'Preface' to *Troilus and Cressida, or Truth Found Too Late,* in *The Dramatic Works of John Dryden,* ed. Montague Summers 6 vols. (New York: Gordian, 1968), 5:12.

12 Joseph Roach, *Cities of the Dead: Circum-Atlantic Performance* (New York: Columbia University Press, 1996), 44–8.

13 R. Duke, 'To Mr Dryden on his play, called Truth Found Too Late,' in Dryden, *Dramatic Works,* 5:30.

14 'Prologue,' *Troilus and Cressida.* In *Dramatic Works,* 5:29. All subsequent references to Dryden's play will be noted parenthetically in the text.

15 'The Grounds of Criticism in Tragedy,' in *Dramatic Works,* 5:27.

16 James, *Shakespeare's Troy,* 117.

17 For studies of the barbarous English national character see Margaret Trabue Hodgen, *Early Anthropology in the Sixteenth and Seventeenth Century* (New York: Harcourt Brace, 1962); Richard Helgerson, *Forms of Nationhood: The Elizabethan Writing of England* (Chicago: University of Chicago Press, 1992); Mary Floyd-Wilson, *English Ethnicity on the Early Modern Stage* (New York: Cambridge University Press, 2002); and my article 'A Degenerate Race: English Barbarism in Behn's Oroonoko and The Widow Ranter,' *ELH* 69:3 (Fall 2002).

18 I have explored this thesis in 'A Degenerate Race.' For an additional reading of *Oroonoko* that locates it in an occasional and polemical context, see Richard Kroll, 'Tales of Love and Gallantry: The Politics of *Oroonoko*,' *Huntington Library Quarterly* 67 (2004), 573–606.

19 John Dryden, *Absalom and Achitophel* in *The Poems and Fables of John Dryden,* ed., James Kinsley (London: Oxford University Press, 1967), 191.

20 Giorgio Agamben, *Homo Sacer: Sovereign Power and Bare Life,* trans. Daniel Heller-Roazen (Stanford: Stanford University Press, 1997), 123.

21 Ibid., 123–5.

22 Dryden, *Grounds of Criticism in Tragedy,* 16.

23 There is a striking contrast with the austere Shakespearean precedent: see William Shakespeare, *Troilus and Cressida* (London, 1601), 4.2:59–71.

24 For discussion of the controversy surrounding this scene, see Eugene Waith, *Ideas of Greatness: Heroic Drama in England* (New York: Routledge, 1971).

25 Dryden, *A Defence of an Essay of Dramatick Poesy*, prefixed to the 1667 publication of *The Indian Emperour*, in *The California Edition of the Works of John Dryden*, ed. E.N. Hooker, H.T. Swedenberg, et al., 20 vols. (Berkeley: University of California Press, 1956–2000), 9:12. See also his later theory of tragedy in the unpublished *Heads of an Answer to Rymer* (1677). *California Dryden*, ed. Samuel Holt Monk (Berkeley: University of California Press, 1971), 17:186.

26 William Davenant, 'Preface' to *Gondibert* (London, 1650).

27 See Kathy Eden, *Poetic and Legal Fictions in the Aristotelian Tradition* (Princeton: Princeton University Press, 1988); Wesley Trimpi, *Muses of One-Mind: The Literary Analysis of Experience and Its Continuity* (Princeton: Princeton University Press, 1983); and Martha Nussbaum, 'Equity and Mercy,' *Philosophy and Public Affairs* 22 (1993).

28 In *Minor Poems*, ed. Norman Ault and John Butt, *The Twickenham Edition of the Poems of Alexander Pope*, ed. John Butt, 11 vols. (London: Methuen, 1939–69), 6:404.

29 For a description of Pope's comments on the Brutus epic, see Miriam Leranbaum's *Alexander Pope's 'Opus magnum' 1729–1744* (New York: Oxford University Press, 1977).

30 BL, Egerton MS 1950.

31 Owen Ruffhead, *The Life of Alexander Pope* (London, 1769), 414.

32 In Alexander Pope, *An Essay on Man*, ed Maynard Mack, *Twickenham Poems of Pope*, 3:163–4, ll. 361–8.

chapter five

# Canon versus Survival in 'Ancient Music' of the Eighteenth Century

WILLIAM WEBER

I have been involved for some time in writing about patterns by which old music was performed in the eighteenth and nineteenth centuries. In my book *The Rise of Musical Classics in Eighteenth-Century England* (1992), I argued that that country was the first to establish a performing repertory one can call 'canonic.'[1] By the 1790s, works of some age appeared on concert programs far more often in the British Isles than anywhere else in Europe, and a set of values became accepted by which such works might be revered. An identifiable repertory of works from the late sixteenth century to the middle of the eighteenth century, from Thomas Tallis to George Frederick Handel, emerged that constituted what is usually called a canon of great works. The canon emerged within a context of post-revolutionary political crisis and social change, forces that dislodged long-standing traditions and altered musical culture fundamentally. The proponents of ancient music, most of them high-church Tories, defined it as a moral cause to be defended against the bad taste found in songs written for the home or operas done at the King's Theatre. The English, indeed Britons as a whole, established their musical life upon a principle found nowhere else in the Western world at the time.

But during the last few years, I began to have doubts about the ways I have used the term 'canon.' While I made no great claims for how consistent it was or how firm an authority it claimed, I sensed more variety within it than I had before, and recognized that it did not last very long. Music historians had in the meantime come to talk about canon a lot more, and I found that the terms they used were disturbing because they usually took for granted that 'The Canon' was born at some point and still exists today. I found that kind of thinking did not recognize the complexities of what was going on in the performance of 'ancient music' around 1790. The result has been for me to rethink the concept of canon and look for alternatives to it.

First, let us scan through the main stages by which repertories of old music evolved in England. Around 1700, the term 'ancient music' became widely used. Taken to mean music composed before around 1625, it was the first term established to denote a performing repertory of old works. In 1726, an Academy of Antient Music began at the Crown & Anchor public house in London, a club made up chiefly of high-ranking church musicians, whose repertory usually contained a piece composed by Giovanni Palestrina in the late sixteenth century. A program of the Academy from 1746 is found in appendix 2, but since it was a highly specialized type of event, I have also included a pair of programs typical of what was done in concerts at the time, all to be discussed below. But then, in 1776, a public series began called the Concert of *Antient* Music – almost always with the anachronistic 't' – that redefined the term 'antient music' much more broadly to mean anything more than twenty years old, and focused chiefly upon the music of George Frederick Handel. This was a very different sort of canonic repertory, one that commanded a much broader public, and starting in 1785 drew the patronage of the king. Still, while ancient music became prominent within English concerts, it did not yet hold a central authority within musical taste.

Taste kept changing quickly, for at the turn of the nineteenth century there began a powerful movement privileging the music of Haydn, Mozart, Beethoven, and a variety of other composers that spread all across the Western musical world. It was called 'classical music' by around 1830. English performing traditions influenced the new movement significantly, chiefly through the oratorios of Handel and the *concerti grossi* of Arcangelo Corelli, but the repertory of ancient music was largely superseded by that of classical music. One sees an example of what emerged in the program of a concert by the Oxford Musical Society in 1832 in appendix 5.

The great variety in content and format found in these five programs frames the issue in this essay. The pervasive presence of Handel's music, and the recurrence of different pieces by John Travers, a learned organist of the time, indicates something of what one might call a canon. But looking at the profusion of composers and works on the programs demonstrates how much tastes varied from concert to concert, and how quickly they changed over time. What then is to be done with the concept of canon within this context?

I particularly welcome the chance to look back at the subject. The proponents of ancient music defined their cause in moral terms, as a means by which to raise the level of musical taste in England at the time. They looked with horror at what the growing music-publishing business was doing in publishing songs written to risqué texts for use in Covent Garden or Drury Lane and then performed in homes. They also had serious doubts about the introduction of Italian opera to London, since that threatened the future of English composers in the musical theatre. In 1711 Arthur Bedford, one of Jeremy Collier's colleagues in the campaign against the theatres, published a book called *The Great Abuse of Musick* that set up music of the sixteenth century as a model by which to bring taste back to former high standards. John Hawkins's *General History of the Science of Practice of Music*, published in 1776, had a similar agenda. If I might borrow from Professor Novak's introduction, we might designate each book a *project* calling for the *retrieval* of music of the past to *improve* musical life in the present.

Yet, in looking at the subject now, I am interested less by projects such as Bedford's than I am by the simple fact that certain pieces of music survived so long. As we shall see, to perform a work over a hundred years old was an extraordinary thing to do at this time. Moreover, the great majority of the old works still in use were not 'retrieved' from the past, for they instead had been preserved in some area of musical memory. A set of quiet traditions had persisted by which old works had survived in one fashion or another, even though little was said about them in print. We shall see that such practices existed to a special extent in England, but can be found in France and Germany as well. Opportunists – Bedford, the leaders of the Academy, and the king himself – took advantage of these traditions and thereby sent them in new directions. If we are to understand how old works became central to musical taste by around 1850, we have to look into the practices and mentalities that maintained them.

It is fruitful heuristically to turn to the idea of *Nachleben*, usually translated as 'survival,' that has been voiced by the art historian Aby Warburg and his interpreters. Early in the twentieth century, concern arose that the concept of the Renaissance was giving art history a rigid teleological definition. Warburg initiated an alternative philosophy that was continued by Edgar Wind and Gertrude Bing, also art historians. They were deeply influenced by anthropology and developed a methodology for analysing not only traces of the past within art works but also the traditions by which they were perceived and revered on a long-term basis.

These ideas have been taken up recently by the Frenchman Georges Didi-Huberman and other French scholars, some in the field of history.[2] Didi Huberman's book *L'image survivante* spells out the concept of history that has emerged within this school. Instead of seeing a style be born, rise to greatness, decline, and die, these authors propose a multi-layered survival of elements derived from different periods, antiquity most important of all, that exist as 'latencies' or 'living fossils' each within works of art and their interpretation. As Didi-Huberman interprets it, 'Survival entails a complex set of operations in which forgetting, the transformation of sense, involuntary memory, and unexpected rediscovery work in unison.'[3]

By this reckoning, time is by nature impure. It is a collection of residues that are essentially anachronistic but participate within the present as memory, indeed as a set of spectres, *fantômes*, that influence perception in ways not easily defined. Didi-Huberman sums it up: '*Nachleben* is impure in much the way *Leben* itself is. Both are messy, cluttered, muddled, various, haphazard, retentive, protean, liquid, oceanic in scope and complexity, impervious to analytical organization.'[4]

This notion of survival can be helpful, if for nothing else than to serve as counterpoise to the rigid canonic assumptions found in musical culture since the middle of the nineteenth century. Today we always think in terms of classics when we think about music; it is hard for us to imagine thinking musically without this overdetermined set of aesthetic notions. Work on the problem of canons in music has generally looked back to what evolved during the nineteenth century from the perspective of what remained within canonic repertories. A danger in thus looking backwards, as I did in my own book to a certain extent, is that one sees only what purportedly led up to the present rather than what existed as a whole in the earlier culture. Another danger is that one can easily take Whiggish delight in deriding the casual social atmosphere perceived to be going on in the opera of the eighteenth century, as has become popular to do recently.[5]

The Warburgian conceptual vocabulary can also help us understand how time and memory functioned within musical life during the seventeenth and eighteenth centuries. The concept of temporal impurity helps us come to terms with the lack of written-out codes of historical or aesthetic nature until at least the end of the eighteenth century, and in many respects not until 1850. We find ourselves very much at sea without such indicators by which to think about music history. For that reason, we need to set aside our presumptions about classics, as Warburg and his followers did in reassessing the concept of renaissance. In 1934 Edgar Wind proposed what he called 'a definite method of approach' by which to study the survival of antiquity. In introducing his *Bibliography of the Survival of the Classics*, Wind stated:

> When we speak of 'survival of the classics,' we mean that the symbols created by the ancients continued to assert their power upon subsequent generations; – but what do we mean by the word 'continue'? Is their significance constantly retained? Or is it not rather forgotten at times, regained and transformed at others? And what are the conditions, what are the effects of 'forgetting' and 'remembering'?[6]

Wind's agenda can guide us in deconstructing our assumptions about music's past and in defining how pieces moved from latent to active roles within musical life, or indeed the reverse, either coming or going within concert programming. The very term 'revival' takes on a new meaning once we make use of his approach: we see that performance of a work after a long period of disuse emerged out of a complex set of traditions and memories.

I would, however, qualify the approaches stated above in two regards. First, what seems haphazard in the survival of old music on one plane, indeed in one period, seems coherent on another, since cultural practices articulated the study or performance of old works. We shall see how that was done in contrasting ways within academic and public contexts. People had a clear enough sense of how spectres of old works related to new ones because opera and concert programs followed conventions that gave them meaning. Second, the survival of music often led to the attribution of cultural authority to it. While that did not by any means involve participation in a formal canon, it did endow a work's memory or its performance with cultural power that could be manipulated in different ways. Indeed, attribution of authority to music often carries with it a claim to authority by the persons doing it. In such a fashion did an empowered musical intelligentsia emerge in close relationship with the musical classics.

The most basic principle within musical culture before around 1800 was the inherent superiority that was perceived for new music over old. The French referred to this principle as 'les progrès de la musique,' that is, the process by which new works are assumed to supplant the old in the 'progress' of time. Note that this idea stood entirely separate from the Enlightenment notion of progress; it denoted a truly ancient mentality of musical culture. Rhetorically, when a work had aged, it customarily received derision for being anachronistic, but at the same time it might also be afforded a certain respect for its crafting. Such respect indicated that a piece of music was entering into the residue of the musical past that in some fashion remained latent within memory. It is important to recognize how rapidly time passed within musical culture; normally within thirty or forty years a style had become anachronistic, and the shape of programs from the earlier time would seem muddled or haphazard to the new generation. Admittedly, by the 1760s some commentators – Charles Avison most prominently, in his *Essay on Musical Expression* (1752/3) – arrived at a better informed and more nuanced historical sensibility than had existed at the start of the century. But the status, indeed the authority, of older works remained in dispute until 'classical' music became firmly established around 1850.[7]

Let us remember that, compared to the plastic arts, music possessed considerably different traditions with regard to the persistence of old works. First of all, no comprehensible artefacts were extant for the music of either Greek or Roman antiquity. Therefore, music lacked the most fundamental basis for an evolving canon of great works. Second, new music took primacy over the old because many works were composed to celebrate events, reducing their value considerably once the event occurred. That the performer was often also composer likewise helped limit the retention of pieces within repertories. Finally, little interaction went on between the various forms of learning within musical composition, philosophy of music, and the art of letters. Few amateurs understood much about learned composition, chiefly polyphonic works for sacred or academic use, even if they might enjoy hearing the music. By the same token, the philosophy and science concerned with music bore little relationship to musical practice, to either composition or performance, save in regard to matters of tuning. Music therefore lacked the intellectual tools by which to form canonic notions such as evolved around Elizabethan poets or the Great Masters during the sixteenth and early seventeenth centuries.

Musical culture nevertheless possessed a complex set of relationships with the past that were defined more by practices than by formal ideas, aesthetics least of all. Use of the term 'ancient' to denote music of the sixteenth and early seventeenth centuries suggests that music was in the forefront in redefining this term away from its traditionally strict association with Greek and Roman antiquity. To some people it must have seemed odd – or, to recall Didi-Huberman's vocabulary, 'muddled' – to call anything that recent 'ancient.' But no matter: traditions within musical learning retained an acute memory of earlier styles and musical sensibilities, if filtered and transformed by the passage of time. The ambiguous nature of how old music was known and the great variety of patterns by which that happened makes the term 'survival' useful for its neutrality in regard to historical greatness. From this starting point, we can reconstruct how older works persisted and how processes of change shifted them from latent to active roles within musical life.

Music can remain in memory in a great variety of ways. Think of a spectrum at one end of which one hears a piece exactly as it was written, and at the other end a new work that makes reference to an old style. Remember what happens today: many people learn the music of Handel or Beethoven chiefly from hearing references to their styles within a movie or background music. Between the two extremes stands the performance of an old work updated with new instruments, tempos, or even harmonies, as was conventionally done until the principle of authentic performance was accepted late in the twentieth century.

Musical memory operated differently within academic and public contexts. Academic knowledge of old music was rooted in the church, where by long tradition music had been taught in its most learned form. Sacred music tended to be more conservative in style than secular music, partly because it was intimately linked with musical pedagogy. With the rise of the Baroque style around 1600 came a practice called the *stile antico*, a composing style of an academic nature that combined techniques from Renaissance polyphony with aspects of the new monodic style. Admission to the famed Accademica Filharmonica of Bologna required composing such a work; Mozart, for example, demonstrated his learning in such a fashion. It seems doubtful that many of its members sang the music of Giovanni Palestrina; instead, they passed on memory of antique styles in works designed strictly to be studied or performed within the academy. But they also would use an old technique in writing a work in modern style in a place where it served a particular musical or theological purpose.

Public memory of old music was less extensive than the academic, but no less acute in its sensitivity. The main context where lay people would hear old music was in churches where a particular work had remained in use to be sung regularly on a feast day, serving as part of local tradition for that celebration. In some cases foundations were set up to support the singing of a particular mass in honour of a deceased person. For example, Francesco Cavalli, the renowned opera composer of the mid-seventeenth century, left an endowment for a requiem mass he had composed to be sung twice annually in his memory.[8] Works that served as 'residue' of earlier stylistic epochs performed an important function in defining a long musical memory for the general public. Even though such musical works were perceived in terms of ecclesiastical tradition rather than any musical canon, they nonetheless acquired a certain authority by virtue of their prominence within feast days.

The rise of public concerts that occurred in the late seventeenth century brought a major new factor to the shaping of musical memory: the need for performing repertory. As public performances became more and more numerous as the eighteenth century progressed, old works sometimes remained in use when an opera company or a concert series could not easily produce enough works to fill its schedule. The public arena thereby became as important as the academic for the shaping of musical memory. The very nature of musical learning underwent fundamental reshaping in the process, chiefly to accommodate the active study and performance of old works.

The persistence of old works due to the need for repertory occurred most noticeably in Paris, indeed in the *Opéra.* That theatre remained open much longer than anywhere else, all but two weeks of the year, and for that reason works written by Jean-Baptiste Lully for the court of Louis XIV were brought back from his death in 1687 to the final such production in 1779.[9] Something similar happened in the court of Frederick the Great: works by Carl Heinrich Graun and Johann Adolf Hasse stayed in use from the 1740s to the king's death in 1786. John Mangum has demonstrated how much their works dominated the repertory in Berlin, owing both to Frederick's own taste and to financial limitations upon the court in mounting new productions after the Seven Years War.[10] The highly prominent concerts in Leipzig's Gewandaus likewise featured excerpts from these operas until the late 1780s, by which time their style had grown far out of date.

England, however, acquired a far longer and deeper musical memory. The earliest date for works that remained in later performances was around 1550, before which time music was notated in ways so different that few could easily understand it. The rich compositional activity at the Chapel Royal during the late sixteenth century was followed by a quiet but unbroken continuity in performance in a few of the more learned cathedrals and chapels. In fact, in 1641, a volume of anthems and services, all by dead composers, was published by John Barnard, a minor canon of St Paul's Cathedral.[11] During the Restoration, increasing attention was paid to this music as churches returned to it in their haste to resume sung services. A movement grew up around this repertory; between approximately 1690 and 1710, Henry Aldrich, rector of Christchurch, Oxford, held meetings in his rooms where both secular and sacred music of the sixteenth century were performed. The Academy of Antient Music then gave ancient music a more formal and public setting, its members including William Hogarth, remarkably enough.[12]

If we return to the concert programs in the appendices, we can see the complex ways by which old works came and went between musical memory and public performance. Appendix 1 illustrates how short a timespan in the age of works was found typically at concerts held around 1700. The celebration of St Cecilia's Day, given by the Edinburgh Music Society in 1695, offered ten works, all by living composers, the oldest composer being only forty-three years old. This was a club where thirty lay and professional players performed together weekly, following a homogeneous instrumental repertory. Membership was limited to men; women attended such public events as were held. The offering of only one vocal number suggests the learned nature of the institution; a concert oriented towards the general public was expected to have as many vocal as instrumental pieces. At the same time, however, events of learned amateurs almost always included a meal, drinks, smoking, and perhaps the singing of theatre songs, some of them with risqué lyrics. In our day, that kind of sociability is not usually associated with learned institutions. Yet, at the same time, the songs they sang tended to date back much farther than the instrumental pieces, since the melodies were often reused for numerous generations.

A program given by the Oxford Musical Society in 1765 (appendix 3) illustrates the rise of practices by which quite old works were now performed on a regular basis. This was a similar kind of institution mingling lay and professional players, but by this time such a society was focused more upon public events than the Edinburgh club, offering greater variety than what musically learned gentlemen might choose to perform among themselves.[13] This program is chosen as a parallel to the one in Edinburgh, however, in that it has less vocal music than had become the norm by 1765. Two out of the ten works performed were by dead composers, typical for public events of the time. Here we see a work by Corelli and one by Handel, each placed in the middle of each part, suggesting a principle of balance rather than one of historical chronology. In such a fashion did practices of programming lend coherence to what might seem haphazard. Written in styles by then quite out of use, the two pieces must have appeared like musical 'spectres,' evoking what Wild called 'symptoms' of memory. By this time playing a concerto grosso by Corelli was conventional at a music society, partly because his music was widely employed in teaching composition in Britain, even though this was not done anywhere else in Europe. Moreover, like the Corelli, Handel's overture to the opera *Pastor Fido* was still performed partly because it was easier for amateurs to do than the highly virtuosic music of composers such as Vivaldi and Geminiani.

After looking at these two programs we can appreciate how extraordinary the one performed at the Academy of Antient Music in 1746 (appendix 2) was in its time. Here we find an extremely specialized area of musical activity, high musical learning reaching out in complicated paths to Italy, Spain, Portugal, and Germany. Five of the pieces were written in the late sixteenth or early seventeenth century, and four between 1695 and 1746. The recent works are all written either in conservative or wholly anachronistic styles. The Palestrina could probably be traced back to the Sistine Chapel, the one place where his works and those of other dead composers were performed regularly, defined clearly in canonic terms.[14] Victoria and Lobo were from Spain and Portugal, respectively, the former achieving high positions in Rome, Cologne, and Spain.

The madrigals by Morley and Byrd formed part of a tradition that combined learning and conviviality such as is unusual in musical life today. It is also much tougher to trace the routes by which secular music survived than is the case for sacred works; one has to trust one's instincts. Two clubs, the Madrigal Society begun in 1740 and the Gentleman's Catch Club in 1761, clearly grew out of a tradition of singing in homes and taverns that had been more or less continuous since the sixteenth century.[15] The Travers Canzonet for Three Voices formed part of the same tradition by which musically educated gentlemen, and in some contexts women, would sing highly contrapuntal pieces at table while sipping some port.

Finally, the growing influence of old music in formal, public concerts appears here in the pieces by Purcell and Handel. Excerpts from Purcell's operas remained in use from his death in 1695. Even more important, his setting of the Te Deum was performed as the main showpiece at the many music festivals held in cathedral towns in the early eighteenth century. A tradition ensued with the addition of Handel's settings of the Te Deum and his oratorios to 'ancient' music that continued with the nineteenth-century repertory of classical music.

But even though one can trace a canon from the Academy's program down to the present, such music remained strictly a curiosity in its time, something much too old and too specialized for the general listener. In 1733 a pamphlet spoke of the 'Gropers into Antique Musick, and Hummers of Madrigals,' the works of whose 'venerable President' [Pepusch] were 'dress'd up in Cobwebs, and powdered with Dust.'[16] Indeed, one musicologist has questioned whether Renaissance polyphony was valued very highly even in its own time outside the small world of learned cathedral musicians.[17] The Concert of Antient Music, by contrast, was aimed at a much broader public than the Academy. While the latter was essentially a professional society for high-level singers and a few of their patrons, the Antient Concert was an elite concert series of the kind that became conventional in nineteenth-century classical-music life. Its programs were focused upon the music of Handel, excerpts from his operas as well as the oratorios. Usually one piece from the sixteenth or early seventeenth century appeared at every other concert. Yet we still find a clutter of musical ghosts: a madrigal by Thomas Wilbye, a Kyrie by Leonardo Leo (one of the founders of the early classical style in the 1730s), and an aria by Hasse from the Berlin repertory just mentioned.

The 1832 program (appendix 5), with its twenty-four pieces, might seem bizarre, indeed haphazard, today. It was one of the four concerts within the annual festival held by the Oxford Musical Society in June. Almost half of the program was by dead composers, a major change in the balance between ancient and modern. While no Elizabethan work was done here, such pieces did still crop up from time to time in programs of this sort. This program contains four main types of music: ancient music by Handel and Travers; late-eighteenth-century opera; English glees and songs, some quite sentimental; and recent opera, a number by the rising star Giacomo Meyerbeer holding a key role as the penultimate work.

Note the remarkable continuity in performance of music by Travers (c. 1703–58) from the Academy of Antient Music in 1746, the Antient Concert in 1780, and the Oxford institution again in 1832. Travers earned his living as organist in the Chapel Royal, also playing in St Paul's Cathedral; even more important, he was amanuensis to John Pepusch, the president of the Academy and the most prominent learned music teacher in London. Musicians such as Travers served as crucial repositories of musical memory. A professional such as he acquired a knowledge of past styles, musical practices, and repertories, both as a teacher and as an organizer of public musical events. He must have influenced greatly what was done with the memory of older styles, helping make decisions on which pieces would be performed and in what terms musically and socially. A song such as his 'Haste, haste, Nanette,' usually called a glee, linked the musical present with the musical past within British culture, since it was influenced in subtle ways by what such Elizabethan composers as Thomas Wilbye had done a hundred and fifty years before – a very long time in musical terms. Its style was not historicist, designed to refer directly to a past style. The glee was thought to be very much a part of that time, but had relatively little to do with what was being written for the opera or instrumental ensembles at mid-century. Simply writing a secular vocal piece without accompaniment, including some polyphony among the voices, in and of itself linked that style with the Elizabethan era. What seems so different from the modern age is that such music was both fairly learned in style and closely linked with informal sociability. It was often sung over the best port in town, at the meetings of the Gentleman's Catch Club formed by the Earl of Sandwich in 1761.[18]

We today react negatively at seeing the 1832 program, finding it to be a hodge-podge of pieces and historical periods, lacking the clear historicist definition we expect to find in the classical-music concert hall. But its title, a 'Grand Miscellaneous Concert,' had a positive connotation in its time, since it denoted a firmly established set of musical expectations and practices. In technical terms it meant that the concert was not devoted to a single long work, as was done several times a year by local musical societies. On a broader plane, however, it communicated to the public that it was the product of compromises struck between the variety of tastes, genres, and cultural needs that impinged upon the directors of the single such institution found in most places. The musical term 'miscellaneous' arose out of the late medieval practice of copying diverse religious texts together, the culmination in printed form being John Dryden's *Miscellany Poems* of 1684. A critical discourse can be found about the 'miscellaneous' program similar to that regarding the literary genre termed 'miscellany,' led by Swift in *Tale of a Tub* (1704) and Shaftesbury in *Characteristics of Men, Manners and Opinions* (1711).[19] Roger North, for example, declared in the 1720s that what he had heard in concerts 'consisted of broken incoherent parts; now a consort, then a lutenist, then a violin solo, then flutes, then a song, and so piece after piece, the time slipping away, while the masters blundered and swore in shifting places, and one might perceive that they performed ill out of spite to one and another.'[20] He found resolution of these issues brought by 'an absolute Dictator, who may coerce and punish the republican mob of music masters.'[21] Contemporaries therefore saw order in this confusion, particularly among the increasingly diverse periods of history from which pieces now tended to come.

But taste was changing fast in 1832, as we have seen was almost always the case in musical life. Within twenty years the sentimental songs and opera numbers such as we see in the 1832 program moved – one might say were *banished* – from mainstream 'classical music' concerts to 'ballad' concerts of supposedly 'light' music designed for a much broader public. The theatre songs by Thomas Arne and William Shield, thus moved, took on canonic status within ballad concerts, being performed with great respect on a regular basis, all of which formed part of a fundamental change in musical culture. Repertories had expanded to a far greater extent than the principle of miscellany could handle, each in the variety of periods, genres, and tastes. Some parts had to go, and a new form of musical authority had to be built. By 1850 the older pieces that we have seen growing in number in programs had achieved a quite new high status within the more learned region of public musical life. The former so unselfconscious performance of older works underwent a fundamental transformation as it was endowed with canonic meaning and authority. A certain area of musical memory thus took on an active and determining role within musical life such as it had not possessed except within pedagogy and very learned composition.

Despite the inflammatory word 'versus' in the title of this chapter, I find that an accommodation can be reached between survival and canon. Each side has to give a little. On one hand, it is clear that musical practices brought coherence to the clutter by which pieces survived. On the other hand, I find myself speaking less about 'a canon,' even less 'The Canon,' and a lot more about 'canonic' practices and implications. The adjective can save us from falling into teleology.

Thus, we see in what diverse and disconnected ways canonic practices appeared between around 1695 and 1832. What led a cathedral to keep a piece in repertory on a feast day had little to do with what made an opera company bring back an old work on a regular basis, or the Sistine Chapel perform works of sacred polyphony by Palestrina. An incomplete but still highly retentive musical memory stood behind all the different practices, only a few elements of which we can reconstruct. At a few critical points in time, works that remained in memory or in performance coalesced into a repertory that was recognized in its own right. Such a repertory would usually last for a certain period of time, but would disappear under the pressure of social and musical change. The operas in Paris and Berlin each went through such a period when old operas stayed on stage. But even though the repertories ended in both cases, important strands of memory remained that returned within subsequent such movements. In Britain, very different repertories and tastes coalesced, first in the Academy of Antient Music between the 1730s and 1760s and then in the Concert of Antient Music between 1776 and 1848. Each of these institutions and the movements behind them came to an end as a reformulation of music and memory came about. The movement for what was called 'classical music' that arose in the early decades of the nineteenth century has had a longer life-span than earlier repertories, but their contents and the ways by which they were perceived or performed have changed so much that new cultural wholes emerged about every third generation.

We have much to gain from the concept of memory, and the unpredictable ways by which a culture can draw upon it. Francis Haskell and Nicholas Penny spoke similarly in *Taste and the Antique,* their book on classical sculpture collected between 1500 and 1900:

> Sometimes the statues we discuss became famous quickly, at others only after many decades; sometimes fame lasted for centuries, at others for only a few years; sometimes it was universal, at others confined to specific countries; sometimes present-day taste would ... acknowledge the justice of such fame, [but] at others it would be startled by it.[22]

## NOTES

1 *The Rise of Musical Classics in Eighteenth-Century England: A Study in Canon, Ritual and Ideology* (Oxford: Clarendon Press, 1992). See also my 'Consequences of Canon: Institutionalization of Enmity between Contemporary and Classical Music, c. 1910,' *Common Knowledge* 9 (2003), 78–99; 'Canon and the Traditions of Musical Culture,' in *Canon vs. Culture: Reflections on the Current Debate,* ed. Jan Gorak (New York: Garland, 2000), 135–52; 'The History of Musical Canons,' in *Rethinking Music,* ed. Mark Everist and Nicholas Cook (Oxford: Oxford University Press, 1999), 340–59; 'Intellectual Foundations of Musical Canon in Eighteenth-Century England,' *Journal of the American Musicological Society* 47 (1994), 488–520; 'Mentalité, tradition, et origines du canon musical en France et en Angleterre au XVIIIe siècle,' *Annales E.S.C.* 42 (1989), 849–75; 'Intellectual Foundations of Musical Canon in Eighteenth-Century England,' *Journal of the American Musicological Society* 47 (1994), 488–520; '*La musique ancienne* in the Waning of the Ancien Régime,' *Journal of Modern History* 56 (1984), 58-88.

2 Aby Warburg, *The Renewal of Pagan Antiquity: Contributions to the Cultural History of the European Renaissance,* intro. Kurt W. Forster, trans. David Britt (Los Angeles: Getty Research Institute for the History of Art and the Humanities, 1999); Edgar Wind, 'Warburg's Concept of *Kulturwissenschaft* and Its Meaning for Aesthetics,' in *The Eloquence of Symbols: Studies in Humanist Art,* ed. Jaynie Anderson (Oxford: Clarendon Press, 1983), 21–37. For more recent commentary, see Georges Didi-Huberman, 'Artistic Survival: Panofsky vs. Warburg and the Exorcism of Impure Time,' *Common Knowledge* 9 (2003), 273–85 and *L'image Survivante: Histoire de l'art et temps des fantômes selon Aby Warburg* (Paris: Les Éditions de Minuit, 2002); and Carlo Ginzburg, 'De Aby Warburg à E.H. Gombrich: Notes sur un problème de méthode,' trans. C. Paolini, in *Mythes, embleme, traces: Morphologie et histoire* (Paris: Flammarion, 1989), 39–96.

3 Didi-Huberman, 'Artistic Survival,' 275.

4 Ibid., 282.

5 Lydia Goehr, *The Imaginary Museum of Musical Works: An Essay in the Philosophy of Music* (Oxford: Clarendon Press, 1992); James H. Johnson, *Listening in Paris: A Cultural History* (Berkeley: University of California Press, 1995).

6 Edgar Wind, 'Introduction: Theme and Method,' in *A Bibliography on the Survival of the Classics,* vol. 1: Publications of 1931, ed. Hans Meier, Richard Newald, and Edgar Wind, edited by the Warburg Institute (London: Cassell, 1934), viii. See also Wind's 'Warburg's Concept of Kulturwissenschaft and Its Meaning for Aesthetics,' in *The Eloquence of Symbols: Studies in Humanist Art,* ed. Jaynie Anderson, biographical memoir by Hugh Lloyd-Jones (Oxford: Clarendon Press, 1983), 21–36.

7 Howard Irving, *Ancients and Moderns: William Crotch and the Development of Classical Music* (Aldershot, Hants [UK]: Ashgate, 1999).

8 'Cavalli,' in *New Grove Dictionary for Music & Musicians,* 1st ed. (London: Macmillan, 1980), 4:24–34; I am indebted to Michael Talbot on this point. See also Lorenzo Bianconi, *Music in the Seventeenth Century,* trans. David Bryant (Cambridge: Cambridge University Press, 1987), 4–6.

9 See my articles '*La musique ancienne* in the Waning of the Ancien Régime' and 'Lully and the Performance of Old Music in the 18th Century,' in *Congress for the Tricentenial of the Death of J. B. Lully, Heidelberg / St. Germain-en-Laye* (Laaber: Laaber Verlag, 1991). Likewise, see my 'Mentalité, tradition, et origines du canon musical' and 'L'Institution et son public: L'opéra à Paris et à Londres au XVIIIe siècle,' *Annales E.S.C.* 48.6 (1993), 1519–40.

10 John Mangum, 'Apollo and the German Muses: Opera and the Articulation of Class, Politics and Society in Prussia, 1740–1806,' PhD dissertation, University of California, Los Angeles, 2002.

11 *The First Book of Selected Church Musick* (London, 1641).

12 Weber, *Rise of Musical Classics,* 68.

13 Oxford Musical Society programs, Bodleian Library.

14 Jeffrey Dean, 'The Evolution of a Canon at the Papal Chapel: The Importance of Old Music in the Fifteenth and Sixteenth Centuries,' in *Papal Music and Musicians in Late Medieval and Renaissance Rome* (Oxford: Clarendon Press, 1998), 138–66.

15 See the minutes of the two societies, held in the British Library.

16 *Harmony in an Uproar,* incorrectly attributed to John Arbuthnot, in *Miscellaneous Works of John Arbuthnot,* 2 vols. (Glasgow, 1751), 2:34.

17 Jeffrey Dean, 'Listening to Sacred Polyphony c. 1500,' *Early Music* 25 (1997), 611–36.

18 See 'Glee' and 'Catch,' *New Grove Dictionary of Music and Musicians,* 2nd ed., ed. Stanley Sadie and John Tyrell (London: Macmillan, 2000); Michael Hurd, 'Glees, Madrigals, and Partsongs,' in *Music in Britain: The Romantic Age, 1800–1914,* ed. Nicholas Temperley (London: Athlone, 1982), 42–65; and Brian Robins, 'The Catch Club in 18th-Century England,' *Early Music* 28 (2000), 517–29.

19 Elizabeth Pomeroy, *Elizabethan Miscellanies: Their Development and Conventions* (Berkeley: University of California Press, 1973); John Dryden, *Miscellany Poem: Containing a New Translation of Virgills Eclogues, Ovid's Love Elegies, Odes of Horace, and other Authors: with several Original Poems by the Most Eminent Hands* (London: Jacob Tonson, 1684). Other such volumes were called *A Handful of Pleasant Delights, A Small Handful of Fragrant Flowers,* and *A Gorgeous Gallery of Gallant Inventions,* and *The Court of Venus.* I am indebted to Anne Cotterill for these references and her thoughts upon the subject.

20 *Roger North on Music,* ed. John Wilson (London: Novello, 1959), 303.

21 Ibid.

22 Francis Haskell and Nicholas Penny, *Taste and the Antique: The Lure of Classical Sculpture, 1500–1900* (New Haven: Yale University Press, 1981), xiv.

# APPENDIX 1

## Musical Society of Edinburgh
## The Order of the Instrumental Music
## for the Feast of St Cecilia
## 22d November 1695

| | |
|---|---|
| Jeremiah Clerk (1674–1707) | Overture |
| John Barrett (?1676–?1719) | Trumpet Sonata |
| Johann Pepusch (1667–1752) | Violins & flutes |
| Johann Pepusch | Violins & oboes |
| Giovanni Bassani (1650–1716) | Sonata |
| Giovanni Bassani | Songs and motets |
| Arcangello Corelli (1653–1713) | Sonata |
| Lord Alexander Colville (1666–1717) | Sonata |
| Gottfried Finger (1660–1730) | Trumpet Sonata |
| Giuseppe Torelli (1658–1709) | Sonata |
| Unidentified composer | Chaconne |

## APPENDIX 2

Motets Madrigals and Other Pieces Performed at the Academy of Ancient Music April 24, 1746

**Part the First**

| | |
|---|---|
| Giovanni Palestrina (1525–94) | Motet for five voices, 'Angeles Domini' |
| John Travers (1703–58) | Canzonet for Three Voices, 'Old I am, Yet can, I think' |
| Henry Purcell (1659–95) | Fifth Act of the Indian Queen, 'While thus we bow before your Shrine' |
| Duarte Lobo (?1565–1646) | Kyrie eleison for four voices |

**Part the Second**

| | |
|---|---|
| Thomas Morley (1557/8–1602) | Madrigal for four voices, 'Say, Gentle Nymphs' |
| Johann Pepusch (1667–1752) | Magnificat |

**Part the Third**

| | |
|---|---|
| Tomas Luis de Victoria (1548–1611) | Motet for four voices, 'Quam pulchri' |
| William Byrd (1543–1623) | Madrigal for three voices, 'The Eagle's Force subdues each Bird that flies' |
| Handel (1685–1759) | Te Deum [unidentified setting] |
| Anonymous | Non nobis Domine |

## APPENDIX 3

Oxford Musical Society January 7 1765

| | |
|---|---|
| G.F. Handel (1685–1759) | Overture, Pastor Fido |
| Thomas Avison (1709–70) | Concerto grosso |
| Joseph Baildon (1727–74) | Ode to Contentment |
| Carlo Campioni (1720–88) | Trio for violins |
| Johann Wagenseil (1715–77) | Concerto for harpsichord |
| Carl Abel (1723–87) | Overture with Horns |
| J.C. Smith (1712–95) | Song, *The Fairies* (1755), 'Hark how the Hounds' |
| Corelli (1653–1713) | Concerto grosso |
| Domenico Ferrari (1722–80) | Solo for violin |
| John Stanley (1712–86) | Concerto grosso |

# APPENDIX 4

Under the Direction of Lord, Viscount Dudley for the Bishop of Durham.
Monday, March 13th, 1780

## CONCERT OF ANTIENT MUSIC

### Act I

| | |
|---|---|
| 2d Overture | Martini |
| Lady, when I behold | Wilbye |
| Heart, thou Seat of Soft Delight (from Acis and Galatea) | Handel |
| Galatea dry thy Tears (from Acis and Galatea) | Handel |
| 2d Grand Concerto | Handel |
| Se tutti mali miei | Hasse |
| We praise Thee, O God | Handel |

### Act II

| | |
|---|---|
| Haste, my Nannette | Travers |
| Overture in Ariadne | Handel |
| Kyrie eleeson | Leo |

# APPENDIX 5

Oxford Musical Festival, Thursday June 16, 1832
Directed by the Chevalier Sigismund von Neukomm of Vienna

**Part I**

| | | |
|---|---|---|
| 1. | First four Movements of the *Dettingen Te Deum* | Handel (1685–1759) |
| 2. | Aria, 'The Fall of Zion' [from opera with new sacred text] | Paisiello (1740–1816) |
| 3. | Glee, 'Blest pair of sirens' | J.S. Smith (1750–1826) |
| 4. | Recitative/Aria, 'Deeper and deeper still,' *Jephtha* | Handel |
| 5. | Recitative/Aria, 'Farewell, ye limpid streams,' *Jephtha* | Handel |
| 6. | Aria, 'Ho perduto,' Paisiello (1740–1816) replaced by: Aria, 'Ituoi fragmenti palpiti,' *Niobe* (1826) | Pacini (1796–1867) |
| 7. | Quintetto, 'Sento o do,' *Cosi fan tutte* | Mozart (1756–91) |
| 8. | Concerto for Cello, composer performing | Robert Lindley (1766–1855) |
| 9. | Song, 'Let the bright seraphim,' *Samson* | Handel |
| 10. | Song and Chorus, 'Let their celestial concerts all unite,' *Samson* | Handel |
| 11. | Song, 'The Sea,' composer directing | Neukomm (1778–1858) |
| 12. | Duetto, 'Io di tutto' | Mosca (1772–1839) |
| 13. | Chorus, 'To Rome's immortal hero,' new words, from *La Clemenza di Tito* | Mozart |

**Part II**

| | | |
|---|---|---|
| 1. | Overture, *Anacreon* | Cherubini (1760–1842) |
| 2. | Glee, 'Deli dove senza mi' | Dr. Benjamin Cooke (1734–93) |
| 3. | Recitative/Aria, 'Ombra adorata,' *Romeo e Giulietta* (1796) | Zingarelli (1752–1837) |
| 4. | Duet, 'I, my dear, was born today,' by John Travers (c1703–58), replaced by Song, 'May morn,' sung by Mr. Braham | Neukomm |

| | |
|---|---|
| 5. Song, 'Lo! hear the gentle lark,' with flute obbligato | Bishop (1786–1855) |
| 6. Duetto, 'All' idea,' *Il Barbiere di Siviglia* (1816) | Rossini (1792–1868) |
| 7. Song, 'Let me careless,' comic opera *Carnival of Venice* (1781) | Thomas Linley, sen. (1733–95) |
| 8. Song and Chorus, 'Haste thee, nymph,' *L'Allegro, il Penseroso, ed il Moderato* | Handel |
| 9. Aria, 'Batti, batti,' *Don Giovanni*, with cello obbligato | Mozart |
| 10. Song, 'There was once a golden time,' by Mrs. Hill Wilson, replaced by Song, 'The Rover's Bride' | George Lee (1802–51) |
| 11. Scena, 'I Violini' (In which Signor de Begnis will perform the part of 'Il Fanatico,' at the rehearsal of his New Overture.) | Sacchini (1730–86) |
| 12. Duetto, 'Ravvisa,' *Il Crociato in Egitto* (1824) | Meyerbeer (1791–1864) |
| 13. Grand Chorus, 'The Arm of the Lord' | Handel |

# PART II

# IMPROVING THE PRESENT

chapter six

# A Revolution in Political Economy?

STEVEN C.A. PINCUS

'What greater demonstration can the world require concerning the excellence of our national government, or the particular power and freedom of this city, than the Bank of England,' gushed one later seventeenth-century Englishman. This new institution, this bedrock of the so-called Financial Revolution, was so much to be lauded because it, 'like the Temple of Saturn among the Romans, is esteemed so sacred a repository, that even foreigners think their treasure more safely lodged there than with themselves at home; and this not only done by the subjects of absolute princes, where there can be no room for any public credit, but likewise by the inhabitants of those Commonwealths where alone such banks were hitherto reputed secure.'[1] This encomium of public credit is surprising because it comes not from a new bourgeois capitalist or even from a junto Whig, but from a man called 'the archivist and to some extent the myth maker of English republican theory': John Toland.[2]

Toland's celebration of the Bank, suggesting that such an institution was only possible and desirable after the events of 1688–9 had altered England's political arrangements, raises a number of questions about the relationship between political economy[3] and the Revolution of 1688–9. Was there a revolution in political economy in 1688–9? What was the ideological, as opposed to the narrowly political, background to the establishment of the Bank of England? What was the relationship between classical republicanism and the Financial Revolution? What role did 'Atlantic' or broader European and worldwide connections play in shaping the later Stuart polity?

The most powerful and influential account[4] of the relationship between political economy and the Glorious Revolution in recent years has been advanced by John Pocock, one of the most prominent exponents of the civic republican tradition. For Pocock, commercial society, 'the assertion of the commercial order,' was not 'a crucial issue in English ideological debate' at the time of the Glorious Revolution.[5] Indeed the really revolutionary consequences of 1688–9 – the reorganization of 'military, financial, and political structures in order to achieve effective participation in continental and imperial warfare' – were 'not fully foreseen or desired by those who invited' William to England. It was because William imposed these changes from without that he is to be seen as 'a revolutionary actor in the history of the British monarchy.'[6] The Financial Revolution, then, was the result of foreign imposition rather than domestic ideological debate. It was at best an unintended, at worst an undesired, consequence of the events of 1688–9. Unsurprisingly, then, the real ideological energy was expended in responding to the Financial Revolution. While political and social thought took off in new directions as a consequence of the structural changes of the 1690s, these new directions developed along the lines predetermined by James Harrington and his followers. 'The structure of ideas' through which 'Englishmen perceived the Financial Revolution when it came upon them' were 'laid down before 1694,' Pocock insists.[7] The 'changes in perception' necessitated by the Financial Revolution were understood 'through the development of a neo-Machiavellian, as well as neo-Harringtonian, style in the theory of political economy.'[8] Neo-Machiavellian political economy was the vehicle for 'a criticism of modernity.'[9] While Pocock is at pains to emphasize the dialectical nature of the response to the Financial Revolution, he nevertheless points out that neo-Harringtonian social thought was 'the dominant mode of Augustan political thought' and continued to be 'dominant over the intellectual scene of the eighteenth century.'[10] Indeed, in Pocock's view, whatever the differences between the defenders of the Financial Revolution and its critics, 'all Augustan analysts of political economy accept the interdependence of land, trade and credit'; 'both factions share not only the same reading of the economic facts, but the same underlying value system, in which the only material foundation for civic virtue and moral personality is taken to be independence and real property.'[11] Fixed not mobile wealth, land not manufactures, was the only allowable basis for politics in the Augustan mind. Those few historians – and they are very few – who have ventured into the terrain of political economic thought in the 1690s have largely echoed Pocock's conclusions.

The only ideological alternative offered in the scholarly literature to neo-Machiavellian political economy, to country ideology, to neo-

Romanism, is an uncaring and radically individualistic utilitarianism. Where Pocock sees the ideological dominance of country ideology, C.B. Macpherson has described 'the emergence of the principles which were to become basic to liberal democracy.' These principles, which Macpherson describes as a 'well-built utilitarian structure,' were essentially possessive individualism: the 'conception of the individual as essentially the proprietor of his own person or capacities, owing nothing to society for them. The individual was seen neither as a moral whole, nor as part of a larger social whole, but as an owner of himself.' Possessive individualists, then, care for nothing besides the protection of their own property; they have no moral conception of the common good that goes beyond the aggregation of the summation of individual enjoyment of property. It is this proto-utilitarianism that Macpherson sees enshrined in the Revolution of 1688–9.[12]

Against these views I will suggest that a revolution in political economy was very much part and parcel of the Revolution of 1688–9. Just as we now know that a war against Louis XIV was central to the ideological program of the revolutionaries of 1688–9,[13] so also dissatisfaction with James II's political economy and his imperial policies alienated a significant segment of the English population. There was a heated and ideologically sharp debate that began before the Revolution and continued long after it. However, that debate, for which 1688–9 was a turning point, was not centrally about virtue and luxury – those terms do play a role – but about whether England should be a manufacturing or an agrarian society, whether England should base itself on mobile or real property. Historians have missed the theoretical richness and subtlety of this debate because they have chosen to focus on 'imaginative literature'[14] – in Pocock's formulation – or 'the ground of intellectual history' – in David Armitage's[15] – rather than on the mass of ephemeral writings, company records, mercantile and diplomatic correspondence, and on the standing-army controversy and the British context rather than on the much larger and longer debates over the East India Company and the creation of the Bank of England. Historians and literary critics have misconstrued the debate because they have insisted on reading only a few canonical texts divorced from the material, social, and political contexts in which they were produced. Indeed, by focusing so heavily on the reactions to the Financial Revolution and on the cultural productions that make up country ideology, historians and literary critics have made it difficult to understand why these dramatic institutional changes took place at all. By recovering the central terms in the debate over political economy – and by giving equal weight to the proponents of the Financial Revolution as well as its enemies – I hope to show that the interpretative field was no more dominated by neo-Machiavellian or neo-Harringtonian economics than it was by proponents of possessive individualism. Instead, there was a fierce debate between a land-based Tory political economy and labour-centred Whig one.

## I

'The past Ages have never come up to the degree of Projecting and Inventing, as it refers to matters of Negoce, and Methods of Civil Polity, which we see this Age arriv'd to,' observed Daniel Defoe in his *Essay Upon Projects.*[16] Defoe made it clear that by projects he meant endeavours 'of public Advantage, as they tend to Improvement of Trade, and Employment of the Poor, and the Circulation and Increase of the public Stock of the Kingdom.'[17] These were not mere inventions. 'The true definition of a project, according to modern acceptation, is,' Defoe explained, 'a vast Undertaking, too big to be managed, and therefore likely enough to come to nothing.'[18] Projects were political economic schemes to promote the common good only achievable in the modern age. It was for this reason that Defoe could 'trace the Original of the Projecting Humour that now reigns, no farther back than the Year 1680, dating its Birth as a Monster then, though by times it had indeed something of life in the time of the late Civil War.'[19]

Defoe, it turns out, was a remarkably perceptive historian. Since at least the 1650s – just after the civil war – English radicals had embraced the possibilities of commercial society. Marchamont Nedham, Slingsby Bethel, Henry Robinson, and Benjamin Worsley, among others, were all men deeply invested in the radical politics of the Commonwealth and critical in a variety of ways of the Protectorate. All of them defended two propositions. First, they argued that property was primarily a human creation and not a natural endowment. Property was potentially infinite because it was the product of human labour. Manufacturing, not farming the land, was the key to England's political and commercial future. Second, they claimed that a national bank – the first project described by Defoe in his *Essay* – would play a vital and constructive role in promoting national prosperity and in providing a bulwark to national security.[20] Although the bank they hoped for was not established, the economic ideas of these writers were remembered and deeply influential after the Restoration. Many of these same writers, joined by a variety of younger pamphleteers, politicians, and merchants, reiterated their ideas and proposals throughout the 1660s, 1670s, and 1680s.

One of the most eloquent and well-developed defences of commercial political economy was offered in 1685 by Carew Reynell. Although Reynell's party political commitments remain shadowy, he was clearly no friend of absolutism. 'And if his Majesty desires to advance his empire,' Reynell advised, 'it is but granting more privileges to trade, and security to men's persons and properties from arbitrary power and control than his fellow princes; and he shall not fail to draw to him all the hands, hearts and purses of the neighbouring nations.'[21] Reynell's political beliefs were determined by his economic convictions. And the touchstone of his economic thought was that 'England is properly a nation of trade.'[22]

Reynell believed along with almost everyone in the later seventeenth century that 'trade and populousness of a nation are the strength of it.'[23] However, the basis of that strength, of trade and populousness, according to Reynell, was labor not land, manufacturing not raw materials. 'It is the manufacturers of a commodity that is in general sale, that employs people and produces the great profit,' he explained, 'although the original materials are not in the country, as silks for example, the making of which employs abundance of people, and with them brings in other things by exportation.'[24] 'It is manufactures must do the work,' he enthused, 'which will not only increase people, but also trade and advance it. It saves likewise money in our purses by lessening importation, and brings money in by exportation.'[25] Manufacturing set in motion a process that rendered property infinite; trade was no longer a zero-sum game.[26] 'Where abundance of manufacturing people are, they consume and sweep away all country commodities, and the wares of ordinary retail trades, with all sorts of victuals, wearing apparel, and other necessaries, and employ abundance of handicraftsmen, in wooden and iron work for tools, and instruments that belong to their trades, and so maintain and increase abundance of husbandmen, retailers and artificers of all sorts,' Reynell detailed, 'and they again increasing, take up more manufactures, and so they thrive one by another, ad infinitum.'[27] 'Though we are a nation already pretty substantial,' Reynell concluded, 'yet it is easy for us to be ten times richer.'[28]

While Reynell was confident that a massive increase in English wealth and consequently English power was attainable, he did not think it would happen of its own accord. He deemed it essential that 'the confusion of trade [be] taken away' and that 'the mysteries of exchange were more publicly known.'[29] He hoped thereby to generate a 'public spirit' that 'gave countenance to brave actions and industrious men, and minded the business of trade and populacy, as much as we do pleasures and luxury.'[30] More important even than a more commercially informed public, however, was a more commercially inclined state. The government action that Reynell called for was not only an elimination of detrimental laws and customs but also the positive creation of state agencies to advance trade. His program had aspects of both negative and positive liberty. Reynell hoped that 'such laws might be made and contrived for the encouragement of trade and manufactures' of which 'the chief things that promote trade and make it flourish are that it be free, naturalization, populacy, [religious] comprehension, freedom from arrests, certainty of property and freedom from arbitrary power, small customs, all conveniency and advantages for trading people: loans of interest, public places of charity for all wanting and distressed people, and also employments ready for all persons that want it.'[31] Reynell, though he was a friend to manufactures and to banks, was no possessive individualist or proto-utilitarian. 'The happiness and welfare of all people arises by having or acquiring, through some industry or other, such conveniency of livelihood, as may not only keep them from want and poverty, but render them pleasant and sociable to one another,' Reynell elaborated; 'this holds both in private persons and families, and also in bodies politic: that they may best be able to grow and flourish, at least bear up against the malignity of enemies and adverse fortune.'[32]

Indeed, Reynell was careful to distinguish in his treatise between trades that promoted the public good and those that did not. Significantly in the ideological context of the 1680s, he singled out the East India trade as particularly deleterious because 'to the East Indies we carry nothing but ready money, and bring in again nothing worth anything but spices.'[33] For Reynell the goal of trade was to bring in raw materials not readily available at home to be manufactured. Since the East India trade did none of that, but only brought in goods for re-export, it potentially benefited the private merchant but not, on balance, the nation.

Reynell's voice may have been an unusually eloquent one, but he had a plethora of ideological fellow travellers. Many besides Reynell were committed to the notion that labour, not natural endowment, created property, that manufactures, not land, was the key to wealth and power. Richard Blome, an expert on the West Indies, endorsed the view 'which is agreed on all hands,' that it is the 'labour' of the lower orders 'that improves countries, and to encourage them is to promote the real benefit of the public.'[34] The violent francophobe William Carter observed that 'where a nation is not rich in mines of gold and silver, it is not capable of being enriched any other way, than by its manufacture.' This premise led Carter to conclude that it was the government's interest to promote manufactures. 'If it be from our manufactures alone that the riches of this nation comes, and if it be from our manufacture chiefly that our shipping is employed, and our mariners bred, if it be from our trading alone, and from the riches which our trading brings in, that his Majesty's customs are raised, and that our fleets have been hitherto built and maintained, and the dominion of the seas hath been preserved,' Carter reasoned, 'then it is and must be from our manufacture only that our bullion hath been brought in, and that the rents of our nobility and gentry doth depend and are sustained. And therefore it must be granted me, that there is no higher interest in the nation.'[35] John Locke, who was one of the earliest supporters of and investors in the Bank of England, was sure that 'if we rightly estimate things as they come to our use, and cast up the several expenses about them, what in them is purely owing to nature, and what to labour, we shall find that in most of them 99/100 are wholly to be put on the account of labour.' No wonder he was convinced that for states 'the honest industry of mankind' and 'numbers of men are to be preferred to largeness of dominions.'[36]

The notion of a public bank had equally wide support in the 1670s and 1680s. An English national bank was already the subject of hot dispute 'in public coffeehouses' on the eve of the Exclusion Crisis.[37] 'If it shall please God once to raise a Bank in London of six hundred thousand pounds fund and anchorage,' thought Andrew Yarranton, who sensibly looked to the Earl of Anglesey for support, 'out of such a Bank will sprout out many lumber houses and smaller banks to quicken trade ... I could write a whole volume of the advantage it would be to our English trades, the growth and manufactures of our kingdom.'[38] In 1683, recalled the political economic writer Adam Anderson, 'Dr. Hugh Chamberlain, a physician, and one Robert Murray, both great projectors, made a mighty stir with their scheme for a bank for circulating bills of credit on merchandize to be pawned therein, and for lending money to the industrious poor on pawns at six percent interest – but it came to nothing.'[39] Sir John Lowther, who was notoriously critical of James II for his failure to put a halt to Louis XIV's aspirations for universal dominion, described a public national bank in his 'rules for increase of trade.'[40] Even the moderate Tory author of the 1685 election pamphlet *The Mischief of Cabals* advocated the creation of a public bank. He thought only a bank would allow England to sustain a war against Louis XIV – a war that he insisted needed to be fought – and he concluded that a bank would aid in the 'circulation of money' that was 'as necessary to the body politic as that of blood to the natural.'[41]

At the accession of James II, then, there was an understanding of political economy which claimed that political power was dependent not on land but on manufactures. This reading of the economy insisted that wealth was potentially infinite, limited not by the size of the kingdom's possessions but only by the industriousness of the population. Since the key to promoting manufactures, the efficient deployment of labour in creating wealth, was the circulation of money to the productive parts of the economy, many continued to advocate the notion of a public national bank, a notion first espoused by the radicals of the 1650s. While these later seventeenth-century merchants, politicians, and thinkers were very much advocates of commercial society, they were not possessive individualists but promoters of the good of the national community. They were *political* economists, not economists.

## II

James II chose Sir Josiah Child rather than John Locke or Carew Reynell as his economic adviser. He chose to understand property exclusively as land and finite rather than as mobile and infinite. Instead of supporting the creation of a national bank, James II supported the exclusive monopoly privileges of the English East India Company.

Sir Josiah Child, whom Defoe singled out in his preface as an evader of the post-revolutionary land tax,[42] was long one of the most important merchants in the kingdom. He successfully took control of the East India Company in the last few years of Charles II's reign, in the years of the Tory reaction (1681–5). In 1682, 'on a sudden,' Sir Josiah Child 'forsook all his old friends that first introduced him, with great difficulty into the Committee, and afterwards raised him to the honour of Governor, throwing them totally out of the management' – Whig friends like Sir Samuel Bernardiston, Thomas Papillon, Sir John Banks, Major Robert Thompson – 'betaking himself to new counselors that were very ignorant in this trade.'[43] These new counselors, it turns out, were a motley collection of Catholic and Tory courtiers.

Unsurprisingly, given the nature of the coup he had engineered, Josiah Child and the East India Company soon became favourites at James II's court.[44] Bruce Carruthers has recently shown that the East India Company 'became closely tied to James II's monarchy,' confirming

Gary De Krey's assessment that 'Child and the East India Company's directorate ... wedded themselves to the Tory position of support for the crown.'[45] James himself was said to have at least £10,000 worth of stock in the Company.[46]

Josiah Child, then, succeeded in seizing almost total control of the East India Company with the aid and support of James II and his courtiers. Just as Child and the East India Company provided unwavering financial support of the king, so James II appears to have embraced Child's economic policies and aspirations. What then was Child's – and by extension James II's regime's – understanding of political economy?

Child was committed to the notion that property was natural, not created by human endeavour, and hence necessarily finite. Possession of land was then necessarily the basis of political power. Since dominion was based in land, Child was certain that commerce was merely the exchange of the growth of that land. 'The principal advantage and foundation of trade in England is raised from the wealth which is gained out of the produce of the earth,' Child contended.[47] Where Reynell had seen a series of knock-on effects from manufactures leading to a possibility of unlimited economic growth, Child saw a finite economy totally 'derived out of this principal stock of good husbandry.'[48] Since no wealth was created by human labour, international trade was necessarily a zero-sum game: 'whatever weakens' Italy, France or Holland 'enriches and strengthens England.'[49]

Child's basic premise that property can only be understood in terms of land, and that therefore trade was necessarily a vicious international competition for limited resources, led him to enunciate traditional classical republican and country ideology concerns. He was convinced that 'luxury and prodigality are as well prejudicial to kingdoms as to private families.'[50] Because land and not exchange or manufactures were the real basis of England's wealth and power, Child urged 'the gentry of England' to leave London and its 'wicked course of life' and return 'to their own countries' to 'betake themselves to a way of husbandry.'[51] Finally, Child thought that the government 'for promoting the credit, and securing those privileges which the land is justly entitled to' should replace 'public taxes that are laid on the land' with an 'excise,' thereby 'easing those few that are the proprietors of land, to lay it on those many that raise their estates out of the produce of the land.'[52]

Child's understanding of political economy, an understanding certainly shared by James II and his court, had significant implications for the organization of the East India trade in particular and for English foreign and imperial policy in general. The notion that trade was finite and that international competition was necessarily fierce, led Child and his ideological fellow travellers to insist that foreign trade be conducted in monopolistic fashion. Competition among English merchants could only be disastrous for England. 'Our affairs in India had been in a wonderful prosperous condition in every place but for the interlopers who unite interests with the Dutch,' Child complained to Secretary Middleton.[53]

These views, finally given the status of law in the King's Bench decision of *East India Company v. Sandys*, paved the way for the development of territorial empire in India. Not only did Child's – and by extension the East India Company's – ideological commitment to an agrarian political economy demand exclusive trading privileges, it also demanded that the Company have property on which to base their trade in India. The East India Company was convinced, as they hoped were 'all English men that have any love for their native country, since this matter has now been beat so thin in debates as well before his Majesty as almost in all other public places,' that 'the English must learn to raise revenues to support their power and increase their strength in India, or give up the whole trade of India to the Dutch in a short time.'[54] To achieve these ends, James II granted to the Company 'all the powers we can possibly desire and all that the Dutch have or can pretend unto.'[55] This allowed the Company to transform itself from 'a parcel of mere trading merchants or peddlers' into a 'formidable martial government in India.'[56]

These were not the flights of fancy of an overly ambitious merchant and his faction in London merely dreaming of empires in the east. A fledgling Indian empire quickly took shape – growing to a population in the hundreds of thousands.[57]

Given the commitment of Child, and by extension both the East India Company and James II's court, to a land-based zero-sum political economy, and given the company's newfound commitment to territorial empire in the East Indies, it is not surprising that England and the East India Company pursued a belligerent policy towards their imperial competitors – the Dutch and the Mogul Empire.

Significantly, there is a good deal of evidence to suggest that this attitude was shared by James II and his court. Just as English public opinion outside the court was whipping itself into a frenzy of francophobic excitement, the East India Company along with James II and his court were demanding a war against the Dutch – a war justified because commerce was integral to power, and depended on imperial landed possessions. In this ideological vision – a vision based very much on the same economic assumptions advanced by Machiavelli and Harrington – trade was a zero-sum game, and England's competitors for trade were the Dutch, not the French. Indeed, Child had long advocated making common cause with the French in the East Indies against the Dutch – a call for a coordination of efforts to which James II appears to have responded.[58] The tension between the English and the Dutch over the East Indies was no state secret. In early 1686, for example, Robert Harley and Roger Morrice both heard that the East India Company's fury over the Dutch seizure of Bantam 'will procure a war with the States.'[59] By 1687, it was known, and indeed advertised, that James II had offered his 'royal assistance' in raising 'some thousands of men' to advance the Company's cause in India.[60] No wonder Viscount Weymouth thought tensions in the East Indies 'may occasion a war.'[61] In this context, then, William's invasion of 1688 needs to be seen not so much as an act of aggression as a pre-emptive first strike.

The East India Company did begin a war in India. But it proved to be a war against the Mogul Empire rather than one against the Dutch. Despite every effort by the Company and by the court to publicize English successes, it was soon known on the Exchange and throughout England that the war was going badly. 'Sir Josiah Child is preparing a relation of his late successes against the great Mogul,' one London newsletter-writer sneered in the summer of 1688, 'yet tis thought his rhetoric will scarce longer gain belief since our actions [stock prices] fall to the lowest ebb.'[62] 'Most men conclude that the East India trade is all lost' as a result of this ignominious war, wrote another.'[63]

Not only did Child alienate his former allies in the East India Company, such as Thomas Papillon and Sir John Banks, but his complicity with James II's regime also managed to upset both Whig and Tory merchants. The exclusive nature of the trading rights he secured, and James II insisted upon, deprived merchants all over the kingdom, and the entire Jewish community,[64] of access to trade. No wonder the merchant community so spectacularly turned against the regime. No wonder there was a 'great grumbling in the City against a certain great East India merchant whose first name rhymes with Goliah.'[65] No wonder the merchant community poured money into William of Orange's coffers in 1688. In August it was said that 'near £200,000 have been remitted' from London 'within a month or six weeks' time.'[66] One diplomat who was taken to see William's treasury in the summer of 1688 reported that there were 'such prodigious heaps of English gold as by his computation less must be left in the kingdom.'[67]

James II, then, did embrace a commercial policy – the policy espoused by Sir Josiah Child and the East India Company. It was a policy that was at odds with the radical commercial tradition – the tradition that property was created by human endeavour and that banks could do much to increase the nation's wealth. The implications of that policy, alliance with France and war with the Mogul Empire and the United Provinces, accorded nicely with the king's own foreign-policy predilections, but put him at loggerheads with the vast majority of the political nation. James II thus alienated not only the East India Company's traditional enemies, but even those powerful merchants within and allied to that company who rejected Child's commercial and imperial vision. The result was that England's merchant community actively supported William's plan for invasion, and provided a key financial prop to the regime in its critical early months. Questions of political economy had played a crucial role in bringing about the Revolution of 1688–9. 'Economic arguments' did not have to wait until 'after the Glorious Revolution,' as David Armitage has maintained, to 'become political and constitutional argument.'[68] Arguments about political economy played a role in generating the ideological energy that erupted in 1688–9; they were not an unintended consequence of those events.

## III

The revolutionaries of 1688–9 not only dislodged James II and his political regime, they also ended Josiah Child's stranglehold on economic ideology. A cacophony of different voices soon began to be heard. The ideology of the radicals of the 1650s, an ideology that had gained a following among those more politically moderate than the brave Commonwealth critics of Oliver Cromwell, quickly re-emerged.

The newly invigorated debate over political economy, though initiated by an English people very much alive to the power of liberty, did not rely upon the classics or their Renaissance and republican popularizers. The ancients were simply no longer relevant in the realm of commerce. 'Livy and those ancient writers, whose elevated genius set them upon the inquiries into the causes of the rise and fall of governments, have been very exact in describing the several forms of military discipline, but take no notice of trade,' observed Nicholas Barbon, 'and Machiavel, a modern writer, and the best, though he lived in a government, where the family of Medicis had advanced themselves to the sovereignty by their riches acquired by merchandizing, doth not mention trade, as any way interested in the affairs of state.' The military revolution of the sixteenth and early seventeenth centuries, however, had necessarily changed the relationship between trade and politics. 'Until trade became necessary to provide weapons of war,' Barbon correctly perceived, 'it was always thought prejudicial to the growth of empire, as too much softening the people by ease and luxury, which made their bodies unfit to endure the labour and hardships of war.'[69] Times had changed. So, argued another commentator, 'whatever low conceits Aristotle or some other pedants may have had of merchandize in old times,' they were irrelevant given the 'dignity' that trade 'has long since arrived to.'[70]

Where James II's government had embraced a land-based, zero-sum understanding of wealth, the notion that wealth was potentially infinite and created by human labour was now everywhere to be found. Manufactures, rather than land, were seen by many to be the key to England's future strength and prosperity. 'It is manifest by experience,' claimed William James, echoing the views expressed by Carew Reynell half a decade previously, 'that where a manufacture and much people are settled in any part of the nation, there the lands are not only occupied, but yield the greatest rents, and the fruits thereof the greatest price.'[71] Sir Francis Brewster was certain that those nations that combine natural resources 'are more to be feared, than those that abound with the blessings of nature, but want that of industry.'[72] John Cary knew it to be 'the great interest of England to advance its manufactures.' It was through manufactures that 'we not only employ our poor, and so take off that burden which must otherwise lie heavy on our lands, but also grow rich in our commerce with foreign nations, to whom we thereby sell our product at greater prices than it would otherwise yield, and return them their own materials when wrought up here, and increased in their value by the labour of our people.' So committed was Cary – Locke's favourite writer on political economy – to an economy focused on labour rather than land that he barely found space in his lengthy tract for any mention of luxury. When he did address the issue it was merely to note that luxury had given rise to the now useful disciplines of physic and natural philosophy.[73]

Given Child's success in purging the Company of his most powerful rivals, the failure of the Company to generate profits for its stockholders, and the unpopular ideological orientation of Child's political-economic program, one would expect the East India Company to come under fire immediately after the Revolution. And it did. One of the Company's defenders complained,

> What great and indefatigable industry hath been employed? What arts and devices made use of, to blast the reputation of the present East India Com-

> pany, is notoriously evident to all, who either give themselves the trouble of listening to those calumnies, daily inculcated in all noted coffee-houses against them, or to the reading of those prints exposed publicly, and delivered gratis in the said coffee-houses to all such as will accept them.[74]

Despite repeated and well-informed predictions of the Company's demise,[75] however, it survived, ultimately receiving statutory legitimization in 1698. The Company's enemies were not defeated, they just shifted their ground.

While the East India Company, its policies, and its influence generated a good deal of furor in James II's reign, after the Revolution the major issue in political economy quickly became that of financing the war. Within a year of James II's flight the tone of the parliamentary debates over finance had become desperate, not to say hysterical. The difficulties of fighting Europe's greatest power on two fronts while relying on land-based conceptions of property were readily apparent. William III and his government were faced with two alternatives: scale back the war against Louis XIV or adopt the full radical implications of the alternative understanding of political economy. They chose to fight.

The starting point of this reorganization was the widely accepted view that finance, not land or virtue, was the sinews of war. By February 1689 the lower house already accepted without argument the claim that 'money is the sinews of war.'[76] 'Trade and navigation,' Sir Dalby Thomas pointed out, 'are undeniably our glory and strength, as well as the only fountain of our riches.'[77]

This understanding of the relationship between wealth and power, so foreign to the writers of classical antiquity and the Italian Renaissance, was a direct result of the changing circumstances of European warfare in the seventeenth century. James Whiston offered an almost identical analysis, adding only that 'the discovery of the wealth of the Indies' as well as technological innovation had contributed to making war 'rather an expence of money than men.' Therefore, he thought that nothing was 'more conspicuous than the advantages of trade' in making 'a prince and people happy at home and formidable abroad.'[78]

The exponents of neo-Machiavellian economics have noted the ubiquity of the notion that money is the sinews of war, without coming to grips with the radical implications of this transformation of political discourse. If wealth was a prerequisite for political survival in the modern world, then concerns about luxury would necessarily be secondary. Indeed, it is remarkable how little the concern about luxury occurs in the fierce political economic debates of the early 1690s. When the term did re-emerge in the later 1690s – the period to which Pocock has devoted almost exclusive attention – it did not loom as large or play the decisive role that it had played earlier. Even then, as Istvan Hont has shown, few followed a neo-Machiavellian economic path. Even the Tory polemicist Charles Davenant, one of Pocock's prime examples, thought the dangers from poverty far outweighed the risks from luxury. Even he believed that it was impossible to return to the simpler world of ancient virtue.[79]

It was in this ideological context, then, that the proponents of labour- and manufacturing-based political economy faced the problem of financing the war against France. Their preferred solution was the creation of a national bank. Proposals and blueprints for banks were to be found everywhere in England in the wake of the Glorious Revolution. 'Banks have not wanted both pleaders and writers in their behalf,' understated one pamphleteer.[80]

Towards the end of the tense session of 1694, Parliament passed legislation creating the Bank of England. The main proponents of the Bank were ideologically predictable. William Paterson and Michael Godfrey, 'a person of great parts and industry, and well known at court and in the House of Commons,' were joined by 'some of the late solicitors against the East India Company.'[81] The initial subscribers to the Bank include a wide variety of Child's East India Company critics and political Whigs including John Cholmley, Sir John Chardin, Sir Robert Clayton, Sir James and Abraham Houblon, John Ward, Thomas Pitt, and John Paige.[82] While the proponents of the Bank had a tough fight in Westminster, they had easier going in the City and the nation at large. William Paterson recalled that the initial subscription to the Bank 'was completed in about ten days' and 'the body of the nation' soon proved 'willing generously to venture their money.'[83] Those who could not stomach Child's land-based zero-sum political economy – MPs, merchants, and the nation at large – turned their hopes away from the creation of a new East India Company towards the creation of a national bank.

The proponents of bank schemes and the defenders of the Bank of England made clear the ideological assumptions that undergirded their enthusiasm.[84] The most eloquent proponent of the Bank was H. M., the author of *England's Glory*. The Bank, H. M. pointed out, could provide 'ready money' in case of 'a sudden emergency' to 'equip Armadas, supply armies, levy soldiers.' When 'there is leisure for deliberation, and ... the Parliament and King judge it requisite, these banks may be in a capacity to supply the crown with whatever money it needs at a reasonable rate.' This ability to support the government financially would 'tend to the overthrowing our enemies by sea and land.'[85] The Bank would also promote wealth and power. 'Money in a nation in trade, is like blood in the veins, if it circulates in all parts, the body is healthy; if it be wanting in any parts, it languisheth,' noted H. M. Banks guaranteed the circulation of money, which will 'beget trade and people, and they will beget riches ... Riches are the conveniencies of a nation: but trade and people are the glory and strength of the kingdom.'[86] H. M. was sure that a bank would initiate the same sort of infinite knock-on effects discussed by Carew Reynell, because with a bank 'people will increase, for trade will bring in people as well as riches to the nation: where trade is, there will be employment; where employment is, there will people resort; where people are, there will be consumption of all commodities.'[87] Luxury, thought H. M., was not a serious problem. 'When the people's yolk is lined with peace and plenty,' he noted, 'it will make them cheerful under it, and not desirous to shake it off. If some few should surfeit and grow wanton, the generality of the people (being content in their condition) would certainly keep them in awe.'[88] For all these reasons the optimistic H. M. could report that 'all men are satisfied a bank will be very ad-vantageous to a nation, especially to a trading people.'[89]

The defenders of the new political economy, the ideological descendants of the 1650s radicals, also drew radically different macropolitical conclusions from Josiah Child and those committed to a land-based zero-sum political economy. Where Child had thought that England's great enemies were the Dutch because they were competitors for commercial empire, the defenders of the new political economy saw French land-based and Europe-centered imperialism as the greatest threat to England. While proponents of agrarian political economy either became Jacobites or lukewarm supporters of the war against France, the defenders of manufactures were its most enthusiastic supporters. One opponent of taxes on manufactures bragged that 'no man in England is more thoroughly sensible than I am of the necessity of carrying on this war.' The stakes could not have been higher. He was sure 'that the liberty of this nation, the preservation of this government, and the security of the Protestant interest throughout Europe' depended 'upon the success of it.'[90]

The Glorious Revolution, then, produced, and by many was intended to produce, a revolution in political economy. James II had devoted himself to an economic understanding that posited property as a natural endowment. In this understanding the world's wealth was necessarily finite, and empires were created by taking land from another state. The

East India Company was thus ideally poised to create a vast new English imperial dominion by warring with the Dutch and the Mogul Empire. After the Revolution an alternative understanding of political economy gained dominance, one that understood England to be a commercial rather than an agrarian society. The proponents of this view, many of whom had been supporters of the East India interlopers, thought that wealth was created by human endeavour and thus potentially infinite. While these political-economic thinkers, writers, and actors initially focused their energies on breaking up mercantile monopolies, in particular that of the East India Company, they quickly shifted their emphasis to creating a national bank. The revolution in political economy was thus a necessary prerequisite to the Financial Revolution. Along with the commitment to a national bank came a commitment to drawing upon merchant knowledge in the creation of national policy and a notion that tax policy should be revised so as to benefit the most valuable area of the economy, the manufacturing sector. In this view, one's true competitors were not states like the United Provinces that specialized in the carrying trade, but large countries like France that could compete with England in manufacturing, and were doing so by pernicious political means. Unsurprisingly, the proponents of this new political economy were far more concerned with the creation of wealth than they were with the corrosive effects of luxury. They were not obsessed with anchoring political personality in real property. Instead, they thought that circulation of wealth was vital to the maintenance of England's national integrity and identity.

## IV

The attraction of civic republican political economy has always been as a profoundly moral alternative to amoral possessive individualism or uncaring utilitarianism. While proponents of neo-Harringtonian political economy have prided themselves on the dialectical nature of their historical enterprise, they have taken the arguments in favour of the Bank of England as read rather than as topics to be explored. At the same time, historians of the Glorious Revolution have focused so narrowly on traditional constitutional and religious issues that they have failed to note the vicious and important debates over political economy that both preceded and deeply influenced the events of 1688–9. Those historians have failed to note the revolution in political economy that was an intended outcome of many of the not-so-reluctant revolutionaries of those years.

The controversy in political economy from 1685 to 1696[91] was not between possessive individualism and neo-Harringtonianism, nor was it between mercantilism and liberalism. Polemicists on both sides of the debate were committed to promoting the common good rather than the profit of private individuals. Polemicists on both sides were sure that the common good, or national interest, was something conceptually distinct from the summation of the private interests of England's population. Similarly, both the defenders and the opponents of the Bank were committed to the notion that the government needed to intervene in the economy. The seventeenth-century advocates of free trade were fierce opponents of exclusive foreign trading concerns created by royal prerogative, but were also great supporters of the creation of the Bank of England. There was broad agreement about the necessity of state intervention in the economy. There was, however, no mercantilist consensus – as so many scholars have assumed – that trade was a zero-sum game.

Nor was political economy merely, as David Armitage has recently asserted, a discipline that 'provided the means to describe and explain the relationships among the three kingdoms.' Political economy was not primarily about 'the triangular relations between England, Ireland and Scotland.'[92] The debate about political economy was about England's economic identity and its relations with Europe and the Indies East and West. Debates about the proper relationship with Ireland and Scotland certainly played a role. But those discussions took place in the broader context of discussions of England's relations with France and the United Provinces, with the East Indies and the Americas.

This approach to the subject reveals that the debate that so deeply influenced the Revolution of 1688–9 and its aftermath was, in fact, between those who understood property as a natural creation and those who knew it to be the result of human endeavour. Josiah Child and James II's court were committed to an agrarian political economy, a political economy that posited a zero-sum world of commercial exchange. The implications of that commitment were support of territorial imperialism in India and war against the Dutch. After the Revolution, Child and his ideological fellow travellers opposed the creation of the Bank both because it would transform England into a commercial society based on mobile wealth and because that very transformation would turn England into a commonwealth.

Those who opposed this agrarian understanding of political economy had an equally sophisticated understanding of their social environment. They argued that property was potentially infinite since it was the result of human labour. In this vision, manufactures rather than land-based empire was the key to England's wealth and power. They were therefore, on the whole, critical of the East India Company's trade to the extent that it imported finished products that competed with English manufactures, while failing to export any of England's finished goods. They also saw little point in military conflict with the Dutch, since there was by definition enough potential property for both to enjoy unlimited economic success. Most important, they were committed to the creation of a national bank based on mobile wealth, since only such a bank could both finance the ideologically urgent conflict with France and ensure the proper circulation of money. Only the circulation of money, they believed, could guarantee the development of English manufactures. Nowhere in this vision was there a celebration of profit-seeking individualists. Instead, the supporters of this commercial political economy celebrated the English nation both for its participatory politics and for its rich communal culture.

The revolution in political economy effected in 1688–9 had been the work of political as well as economic radicals, of which Daniel Defoe was one. Their political achievement was to create the financial mechanisms that allowed England to carry on wars on a scale previously unimaginable against an enemy that was more powerful and more ambitious than any they had previously faced.[93] They were modern projectors. Their ideological achievement was just as great, for they had forced classical republicans to jettison Harrington's notions of political economy. When Harrington was embraced – and not modified so much as to reverse the political economic message – it was by Tories and Jacobites. That Toland, Molesworth, John Trenchard, and Thomas Gordon could be such enthusiastic supporters of the Bank of England shows the fraying of the republican program. In this sense, Blair Worden is surely right to suggest that in the wake of the Revolution 'the teaching of republicanism became ever less precise. The movement became ever more a language, ever less a programme.'[94] Republican politics and the civic republican political economic vision had been severed, the Whigs embracing one side, the Tories the other.

## NOTES

This essay draws heavily on material from the chapter 'Revolution in Political Economy' in my forthcoming monograph *The First Modern Revolution.* I have presented a related line of argument in David Womersley, ed., *The Culture of Whig-*

*gism* (University of Delaware Press, forthcoming). I am extremely grateful for the comments and criticisms I have received on this work at the Clark Library, the Institute of Historical Research, the University of Chicago, Cornell University, and the University of Michigan. In particular I am grateful for the comments and suggestions of Mike Braddick, Robert Brenner, Alan Cromartie, Mark Knights, Allan MacInnes, Nick Von Maltzahn, Jim Robinson, Susan Stokes, Rachel Weil, and Blair Worden.

1 John Toland, *The Oceana of James Harrington and his Other Works* (London, 1700), 3.
2 J.G.A. Pocock, *Virtue, Commerce and History* (Cambridge: Cambridge University Press, 1985), 233.
3 The term 'political economy' is, of course, a conscious anachronism, but it is an appropriate one. The debates in this period were never about the economy abstracted from the state. I take some comfort in my use in that many other scholars are coming to use the term. I use the term with the capaciousness suggested by Donald Winch: Donald Winch and Patrick K. O'Brien, eds, *The Political Economy of British Historical Experience* (Oxford: Oxford University Press, 2002), 4. Julian Hoppitt describes his next project as being about 'political economies, 1660–1800,' ibid., 3. Debates over political economy were about many things in the period I am describing. I am highlighting one important and, I would argue, essential strand.
4 A much more comprehensive discussion of the secondary literature will appear in my *First Modern Revolution.*
5 Pocock, *Virtue, Commerce and History*, 108; Pocock, *The Machiavellian Moment* (Princeton: Princeton University Press, 1975), 423–4.
6 Pocock, *Virtue, Commerce and History*, 230; Pocock, *Machiavellian Moment*, 424.
7 J.G.A. Pocock, 'Early Modern Capitalism – the Augustan Perception,' in *Feudalism, Capitalism & Beyond*, ed. Eugene Kamenka and R.S. Neale (London: E. Arnold, 1975), 71. I owe this reference to my research student Ben Stone with whom I have had frequent and fruitful discussions of many of the issues addressed in this essay.
8 Pocock, *Machiavellian Moment*, 423.
9 Ibid., 423.
10 Ibid., 426; Pocock, 'Early Modern Capitalism,' 70.
11 Pocock, 'Early Modern Capitalism,' 75.
12 C.B. Macpherson, *The Political Theory of Possessive Individualism* (Oxford: Clarendon Press 1962), 1, 3, 258, 270.
13 Steven Pincus, '"To Protect English Liberties": The English Nationalist Revolution of 1688–9,' in *Protestantism and National Identity: Britain and Ireland*

*c.1650–c.1850*, ed. Tony Claydon and Ian McBride (Cambridge: Cambridge University Press, 1998), 75–104.

14 Pocock, 'Early Modern Capitalism,' 72.

15 David Armitage, *The Ideological Origins of the British Empire* (Cambridge: Cambridge University Press 2000), 3.

16 [Daniel Defoe], *An Essay upon Projects*, ed. Joyce Kennedy, Michael Seidel, and Maximillian Novak, in *The Stoke Newington Daniel Defoe Edition* (New York: AMS Press, 1999), 7.

17 Ibid., 9.

18 Ibid., 13. Here Defoe clearly distinguishes his discussion from the kind of projects delineated by Joan Thirsk in her *Economic Policy and Projects* (Oxford: Clarendon Press, 1978). For Thirsk, 'A project was a practical scheme for exploiting material things; it was capable of being realized through industry and ingenuity. It was not an unattainable dream like the commonweal' (1). Therefore, Thirsk is wrong to accuse Defoe of seeing 'too much novelty in his own day' (9). The sixteenth- and early-seventeenth-century projects that Thirsk describes were of a scale much smaller than those that Defoe dignifies with the name of projects.

19 Defoe, *Essay*, 14.

20 Steve Pincus, 'Neither Machiavellian Moment nor Possessive Individualism: Commercial Society and the Defenders of the English Commonwealth,' *American Historical Review* 103.3 (June 1998), 720–1.

21 Carew Reynell, *A Necessary Companion or, The English Interest Discovered and Promoted* (London: William Budden, 1685), 71–2. The arguments advanced here are a radical condensation of a chapter in my forthcoming book *The First Modern Revolution.* That chapter is in turn based on a reading of every title listed in the Goldsmiths-Kress catalogue of economic literature from 1685–96 as well as a wide range of manuscript material. In this essay I have focused on a small number of texts because of their clarity, extent of vision, and their importance. The same argument could have been advanced by deploying a wide range of materials. I feared that that approach would have made me appear to be the author of the intellectual coherence of the positions.

22 Ibid., sig. A7r, 5.

23 Ibid., 18.

24 Ibid., 17–18.

25 Ibid., sigs. [A1]v–[A2]r.

26 Reynell's argument, which was in fact very widespread, flies in the face of assumptions bv historians and economists that contemporaries had no notions of sustained economic growth: D.C. Coleman, 'Labour in the English Economy of the Seventeenth Century,' *Economic History Review*, n.s., 8.3

(1956), 287–8 (though Coleman does admit that there were some who wished it were not this way); Antony Brewer, 'The Concept of Growth in Eighteenth-Century Economics,' *History of Political Economy* 27.4 (1995), 609–10 and passim. More recently Kenneth Morgan has argued that until 1815 'the political economy of the British empire was underpinned by a mercantilist framework.' Mercantilists all agreed that increased productivity was impossible. 'Mercantilists viewed overseas trade as a zero-sum game.' Kenneth Morgan, 'Mercantilism and the British Empire, 1688–1815,' in Winch and O'Brien, *Political Economy*, 165, 168. I am not the first to notice that this dominant, almost hegemonic interpretation is based on a very partial reading of the evidence. See the important article by Richard C. Wiles, 'The Theory of Wages in Later English Mercantlism,' *Economic History Review*, n.s., 21 (1968), 113–26.

27 Reynell, *Necessary Companion*, 48.

28 Ibid., sigs. A5v–A6r.

29 Ibid., 16.

30 Ibid., sig. A4.

31 Ibid., 6.

32 Ibid., 1–2.

33 Ibid., 13–14.

34 Richard Blome, *The Present State of His Majesty's Isles and Territories in America* (London, 1687), 127.

35 [William Carter], *The Reply of W. C.* (London, 1685), 49.

36 John Locke, 'Second Treatise of Government,' in *Political Writings of John Locke*, ed. David Wootton (New York: Mentor, 1993), 281–2. For Locke's support of the Bank, see John Locke to Edward Clarke, 30 June 1694, and Locke to Clarke, 6 August 1694, in Benjamin Rand, ed., *The Correspondence of John Locke and Edward Clarke* (Cambridge, MA: Harvard University Press, 1927), 395, 397.

37 Andrew Yarranton, *England's Improvement by Sea and Land* (London: R. Everingham, 1677), 20.

38 Ibid., 22–3.

39 Adam Anderson, *An Historical and Chronological Deduction of the Origins of Commerce*, vol. 2 (London: J. White et al., 1801), 564.

40 Sir John Lowther, Notes on Trade, ca. 1680, Cumbria Record Office, D/Lons/W1/63, pp. 1–2.

41 *The Mischief of Cabals*, 7 May 1685 (London, 1685), 35–6.

42 Defoe, *Essay*, 4.

43 *Some Remarks upon the Present State of the East-India Company's Affairs* (London, 1690), 3; Nathaniel Cholmley (Whitby) to Cholmely Stephens, 30 December

1684, North Yorkshire Record Office, ZCG, unfoliated; Cholmley to John Healtfield, [January 1686], ibid.

44 James had seen Child as a political opponent in the 1660s and 1670s. Only after 1678 did his political alliances shift, a shift that was consolidated by his break with Thomas Papillon in 1682. William Letwin, *The Origins of Scientific Economics: English Economic Thought 1660–1776* (London: Methuen, 1963), 21–4, 29, 35.

45 Bruce G. Carruthers, *The City of Capital* (Princeton: Princeton University Press, 1996), 146; Gary De Krey, *A Fractured Society* (Oxford: Clarendon Press, 1985), 24; Henry Horwitz, 'The East India Trade, the Politicians, and the Constitution: 1689–1702,' *Journal of British Studies* 17.2 (1978), 1.

46 Morrice, Entering Book, 1 June 1689, Dr Williams Library, 31Q, p. 560.

47 Child, *A New Discourse of Trade* (London, 1694), 7.

48 Ibid., 8.

49 Ibid., 3.

50 Ibid., Imprim: 24 December 1692, sig. [A4v].

51 Ibid., 27–8.

52 Ibid., 17.

53 Sir Josiah Child to Lord Middleton, 1 September 1683, British Library, Add. 41822, f. 25r.

54 East India Company to General of India and President and Council of Fort St George, 25 January 1688, India Office Library, E/3/91, f. 245r.

55 East India Company to President and Council of Fort St George, 24 February 1686, ibid., f. 54r.

56 East India Company to President and Council of Fort St George, 9 June 1686, ibid., f. 70v.

57 Josiah Child, *A Supplement to a Former Treatise concerning the East India Trade* (n.p., 1689), 7–8; and Om Prakash, *European Colonial Enterprise in Pre-Colonial India* (Cambridge: Cambridge University Press, 1998), 148.

58 Child to Middleton, 6 September 1683, BL, Add. 41822, f. 28v; Representation of the Deputies of Amsterdam, 1686, BL, Add. 41814, f. 119v; EIC to General of India and Council at Bombay, 27 August 1688, IOL, E/3/91, f. 273r.

59 Roger Morrice, Entering Book, 9 January 1686, Dr Williams Library, 31P, 509; Robert Harley to Sir Edward Harley, 9 January 1686, BL, Add. 70013, f. 314r.

60 London Newsletter, 9 July 1687, Folger Shakespeare Library, Lc1831; London Newsletter, 16 July 1687, FSL, Lc1821; Sir John Jacob to Earl of Huntingdon, 12 July 1687, Huntington Library, HA 7148.

61 Viscount Weymouth (Longleat) to Sir Robert Southwell, 28 May 1687, BL, Add. 28569, f. 63r.

62 London Newsletter, 4 August 1688, Add. 4194, f. 291v. For an example of Child's rhetoric, see Child, *A Supplement, 1689*, 9.

63 Newsletter addressed to Earl of Suffolk, 25 July 1688, BL, Add. 34487, f. 17v. See the similar assessment in John Verney (London) to Sir Ralph Verney, ca. 24 July 1688, Bucks. RO, Verney MSS, reel 42. This view was elaborated after the revolution, see *Some Remarks*, 4, 6–7.

64 John Cholmley (London) to John Aelst, 25 November 1684, North Yorks. RO, ZCG, unfoliated; John Cholmley and Ambrose Isted to John Aelst, November 1684, ibid. James II's alienation of the extremely wealthy, if small, London Jewish community bears further investigation.

65 Newsletter from London, 14 August 1688, BL, Add. 4194, f. 299.

66 Daniel Petit (Amsterdam) to Middleton, 17/27 August 1688, BL, Add. 41816, f. 160r.

67 Peter Wyche (Hamburg) to Middleton, 18 September 1688, BL, Add. 41827, f. 90r.

68 Armitage, *Ideological Origins*, 166. Armitage seems to be relying on the work of Terence Hutchinson, *Before Adam Smith: The Emergence of Political Economy, 1662–1776* (Oxford: Basil Blackwell, 1988), 56.

69 Nicholas Barbon, *A Discourse* (London, 1690), sig. A3.

70 *The Character and Qualifications of an Honest Loyal Merchant* (London: Robert Roberts, 1686), 11.

71 William James, *Englands Interest* (London, 1689), 2.

72 Francis Brewster, *Essays on Trade and Navigation* (London, 1695), 1. Brewster, it should be noted specifically argued that there was no necessity of belligerent competition between the English and the Dutch; p. ix.

73 John Cary, *An Essay on the State of England* (London, 1695), sig. [A7], 2, 6 12, 23. For Locke's assessment of Cary's work see John Locke to John Cary, 2 May 1696, in Benjamin Rand, ed., *The Correspondence of John Locke and Edward Clarke* (Cambridge, MA: Harvard University Press, 1927), 472. The *Essay*, Locke says, 'is the best discourse I ever read on that subject, not only for the clearness of all that you deliver and the undoubted evidence of most of it. But for a reason that weighs with me more than both those, and that is that sincere aim at the public good and that disinterested reasoning that appears to me in all your proposals: a thing that I have not been able to find in those authors on the same argument which I have looked into.'

74 N. T., *A Modest and Just Apology for; or Defence of the Present East India Company* (London, 1690), 1.

75 See, for example, John Cholmley to ?, 14 January 1691, North Yorks. RO, ZCG, unfoliated; J. Hill (London) to Sir William Trumbull, 26 October 1691, Berkshire RO, D/ED/C33.

76 *Orange Gazette*, 1 March 1689, p. [2].

77 Thomas, *An Historical Account of the Rise and Growth of the West India Colonies,* (London, 1690), 1

78 Whiston, *A Discourse of the Decay of Trade,* 2–3; Whiston, *To the King's Most Excellent Majesty* [1693], [1].

79 Istvan Hont, 'Free Trade and the Economic Limits to Modern Politics: Neo-Machiavellian Political Economy Reconsidered,' in *The Economic Limits to Modern Politics,* ed. John Dunn (Cambridge: Cambridge University Press, 1990), 64–5. It is surely significant in interpreting Davenant's political economy that he was on the East India Company's payroll for much of the 1690s: Charles Davenant, Receipts of Payment from the East India Company, Bodleian, MS Rawl D747, ff. 194–201.

80 *Some Useful Reflections upon a Pamphlet Called a Brief Account of the Intended Bank of England* [1694], 3.

81 Jerry Squirt, *Some Account of the Transactions of William Paterson* (London, 1695), 4–5; Horwitz, 'East India Trade,' 6–7.

82 *A List of the Names of all the Subscribers to the Bank of England,* [1694]. D.W. Jones has shown that the founders of the Bank of England were precisely the opponents of the Old East India Company: *War and Economy in the Age of William III and Marlborough* (Oxford: Blackwell, 1988), 12–13, 298–9, 334–5.

83 Paterson, *An Inquiry into the State of the Union* (1717), in *The Writings of William Paterson,* ed. Saxe Bannister, 3 vols. (New York: A.M. Kelley, 1968), 2: 64, 67.

84 I have not discussed the 1696 debate over the Land Bank in this essay – that is treated fully in the last two chapters of my forthcoming *First Modern Revolution.* The Land Bank was defended by Child as a last-ditch defence of land-based political economy.

85 H. M., *England's Glory; or, the Great Improvement of Trade in General by a Royal Bank* (London, 1694), 21–2, 31.

86 Ibid., sig. A3r, 11–12.

87 Ibid., 18–19.

88 Ibid., 23–4.

89 Ibid., sig. A4r.

90 *Some Considerations about the Most Proper Way of Raising Money* (London, 1693), 31.

91 I am aware that the supporters of land-based political economy did not withdraw from the ideological battlefield after 1694. In 1696 there was a fierce struggle over the creation of the Land Bank, a Tory/Jacobite alternative to the Bank of England. That bank was unable to raise the necessary subscription. I deal with the episode in the final chapter of *The First Modern Revolution.* In many ways, the land-based political-economic alternative survived even that episode. Land-based political economy was the ideological prop of the South Sea Company and would later justify Robert Clive's actions in India.

92 Armitage, *Ideological Origins*, 148.

93 In this sense I agree with Maximillian Novak's assessment that 'the government itself had turned projector in order to raise money to continue the war.' Maximillian Novak, *Daniel Defoe: Master of Fictions* (Oxford: Oxford University Press, 2001), 115.

94 Worden, 'Revolution of 1688–9 and the English Republican Tradition,' in *The Anglo-Dutch Moment: Essays on the Glorious Revolution and Its World*, ed. Jonathan Israel (Cambridge: Cambridge University Press, 2003), 269.

chapter seven

# 'Wandring Ghosts of Trade Whymsies': Projects, Gender, Commerce, and Imagination in the Mind of Daniel Defoe

KIMBERLY LATTA

One of the most prolific writers on commerce, Daniel Defoe may be our greatest source for social attitudes about emergent capitalism in the early eighteenth century. But his ideas about the imagination were as forward-thinking for poetry as they were for economics, and we can say so without claiming that he innovated dramatically in either arena. In his time, Defoe was a brilliant synthesizer and gifted register of contemporary wisdom on a variety of topics, including spirits. Indeed, ghosts and spectral realities preoccupied Daniel Defoe for reasons very similar to his fascination with the imaginative generation of aesthetic and economic values.[1] He was interested in the relationship between the material and the metaphysical. What I want to show here is that Defoe's economic thought wound inextricably together with his theory of imagination, and that in both subjects he expressed a gendered consciousness. This consciousness opened itself as never before to the positive potential of *projecting*, a word that frequently crops up in seventeenth-century texts as a metaphor for the activity of bringing forth the New in property as well as in thought. Projecting (from the Latin verb *proicere*, which means to cast outward or to throw) was used interchangeably with invention to evoke the sense not only of commercial hucksterism but also of inventive wit in the seventeenth century. In 1647, for example, the translator of *Daphnis and Chloe* wrote 'Daphnis was more of a projecting wit than she.'[2] Projecting was also the term for the final stage of alchemical transmutation, in which the adept cast the powdered Philosopher's Stone (called *the powder of projection*) across a metal in order to transmute it into gold. Projecting thus conveyed a sense of somewhat dubious metamorphosis and the generation of something from nothing, which merged with the very old sense, dating back to 1400, of the word 'to project,' which meant to plan or devise a scheme. Early modern alchemical works 'regularly' compared 'the alchemical process to the gestation, birth and nutrition of a child' and the alchemist to a mother.[3] Although the term in alchemy indicated the metamorphic process in which the seeds of pure essence were heated and 'multiplied' into gold, a process in which something comes to life, in other arenas 'projecting' acquired debased overtones of fraud and became linked to risky ventures, false promises, ephemeral payoffs, fictional futures, and, most especially, excessive generation or production that carries indefinitely on towards the future with no certain goal or foreseeable end point.

Defoe ambivalently embraced projecting, which for him was an act of throwing forward into the future for the sake of moving forward and taking risks, rather than for the sake of returning to a teleological origin – God the Father, truth, the state, the common good – that tradition constructed as the proper end (telos) of all human activity. Although the modern shift away from an Aristotelian ethics of limited, teleological generation and towards a greater acceptance of unlimited, open-ended generation in aesthetics and economics has been studied before,[4] previous analyses have not considered how gendered polarities often structured the rhetoric of economic as well as poetic production. This essay examines the way that Defoe's figures of 'Witt,' 'Credit,' 'Trade,' and 'Projecting' represent creative generation in commerce, as well as in language, and generally follow established oppositional categories that linked reason with masculinity and irrationality with femininity. As ever with Defoe, things get interesting where his oppositions begin to break down and blur together.

I have organized the chapter into two sections. The first, on 'economic projecting,' considers Defoe's contribution to an on-going discursive rehabilitation of strategies for making wealth (such as credit and stock speculation) that were still associated with usury, or the getting of money from money. The second section, on 'imaginative projecting,' traces an analogue to that political-economic debate in late-seventeenth- and early-eighteenth-century discussions about the term 'wit' and the imagination. Just as some liberal writers had advanced the idea that human labour created new wealth, some poets were beginning to challenge the more conservative view of the human imagination as a merely imitative and accumulative faculty by representing it as an autonomous, generative agency. The development of the idea that new wealth could be created seems to have been related to the nascent conceptualization of the imagination as a creative faculty. This paper's organization into sections on 'economic' and 'imaginative' projecting, therefore, is meant merely as an artifice for the sake of explaining concepts that actually remained completely intertwined with one another during this period.

## I. Economic Projecting

There were two schools of thought about the generation of wealth during Defoe's time. One maintained that human labour could in fact bring forth potentially limitless new wealth, while the other claimed that wealth could not be created, only discovered, and that real value inhered primarily in land. Conservative writers endorsed a classical republican ethos and rejected a political economy based on mobile wealth throughout the 1600s and 1700s. Yet as Steven Pincus has argued recently, mid-seventeenth-century merchants, such as Slingsby Bethel and Marchamount Needham, and a number of openly Whig writers after them, 'valued human choice, the human capacity to create wealth, and epochal change in human history.'[5] Opponents of this point of view tended to regard imaginative strategies for generating profits as modern forms of usury.

Where does Daniel Defoe fit into this constellation of attitudes towards the generation of new wealth? The speaker of *An Essay upon Projects* promotes human choice and the capacity to create wealth, as well as a society whose collective members will benefit from allowing individuals to pursue their private aims. Indeed, he defies centuries of tradition by declaring that human ingenuity could bring forth the New: 'Some are apt to say with *Solomon, No new thing happens under the Sun, but what is, has been,* yet I make no question but some considerable Discovery has been made in these latter Ages, and Inventions of Human Original produc'd, which the World was ever without before, either in whole, or in part.'[6] In short, human ingenuity brings forth things that did not exist before. The speaker links this creative capacity to 'the Merchandizing Part of the World,' who, he enthuses, 'may more truly be said to live by their Wits than any people whatsoever.' He also states that imagination lies at the heart of commercial generation: 'All Foreign Negoce ... is in its beginning all project, Contrivance, Invention.' Projectors are not only merchants; they are also 'authors,' who employ 'wit,' 'invention,' and 'genius' in order to create the New (*Essay* 35–6). The speaker would seem, then, to advance a Whig or liberal position that maintains that new wealth can be created.

The speaker's argument that reliable new wealth can be made, not merely found, depends upon the assumption that a difference between 'honest' and 'dishonest' means of making money can be ascertained. While the speaker freely admits that unscrupulous projectors abound, and condemns them in the strongest language he can muster, he also insists that 'the Honest Projector is he, who having by fair and plain principles of Sense, Honesty, and Ingenuity, brought any Contrivance to a suitable Perfection, makes out what he pretends to, picks no body's pocket, puts his Project in Execution, and contents himself with the *real Produce*, as the profit of his Invention' (*Essay* 45; my emphasis). To demonstrate this point, the speaker gives many examples of successful projects that not only realize the desires of their authors for money, but that also 'tend to Improvement of Trade, and Employment of the Poor, and the Circulation and Increase of the public Stock of the Kingdom.' A projector who 'aims primarily at his own Advantage' creates something valid when he builds 'on the honest Bias of Ingenuity and Improvement.' 'Dishonest' projectors are criminal not because they serve their own self-interests, but rather because are unproductive: they fail to yield 'circumstances of Public Benefit added' (*Essay* 36). This argument foreshadows the Mandevillian view that individual avarice could actually promote the common good.

Paradoxically, the speaker's insistence on a discernable distinction between 'honest' and 'dishonest' projecting observes a line of reasoning that follows an Aristotelian logic most commonly adopted by Tory conservatives, who held that no new *real* wealth could be generated. In fact, Aristotelian teleology lies at the root of the controversy over whether or not the usury-like schemes for producing profits were good or bad for the state, as well as the general break with scholasticism that innovators in science, art, and philosophy made at this time. In the *Politics* Aristotle distinguished between 'natural' and 'unnatural' modes of getting wealth according to a teleological ethic of purpose, or ends, and limitation. Aristotle argued that the *telos* or purpose for which wealth was gotten should govern the ethics of its production, pointing out that Solon had been wrong to say that 'no bound to riches has been fixed for man.' On the contrary, Aristotle maintained, 'there is a boundary fixed, just as there is in the other arts: for the instruments of any art are never unlimited, either in number or size, and riches may be defined as a number of instruments to be used in a household or in a state.'[7] Riches are merely an 'instrument' of that art by which the household or state is maintained, and the number of instruments needed for that art is limited. Wealth-getting in retail trade, he argued, served no legitimate noble end, such as the support of the state or the household, but rather only the desire of the merchant to generate more riches. 'There is no limit of the end' of this kind of wealth-getting, which, he said, produces 'riches of the spurious kind.'[8] Property produced for the sake of itself begins and ends with itself and therefore is both unethical and unreal. It contradicts the law of nature that dictates that all generation should begin and end with a *telos* other than itself. This kind of economic production, which Aristotle referred to as *chrematistics*, is autonomous because it rests on no foundation other than itself. It is open-ended because it recognizes no conclusion. Tory writers, like the sixteenth- and seventeenth-century opponents of usury whose arguments they echoed, often accused the men who advanced projects for generating potentially limitless new wealth of observing no other authority than their own greed and of creating nothing but promises. The speaker of *An Essay upon Projects* seems to take up a modified Tory position by marking out a difference between 'honest' and 'dishonest' projecting that is analogous to Aristotle's distinction between limited and unlimited generation. Honest projecting benefits the State, while dishonest projecting serves the interest of none but the projector. It would seem, then, that the speaker approves of limited projecting and disapproves of the unlimited kind.

Nevertheless, maintaining a distinction between the two kinds of projectors ultimately defies the rhetorical powers of this ostensibly Tory persona, who alleges that 'this Age swarms with such a multitude of Projectors more than usual, who besides the Innumerable Conceptions which dye in the bringing forth, and (like Abortions of the Brain) only come into the Air, and dissolve, do really every day produce new Contrivances, Engines, and Projects to get Money, never before thought of' (*Essay* 35). Not only are the 'authors' (and parents) of both abortive and viable projects, of mere cheating promises and public-benefiting dreams, the *same people*, but they are also prompted to come up with these useful and useless schemes by 'Necessity, which is allow'd to be the Mother of Invention' (*Essay* 34). All projects begin and end with the aim of making money, a condition that the pragmatic Defoe refuses to condemn. Aligning economic production with imaginative production and the commonplace of the mind as a womb, *An Essay* suggests that the difference between good and bad projecting is impossible to determine until the schemes have completed their terms and have been born into the world. This is why Defoe's speaker cannot bring himself to repudiate 'thieving' projecting altogether, since 'Success has so sanctifi'd some of those other sorts of Projects, that 'twou'd be a kind of Blasphemy against Fortune to disallow 'em' (*Essay* 38). Defoe dons a Tory mask, then, in order to de-fend a set of interests that are more properly aligned with the Whigs.

In *An Essay upon Projects*, then, the tenuous difference between 'honest' and 'dishonest' projecting amounts to something less like an Aristotelian distinction between limited and unlimited generation and more like the difference between 'truth' and 'lies.' Dishonest projectors begin with the intention of generating nothing other than fictions whose circumstances will line their own pockets and serve no positive, public ends. This ethical standard helps to explain Defoe's subsequent protestations of desiring to ameliorate his reader's morals in literary commodities in which the difference between 'truth' and 'lies,' public service and private interest, are not at all easy to detect.[9] These imprecise distinctions indicate, perhaps, Defoe's deep ambivalence about whether or not *any* kind of projecting actually generated the New. But his willingness to defend such production also demonstrates his break with traditional Aristotelian notions of bounded or limited production and movement towards the justification of autonomous, open-ended generation and the expansion of trade.

Defoe may have promoted trade most often through a Tory persona, as Maximillian Novak has argued, but this persona did not finally endorse a position that we can call mercantilist.[10] Mercantilism was hardly a systematic ideology, as Novak himself admits, yet most mercantilists generally regarded the world's wealth as fixed and limited. For them, trade remained a zero-sum game. Defoe remained politically aligned with the Whigs and a defender of the economic management of the Earl of Godolphin and of the interests of the Bank of England.[11] Those interests depended upon a nation-wide comfort with the idea that controversial, speculative methods for making money – projecting – would lead to an increase in English prosperity in general.

Yet for all his youthful enthusiasm for 'honest' projecting, Daniel Defoe seems to have given up the attempt to defend the virtue of this particular 'Faculty' by the time he embarked on the project of the *Review*. This periodical of political and economic commentary began in 1704 as *A Weekly Review of the Affairs of France* and soon expanded to encompass *All Europe, As Influenced by that Nation: Being Historical observations, on the Public Transactions of the World.* Defoe's *Review* first ran twice, then thrice, a week under various titles until 1713. The departure from the rhetorical feats of *An Essay* to those of the *Review* may have more to do with terminology than philosophy, however, since Defoe wrote the astonishingly regular and lengthy publication in service to the government of Robert Harley, later the first Earl of Oxford, Secretary of State from 1704–8,

Chancellor of the Exchequer from 1710, and leader of the Tory ministry from 1710–14. Although the *Review*'s persona defends other kinds of open-ended, autonomous generation, such as credit, he reserves nothing but scorn for 'Projecting,' who figures in a family history of commerce as a 'Prolifick Monster' giving birth to uncountable children. He tells us that she is the daughter of 'Trade,' herself a woman, who, having 'grown wanton by wealth,' imprudently went a 'caterwauling after novelty' in the streets. There she met 'Alderman Avarice,' who raped the unfortunate, wandering seeker of the New.[12] Avarice then forced himself on his own daughter, 'Projecting,' who soon 'grew big with a Generation of Mischiefs.' Comparing Projecting to the Hydra, the mythical, many-headed beast who grows two new heads for each one lopped off, Mr Review moans that her most horrible offspring is the 'Stockjobber,' a feminine 'Devil of a Monster – with a hundred Hands, and every Hand a hundred Fingers.'[13] The morbid delight conveyed in the colourful description of this prodigy of proliferation, not to mention the allusive, metaphorical profusion he employs, suggests that Mr Review's repudiation of projecting is less than absolute.

Obviously, Defoe's complex representation of projecting as a monstrous mother relies on the commonplace that says the mind is a womb.[14] What is less obvious is that he combines that commonplace with a similar metaphor that identifies money itself as a kind of womb and that figures its generation in terms of a dangerously and sinfully autonomous, fertile woman. Anti-usury pamphlets often likened the 'birth' of money from money in such terms of feminine disorder and procreative excess.[15] The whore, Projecting, and her daughter, Stockjobber, descend from this figurative tradition. Representative of texts that link usury with female vice is Thomas Wilson's *A Discourse upon Usury*, which features a discussion between four men who represent the scope of opinion on usury at the end of the sixteenth century. The most conservative speaker, the preacher Oekerfoe, rails that 'Usurie' is 'the daughter of covetousness, a monstrous daughter I say, of an horrible fowle foster dame.'[16] Letting her loose in the land, he warns, will usher in a complete breakdown of the social order, because 'usurye, the daughter of covetousnes, the mother of mischiefe, and the very hel of evil, overthroweth trades, decaieth merchandizes, undoeth tillage, ... beateth down nobilitie, bringeth dearth and famine, ... and last of al causeth destruction and confusion universallye.'[17] Like Wilson's 'monstrous daughter of covetousness,' Defoe's allegorical Projecting arises from the union of Trade and Avarice, and becomes herself a 'horrible fowle foster dame.' Her numberless, destructive offspring undermine all positive industry, starve the nation, and wreak havoc in general.

Defoe's representation of Projecting as a prodigiously fertile, feminine beast vividly echoes the vision of another conservative denouncer of usury in Wilson's *Discourse.* The Civilian lawyer fulminates, 'What is more against nature, then that money should beget or bring forth money, which was ordeined ... not to increase it selfe, as a woman dothe, that bringethe forthe a childe ...? And therefore Aristotle saithe that such money as bringeth forthe money through usurie is an ugly beast, that bringeth forthe monsters from time to time, suche as are not in nature ... contrary to nature, order, and al good reason.'[18] Like many other early modern writers, the Civilian alludes to the passage in the *Politics* where the philosopher explained that 'this term interest [*tokos*], which means the birth of money from money, is applied to the breeding of money because the offspring resembles the parent. That is why of all modes of getting wealth this is the most unnatural.'[19] Defoe's representation of Projecting and Stockjobber as hydra-like, breeding monsters clearly carried this Aristotelian imagery forward into the eighteenth century, where it merged with a conservative, Tory argument against economic strategies for generating wealth such as credit and the National Debt. Jonathan Swift, for example, represented the undertakers of the South Sea trading company as 'midnight hags / Tormenting Fools be-hind their backs' in 'The Run upon the Bankers' (1720). Swift's hags observe no authority but their own and autonomously conjure false values that blight the nation.

Like Swift's witches, Defoe's whore, Projecting, figures a catastrophically autonomous generation, and seems therefore to promote a Tory argument about the generation of the New. Mr Review breathlessly recounts that, long after Avarice's rape, 'the old Mother teem'd still and brought forth every Hour one Plague or another; but t'would be endless to mention all her Children, that no Tongue can describe, nor no Eye can see the End of.' Recognizing no goal, authority, or end point but only the perpetual satisfaction of her own desire, Projecting figures the chrematistic generation that Aristotle denounced. To invoke Wilson's terms, Projecting, the daughter of covetousness, is also that vice's mother. Projecting would seem therefore to figure a principle of consumption that lies at the dark heart of commercial production, the canker that eats up the abundance that profits generated from profits always harbour, as opposed to the solid, putatively 'real' values found (not made) in land.

Nevertheless, Defoe's representation of Projecting can be understood not as a figure that coherently and consistently promotes a Tory objection to the generation of the New, but rather as an oppositional figure, an 'other' whose existence ultimately underscores the liberal point of view. Projecting represents the negative, usurious kind of autonomous generation that Mr Review strenuously seeks to distinguish from its positive counterpart. In other words, this whore plays a part in a rhetorical strategy that Defoe employed in order to highlight the beneficial virtues of activities of autonomous generation that had been denounced for centuries and was still largely regarded as a species of devouring usury: credit and long chains of interest-bearing debts. The *Review*, thus, carries on the ideological work of *An Essay*, which had demonstrated a discernable distinction between 'honest' and 'dishonest' projecting, or autonomous generation, in commerce.

In his allegories of Trade, the central focus of the *Review*, Defoe presents as confusing a face to the world as he does when he discusses projecting. Mr Review insists that Trade is 'the great Foundation of *English* Wealth' (*Review* 8.30 [1711]), but he also, paradoxically, characterizes Trade as a woman periodically beset by 'wandring Ghosts of *Trade Whymsies*' in her sleep, which cause her to give birth to 'multitudes of Mushrooms ... born to evaporate by Time, and dye in the handling.' Like the 'Abortions of the brain' that the spawn of Projectors bring forth, these ephemera autonomously 'spin out their own Bowels like the Spider' (*Review* 3.126 [1706]). What is Mr Review actually attempting to prove with his portrait of Trade as a woman beset by whims and fleeting, irrational thoughts? Surely such autonomous generation, which, like projecting, begins with nothing more than the *dream* of wealth, can hardly constitute the 'foundation' of the economy? Like the ephemeral 'projects' that prompt the people whom their 'authors' have duped to form committees, set up offices, shares, and books, how could such abstractions such as whims and ghosts yield real, material benefits, real 'Produce,' or live offspring?

Mr Review poses a version of this question to himself and responds suggestively, 'If any Man requires an Answer ... they may find it in this Ejaculation – Great is the Power of Imagination! Trade is a Mystery, which will never be completely discover'd or understood.' Like a woman, 'it suffers Convulsion Fitts, hysterical Disorders, and most unaccountable Emotions ... today it obeys the Course of things, and submits to Causes and Consequences; tomorrow it suffers Violence from the Storms and Vapours of Human Fancy ... and then all runs counter, the Motions are excentrick, unnatural and unaccountable – A Sort of Lunacy in Trade attends all its Circumstances, and no Man can give a rational Account of it' (3.126 [1706]). Identifying Trade with imagination and with femininity, Mr Review suggests that Trade is yet another species of projecting. She generates without the government of masculine reason, and she produces nothing real, only fantasies. And Trade without reason, generating only for the sake of generating, is a kind of unlimited generation (which Aristotle tells us is unwise and which Wilson insists will bring our ruin). He seems once again to be adopting a Tory point of view.

But this conclusion, that Trade should not be allowed to run free of manly control, does not actually answer the question, How do dreams and ephemera constitute the foundation of the English economy? In the event that we might feel inclined to discount Mr Review's association of Trade with female mutability as a statement of denigration on the grounds that a living denizen of the Age of Enlightenment could hardly be expected to esteem such wild irrationality, we should remember that Defoe praised Noah's Ark as a divinely inspired 'Project' that blurred the boundary between madness and reason: 'No question seem'd so ridiculous to the Graver Heads of that Wise, tho' Wicked Age, that poor *Noah* was sufficiently banter'd for it; and had he not been set on work by a very peculiar Direction from Heaven, the Good old Man would certainly have been laugh'd out of it, as a most senseless ridiculous Project' (*Essay* 40). Far from vilifying the vapourish, mutable, and unaccountable faculties of illusion that drive trade, Defoe, at times, actually seems to admire these very elements as part of the alchemy through which ideas give birth to real, new wealth. In short, Defoe links trade with the generation of new wealth and the creative imagination, or *fantasia*, which was already associated with women, wildness, and unpredictability, as well as with usury. I will return to this argument in detail in the next section. The point to get across now is that Defoe did not insist that the economy functioned in a reasonable, predictable manner, or that its production depended upon rational men rationally going about their business.

To demonstrate the argument that Defoe endorsed irrationality and a kind of fertility of the mind that had long been associated with femininity (what are the Muses if not this? or Dante's Beatrice, Petrarch's Laura?) as a means by which new wealth could be generated, we may turn to Mr Review's passionate encomiums to 'Credit, that Incomparable Lady, that Beauty of our Age,' and most fertile and most self-referential agency of wealth generation, who flirts while holding herself out of reach, like a 'coy Lass' (*Review* 3.5 [1706]). Credit, on Mr Review's view, was 'a Beautiful Lady, but the strangest, coy, humorsome Thing that ever was heard of

– made all the world Court her, Jilted them all in their turn, and indeed was never constant to any of her Lovers.' This marvellous, quasi-magical faculty of generativity, Mr Review sings with rapture, 'has paid Interest for nothing, and turn'd nothing into something, Coin'd paper into metal, and stamped a Value upon that what had no value before' (*Review* 7.55 [1710]). 'That substantial Non-Entity call'd CREDIT' allowed England to carry on the war 'with a hundreth Part of the ready Money that it would otherwise require' (6.31 [1709]). Credit has 'made Paper pay Millions instead of Money, doubled and trebled our specie by circulation; ... brought out our hoards, melted down our Plate, sold our Jewels to take Air for Silver, and split Stickes for Gold' (7.55 [1710]). Credit performs alchemical magic, producing gold from dross.[20] Although Mr Review concentrates on the mysteriousness of credit's generative abilities, her 'Power of Transmutation,' he also extols credit as a 'Teeming Womb' and 'Mother of Great Designs' (3.5 [1706]). Thus, he clearly comprehends the imaginative powers flowing through the economy specifically as a reproductive agency associated with maternity, but also with autonomous, open-ended generation.

Given Defoe's obvious enthusiasm for Credit's alchemical powers, we cannot conclude that the portrait of Projecting as a generating demon represents his total or unqualified repudiation of autonomous generation, nor his endorsement of a strictly empirical epistemology. As Terry Mulcaire observes, Defoe's representation of the free market as a woman associates 'it with the freedom of desire and imagination, not the freedom that comes from following divine prescription or natural law.' In fact, 'the figure of an autonomous male property holder, rationally pursuing his economic self-interest in the marketplace that ran according to fixed and rational laws, was less important to early liberal ideology than critics have long believed, while a complex psychological and aesthetic investment in the imagined figure of a desiring and desirable but deeply volatile female identity was much more important.'[21] Rather than assuming that the jilt Projecting and many-handed Stockjobber demonstrate Defoe's alignment with a conservative, Tory position, we need to see that they figure as negative alternatives to other, positive forms of open-ended generation that had to be cleansed of the taint of usury. These alternatives allow him to justify a Whig ideology. Credit possesses usury's extraordinary generativity, but her generation, however erratic, is both lawful and natural because it is essential to the productivity of England: 'Credit keeps Thousands of Families at Work, and gives Bread to Nations,' Mr Review instructs. It 'is the Health of our Body Politick' (*Review* 7.118, 119

[1710]). Like a good mother, Credit keeps money circulating through the English economy and sustains its citizens in hard times.

Anxious to establish some kind of basis for Credit, and in response to the assertions of more conservative economic commentators who claimed that no new wealth could be generated, Defoe insisted that human ingenuity and industry could create property. 'Every poor Man's Labour is an Encrease of the publick Stock,' he claimed. 'Whosoever he be, that improves Art to any Degree that was not known before, is a public Advantage to the General ... every Manufacturer, that makes that at home, which till then was bought from abroad, is a Blessing to his Country, an Advantage to the general Stock, and encreases both Wealth and People' (*Review* 5.107 [1708]). The source of England's prosperity and fertility lay not just in its land, but also in human labour involved in commercial trade, the production of commodities for circulation. Thus, Defoe reconceived acquisitiveness and the desire for money in general as the human power to generate wealth, not just from land given by God, but through divinely bestowed, generative faculties for bringing forth wealth out of the self through intellectual feats, magnificent management, projecting. Open-ended generation, maligned as a grotesque, unceasing, autonomous female reproduction in the anti-usury pamphlets reappears here positively revalued in the character of the industrious merchant and the Lady of the Exchange.

Nevertheless, just because Defoe moved away from the Aristotelian paradigm that governed more traditional thinking about economic generation during this period and advocated the claim that money could indeed give birth to money does not mean that he abandoned the paradigm altogether. Defoe remained conflicted about autonomous, unlimited generation, and this explains why it is difficult to interpret his representations of Projecting, Stockjobber, and Credit. On the one hand, we can think of Projecting and Stockjobber as *confirming* the Aristotelian paradigm insofar as they can be understood as examples of open-ended generation over against Credit, which is offered as an example of bounded, or limited, generation. Mr Review seems to be trying to pursuade us that Credit, whom the various merchants and parliamentary leaders woo in pursuit of marriage, and who often, in fact, does marry and set up shop with her favourites, is the sort of creature who settles down and can be encompassed by the legal, emotional, and mythical boundaries of heterosexual wedlock, which would technically render her a *femme couvert*, a woman enclosed and *limited* by her husband's identity.[22] Such a situation would bring Credit's unpredictable and irrational fecundity into the management of rational masculine government in the household, which is where Aristotle located the 'art of getting wealth.' Indeed, Aristotle insisted that the husband should rule his wife just as the soul rules the body and as 'the mind and the rational element' rule the passions. As he further explained, 'the equality of the two or the rule of the inferior is always hurtful ... the male is by nature superior, and the female inferior; and the one rules, and the other is ruled; this principle, of necessity, extends to all mankind.'[23] Lady Credit might be understood along Aristotelian lines as the wife and Parliament, or the nation, as the husband whose *telos* governs her fertility, the 'mind and rational element' that orders her procreativity. The bad mothers Projecting and Stockjobber establish a discursive space against which the positive term may be allowed to emerge as a putatively reliable, bounded generation. Still, Projecting and Stockjobber are necessary, even inextricable, from Credit insofar as it would not be possible to understand the legitimacy of the latter without the illegitimacy of the former.[24]

On the other hand, the opposition between the negative and positive terms remains and may also be understood as one that breaks with the Aristotelian paradigm. As Mr Review freely admits, the 'coy lass' Credit escapes all government, marital or otherwise, that monarchs, merchants, and Parliaments would apply to her.[25] And what actually distinguishes one category of getting money from money, or usury, from the other? Mr Review musters up a strong argument in favour of Credit as a limited generation based on 'real' values, such as land and honourable dealing, and against Projecting as an unlimited generation based on something ephemeral. But he soon argues himself into circles, as it were. 'Land then,' he argues, 'is the first Fund, the Original of Trade, that must be confess'd; but Trade is the Fund of Improvement, even to land' (*Review* 8.22 [1710]). Trade, and the growth that, he argues, is based upon solid, intrinsic values, finally depends on a dialectical relationship between land and trade, materiality and the process of exchange, in which seemingly metaphysical, artificial values – exchange values – are generated. He also argues that 'Nothing but punctual honourable dealing can restore Credit, *not that*, without a Series, a continued Practice of such dealings. How to do that ... I must own is the Miracle' (*Review* 3.5 [1706]). What is desired is an endless *multiplication* of wealth that can only take place through a deferral of closure, the indefinite postponement of circulation's end. As he himself admits, once the 'last Act of Trade' is reached, when people exchange credit for goods that they need and intend to consume ('Food or Cloathes') multiplication stops, credit dies, and trade withers: 'here is no rolling on' (6.33 [1709]). The only *real* fund of value is neither labour, nor land, but rather *circulation*, and a continual process of projecting values through open-ended generation. 'Public Credit supports Power, Power Supports Commerce, and Commerce Credit; and Endless Circulation Runs through these Things. Their Affinity, Connection, and Dependence is such, that they ever Influence one another, and rise and fall together' (7.118 [1710]). Credit, therefore, like projecting, constitutes a genuine, if utterly incomprehensible, source of the wealth upon which the future of English happiness depends. Trade is a mystery.

Thus, both Credit and Projecting appear in Defoe's prose as virtually indistinguishable faculties of autonomous generation, one of which is defined negatively in order to justify the other. Like Hegel's master and slave, the elevation of the former depends on the degradation of the latter. We must therefore conclude that, while Defoe remained ambivalent about the ethics of 'projecting' and open-ended proliferation, he conclusively broke with the Aristotelian paradigm by justifying the kind of unlimited production that Aristotle attacked. This justification also allowed him, paradoxically, both to confirm and challenge Aristotle's argument that a well-ordered society is characterized by the rule of the rational over the passionate, and of men over women. The next section develops this last point.

## II. Imaginative Projecting

We can trace an analogue to the political-economic debate about the possibility of generating new wealth in the late-seventeenth and early-eighteenth century to contemporary discussions about poetic making and imagination. Defoe located human creativity in the inextricable spheres of wealth production and human reproduction, a convergence already present in anti-usury discourse, as we have seen. During this period, it made sense to him to locate wit, a word used interchangeably with fancy, imagination, and genius at this time, in yet another allegorical genealogy of commerce. Not surprisingly, this 'Witt,' a 'meer jilt' and a 'common whore,' resembles both Credit and 'Projecting.'

The term 'wit' derives from the Anglo-Saxon *witan*, to know, and originally denoted the mind and the senses. In the late seventeenth century 'wit' also connoted rhetorical quickness and talent but also, in some cases, lunacy. Although a single definition for 'wit' is notoriously difficult to articulate during this period, the term seems to have undergone a discursive re-evaluation similar to that of the concept of usury. The *OED*'s fifth substantiation of 'wit' demonstrates that, in the early modern and early Enlightenment periods, wit connoted innate, organic creative talent, akin to later nineteenth-century definitions of genius.[26] In Defoe's time, writers disagreed with one another about the question of whether or not 'wit' was a faculty for generating the New, or whether it only served to accumulate images already in existence. Thomas Hobbes, for example, argued that the imagination is nothing more than the combination of ideas stored in memory. He allied imagination with 're-conning' (as in *connaissance*, recognition): 'Sometimes a man knows a place determinate, within the compasse whereof he is to seek; and then his thoughts run all over the parts thereof, in the same manner, as one would sweep a room, to find a jewell; or as a Spaniel ranges the field, till he finds a sent; or as a man should run over the Alphabet, to start a rhyme.'[27] Dryden famously defined 'wit' with the same hunting-and-gathering metaphor, and John Locke also spoke of wit as 'the assemblage of *Ideas*, and putting those together with quickness and variety ... to make up pleasant Pictures, and agreeable Visions in the Fancy.'[28]

Hobbes's representation of the imagination as the movement of thoughts, which he divided into 'unguided' and 'regulated' categories, corresponds in part to Aristotle's distinction of the art of getting wealth into unlimited and limited categories. Hobbes argued that 'in the imagining of any thing, there is no certainty what we shall imagine next; only this is certain, it shall be something that succeeded the same before, at one time or another.'[29] Unguided imagination is that 'trayne of thoughts, or Mentall Discourse' that 'wanders' without any certain goal, or without the governance of any 'end and scope of some desire.' Hobbes gave as an example of such dangerous and 'wild ranging of the mind' the procession of thoughts that led to the betrayal of Christ, and thereby suggested that all unguided thought will lead to baneful results. Thoughts should be limited, or regulated, by a conscious 'end.' 'Invention,' Hobbes said, is the train of thoughts 'governed by designe,' as well as 'Remembrance' and the spaniel ranging over the field of memory with a certain purpose.[30] Unguided, wandering, irrational linguistic signification was, alternately, something like the proliferation of false values in money. 'For the errours of Definitions,' he explains, 'multiply themselves, according as the reckoning proceeds; and lead men into absurdities, which at last they see, but cannot avoyd, without reckoning anew from the beginning, in which lyes the foundation of their errors. From whence it happens, that they which trust to books, do as they that cast up many little summs into a greater, without considering whether those little summes were rightly cast up or not.' Hobbes also compared the 'abuse of language' to the accumulation of imaginary values. 'Words are wise mens counters,' he warns, 'they do but reckon by them: but they are the mony of fooles.'[31] False sums in words come about from the lack of a proper, bounded end for its proliferation. Without that *telos*, Hobbes insisted, one cannot regulate the proper relationship between signifier and signified, and then meaning proliferates wildly and incoherently. As Aristotle said of chrematistic production, such unlimited association generates nothing but 'riches of the spurious kind.'

The polarity that Defoe sets up in the *Review* between the twins 'Witt' and 'Invention' would seem to reiterate Hobbes's contrast between wandering and focused, open-ended and bounded, thinking. As Mr Review tells the story, the twins are the offspring of Necessity, but only one of them amounts to much. The boy, Invention, gets married and has scores of useful children:

> the names of some of there were as follows, INDUSTRY ..., INGENUITY, HONESTY ... besides several others – And these all partaking of the same Temper with their Father, set all the World to Work: had the Example been follow'd, and had not *Necessity's* other Branch, Calld *Wit*, and her Numerous Progeny of Bastards destroy'd it, as fast another Way, they [would have] laid such Stores up, and such Foundations of Wealth and Prosperity, that the Name of their Mother POVERTY was in a fair Way of being forgotten in the World, and all the Misery Mankind seem'd to be born to look'd as if it might have been prevented.

Invention observes steady goals: work, riches, and general prosperity. His generation, in both senses of the word, is productive of real values. Witt, by contrast, 'indeed prov'd a mere Jilt, turn'd common Whore, and her Numerous, tho' Spurious Race, has filled the World with Fops and Beggars, who like the Drones in the Hive, starve and help to undoe Mankind' (*Review* 8.38 [1711]). She proliferates wildly, without reason or aim, and therefore constitutes an open-ended generation that not only fails to support the household and the state, to invoke Aristotle's terms again, but that also de-generates whatever wealth her brother Invention brings into the world. Witt, Mr Review complains, 'fills the World with Misery, Poverty, Woe, and Wickedness' (ibid.). The allegory therefore seems to confirm both the Aristotelian and the Hobbesian prohibitions of unlimited generation of wealth and thought.

A key difference divides Defoe's 'Invention,' however, from Hobbes's definition of that concept as a train of thoughts 'governed by designe.' While Hobbes regards it as an essentially barren agency that observes a logical progression of what 'succeeded the same before,' Defoe defines Invention as a generative faculty for bringing forth the New and other agencies of fabrication. Invention fathers Industry, Ingenuity, and Honesty, who in turn generate 'Ploughman,' 'Grazier,' 'Miner,' 'Gardiner,' 'Dary,' 'Handicraft,' 'Wright,' 'Smith,' 'Tinker,' 'Engineer,' 'Manufacture,' 'Barter,' and 'Trade.' While Hobbes defines 'invention' according to its Latin sense (*invenire* means to find or discover), Mr Review renders the term according to an altogether different word, *genare,* to generate or bring forth. Since Invention's offspring also include 'Money,' 'Coin,' 'Merchant,' 'Shopkeeper,' 'Credit,' and 'Projector,' who, in this particular genealogy, is a 'Famous' *male* 'Progenitor' (*Review* 8.38 [1711]), Invention may be understood as yet another faculty for (be)getting money from money, which Aristotle claimed was 'unnatural.' Yet insofar as he and his offspring produce in a rational, comprehensible manner that benefits the state, Invention figures a *relatively* limited generation. Mr Review defends this kind of imaginative generation over against 'Witt,' which he associates with wild proliferation, irrationality, and femininity.

Defoe's association of wit with wildness, incoherence, and women carries forward ancient ideas that linked imagination and figurative language with passionate disorderliness in need of reason and restraint, often in gendered terms.[32] John Milton, for example, believed that no authentic, praiseworthy, or sacred art could issue from the wild fertility of 'Fancy' let loose from the governance of Reason, her 'Chief.'[33] Thomas Rymer explained that '*Fancy* leaps and frisks, and away she's gone, whilst *reason* rattles the Chains and follows after.'[34] And Thomas Shadwell, admiring those who 'subjected wit to the government of judgment,' argued that none of the 'very fanciful Plays, admired most by Women, can be so good a Play as one of *Johnson's* correct and well-govern'd Comedies.'[35] Alluding, perhaps, to the story of Phyllis astride Aristotle, Swift suggested that once a 'Man's Fancy gets *astride* on his Reason,' then 'common Understanding, as well as Common Sense, is Kickt out of Doors.'[36] Locke linked metaphor, allusion, and simile with irrationality and femininity, as well as imagination with madness: '*Mad men,*' he intoned, 'have not entirely lost the faculty of reasoning,' but 'having joined together some *Ideas* very wrongly, they mistake them for Truths.'[37]

Defoe similarly drew upon gendered polarities of masculine reason and feminine passion to describe the imagination in *The Pacificator* (1700):

> *Wit* is the *Fruitful Womb* where Thoughts Conceive,
> *Sense* is the *Vital Heat* which Life and Form must give
> *Wit* the *Teeming Mother* brings them forth,
> *Sense* is the *Active Father* gives them worth.[38]

Much like many commentators during this period, Defoe's speaker stressed the need for harmony between two relatively equal powers, wit and sense, and argued that the latter organized and shaped the former. If he departed from convention by granting generative powers to wit in *The Pacificator*, he also upheld the Aristotelian theory that only male agents had formative power, and that women passively contributed mere matter to their children.[39]

The whore, 'Witt,' whom we meet in the *Review*, however, demonstrates none of the redeeming qualities of her predecessor, 'wit,' who is happily married off to Sense in *The Pacificator*. Like effervescence, froth, or 'trade whimsies,' this Witt subsists on the level of the literary alone, corrodes whatever sense might dwell beneath her, and then evaporates. Moreover, the hordes that pour out of her body consume everything in sight. They most vividly correlate to Projectors, whom Mr Review denounces as 'Locusts and Caterpillars': 'These Vermin fly-blow'd the Nation, and bred Moths even in our Parliament House, which eat up all our Funds, canker'd our Money, and in short threw the whole Kingdom into Convulsion Fitts' (*Review* 6.31 [1709]). Witt's offspring are not only *like* those of Projecting, they *are* projectors, who leave 'nothing but Cobweb, and a tangl'd Husk of Emptyness in the Fingers of those Fools, that were deceiv'd with the Appearance' (*Review* 3.126 [1706]). To cite the terms from the *Pacificator*, Witt's womb teems without Sense, and she brings into the world nothing but ephemera because she lacks the masculine 'vital heat' that gives life and shape. In order to become something viable, in order to become art, Witt's matter (and *mater*, mother) must encounter masculine, form-giving agency. This Aristotelian notion correlates to the principle of limited generation, in so far as Witt's reasonless generation constitutes a dangerously autonomous proliferation, a wandering proclivity not unlike that of Trade, who disastrously encounters Avarice and dooms humanity to poverty.

Mr Review also reiterates the association of femininity, imagination, and irrationality, as well as the Aristotelian-Hobbesian principle that unlimited generation leads to false values, when he illustrates Trade as a woman whose imagination escapes the management of reason in sleep:

> Multitudes of Mushrooms have obtain'd upon the World, whose Birth was the Produce of meer Vapour and Exhalations; who, as they sprung up in the dark Midnight Moments of Trade, when her Eyes were shut, and when she was as it were doz'd with Dreams, and hagrid with wandring Ghosts of *Trade Whymsies*; so they were born to evaporate by Time, and dye in the handling, that by the Nature of them were destin'd to dissolve like a Cloud, and spin out their own Bowels like the Spider, that had nothing material in them, but being merely imaginary in their substance.

This portrait recalls the famous passage in the *Timaeus* in which Plato decries *fantasmata* that arise in the mind asleep or under divine inspiration, a notion that Cicero developed into a doctrine that held that divine *furor* gave rise to poetry.[40] Paradoxically, the colourful, figurative portrait of Trade as dreaming woman and the allegory of Trade debauched by Avarice *exemplify* the allusive, ungrounded signification (metaphor building upon metaphor) that they ostensibly denounce. Mr Review employs imaginative, open-ended proliferation in language, or wit (or projecting!), in service to the stated aim of enlightening his countrymen and countrywomen about what terrible things happen when the economy throws off the management of reason and allows imagination to wander (a scene he also imagines as Trade going a-caterwauling after Novelty). In other words, Defoe's speaker hypocritically employs metaphor and allusions, the tools of the poet under the possession of the *furor*, in order to denigrate unlimited generation as a kind of madness, which is further compared to feminine promiscuity.

We can say with certainty that the feminine figures of 'Witt' and 'Trade' demonstrate Defoe's anxieties about the wandering, autonomous properties of allusion, metaphor, and unlimited signification in general, a nervousness he displays also in his economic theory. In both registers (of imaginative and economic generation), which were obviously inextricable from one another in his world view, Defoe falls back on gendered oppositions that align men with reason and order and women with madness and chaos. Apparently, he does so in order to distinguish between that sort of 'projecting' that he could stomach, and that which he could not abide. He abandons these polarities temporarily when he celebrates Credit – that overtly feminine agent of self-referential generation and signification – as an agency that gives birth to the New entirely through her own alchemical power. But even Credit's reputation undergoes continual attack, suggesting that this 'lady' remains difficult to distinguish from a tramp.

In the *Review*, then, Defoe only imperfectly extracts the category of 'Credit' from what had been denounced in economics as usury, which appears in the form of Projecting and Stockjobber. This is so even though his speaker defines credit as a legitimate form of open-ended generation that fits with the notion that 'God, in the Order of Nature ... made Trade necessary to the making the Life of Man Easy' (*Review* 1.54 [1713]). Similarly, by steeping the category of 'Witt' in ancient platitudes about feminine wildness and lack of formative agency, he creates in opposition a *new* category for imaginative generation (invention) that shares all of the older term's prodigious fecundity and none of its vices, a vision of the fertile imagination whose production is orderly, reasonable, and comprehensible within an Aristotelian teleology. In other words, he comes up with a strategy for cordoning off the dangerous fecundity that he associates with Witt *and* 'the Faculty of Projecting' (as his speaker puts it in *An Essay*) by developing a more positive imaginative faculty, a vital power of bringing forth the New. He, confusingly, sometimes calls it 'Projecting' and at other times 'Invention,' and he usually (though not always) genders this faculty masculine. His organizing principle appears to arise from a gendered polarity that arranges reason over madness, sense over non-sense, 'real' fecundity over barren generation, and men over women.

But the fact remains that Defoe employed 'wit,' and a good deal of metaphor, allusion, and other forms of eloquence, in order to denounce open-ended signification; self-reflexively committing the very crime he condemned in order to condemn it; projecting copiously in order to rein in projecting! Defoe's ambivalence about autonomous generation stemmed from an ancient philosophical resistance to unlimited generation, yet he also clearly understood poetic and economic imagination to be part and parcel of the problem that afflicts every child of Necessity: getting on in the world. This astonishingly prolific author had constant financial difficulties and coped with the ever-present possibility of ruinous insolvency by generating without end. And when Mr Review admits that 'writing upon Trade was the Whore I really doated upon, and design'd to have take up with, but she is ravish'd out of my hands' (*Review* 1.106 [1713]) in the last issue of the *Review*, he confesses his inability to resist what he experienced as a self-indulgent desire to pursue what could not be captured in language. He does this through an unceasing signification directed blasphemously *not* towards a certain end, but rather towards one that he indulged for the pure pleasure of signifying.

We may understand Defoe's fascination with Trade in terms of a sexualized masculine desire that longs for mastery but also to reproduce itself, a dream of governing creativity, and a fear that, once the dilating womb of discourse or commerce opens up, its gates will never shut again.[41] In service to a patriarchal aesthetic economy, Mr Review clutches to his breast a dream of autonomous generation (Credit, imagination, Projecting, Witt) always threatening to escape its government. He also fantasizes about that source of value as a woman impervious to masculine reckoning. 'No Man,' he writes, 'can give a rational account of it.' Only in this sense of a *desire* to impose order on autonomous, open-ended proliferation can I agree with Maximillian Novak's astute comment that, while 'he liked to flirt with his "coy mistresses" – credit, paper money, and land banks,' Defoe remained 'at heart, a most faithful husband to solid mercantilist ideals of intrinsic value and the balance of trade.'[42] Because mercantilists generally believed that there was a limited amount of wealth in the world, and Defoe clearly dreamed of schemes by which more could be created, he challenged the idea that projecting remained nothing more than fantasies for generating what would always be purely fictional values. In other words, he actually valued wit's fantastic *fantasmata* and insisted that they were real and valuable.

How might the word 'wit,' which originally meant 'the faculty of knowing or perceiving, the mind,' have taken on the seemingly opposite sense of a faculty for wild, irrational, and unlimited generation of ephemera, illusions, and abstract values? This transition is comprehensible within the context of sceptical philosophy, particularly that of Descartes, who contributed to the idea of the imagination as an essential, active, and creative aspect of the mind.[43] In 'Cogito and the History of Madness,' Jacques Derrida reads Descartes over Foucault's shoulder to show that the founding moments of modern philosophical discourse, a language that would speak rationally and tell the truth about things, can be understood as an effort to cordon off a part of the discursive totality felt to be at odds with reason.[44] Derrida argues that the discourse of reason is finally impossible to separate from the discourse of madness, and that the project to think ourselves beyond the division of reason from madness is itself insane and impossible, although we nonetheless remain engaged in it. My research suggests that early-eighteenth-century aesthetics developed as part of a larger, gendered Enlightenment project to define 'reason' over against its various antitheses, including 'imagination,' and that this development is related to an analogous intellectual movement in which some forms of autonomous and unlimited commercial generation were defined in opposition to usury. In this emergent, discursive formulation, woman was to man as imagination was to judgment, and therefore gendered distinctions became central to eighteenth-century epistemology, commerce, and aesthetics.

Daniel Defoe insisted on the polarity between reason and madness as factors leading to 'honest' and 'dishonest' projecting, even while he sought a language with which to defend ceaseless dilation and what appeared to him to be a wild, interior fecundity generating beyond all the limits that human reason would impose upon it. Can we say, then, that, when he celebrated this projecting 'whore' in himself, he acknowledged cultural stereotypes that denied *materia* (from *mater*, mother) but transcended them by embracing a more fluid, complexly gendered sense of self? Possibly. Defoe accepted his immersion in an essentially irrational economy. He saw that imagination could lead to chaos, but also that government could lead to tyranny. What he looked for was a balance, a way of dwelling in the world. If more conservative writers had argued *ex nihil nihil fit*, Defoe optimistically insisted that women and men constituted *somethings*, however flawed, from which the New and the Real could issue.[45]

## NOTES

1 For an important essay concerning Defoe's interest in ghosts as *fantasma* crucially sustained in the literary medium, see Jayne Lewis, 'Spectral Currencies in the Air of Reality: *A Journal of the Plague Year* and the *History of Apparitions*,' *Representations* 87 (2004), 82–101.

2 Thornely, trans. *Daphnis and Chloe*, 113, cited in *OED*.

3 Gareth Roberts, *The Mirror of Alchemy: Alchemical Ideas and Images in Manuscripts and Books from Antiquity to the Seventeenth Century* (Toronto: University of Toronto Press, 1994), 20.

4 See, for example, Michael McKeon's 'The Origins of Aesthetic Value,' *Telos* 57 (Fall 1983), 63–82.

5 Steven Pincus, 'Neither Machiavellian Moment nor Possessive Individualism: Commercial Society and the Defenders of the English Commonwealth,' *American Historical Review* 103 (1998), 708. See also his forthcoming book, *The Glorious Revolution and the Origins of Liberalism.*

6 *An Essay upon Projects* (1687) in *Political and Economic Writings of Daniel Defoe,*

ed. W.R. Owens (London: Pickering & Chatto, 2000), 34. All subsequent references to this edition will be made parenthetically in the text.

7 *Politics*, book 1, §8, 1256a–b, in *The Complete Works of Aristotle: The Revised Oxford Translation*, ed. Jonathan Barnes, 2 vols. (Princeton: Princeton University Press, 1984), 2:1997.

8 *Politics* 1, §10, 1258a–b; in *Complete Works of Aristotle*, 2:1997.

9 See, for example, the preface to *Roxana* (1724), in which the Relator asserts that 'this Story differs from most of the Modern Performances of this kind ... in this Great and Essential Article, Namely, That the Foundation of This is laid in Truth of Fact; and so the Work is not a Story, but a History.' *Roxana*, ed. David Blewitt (London: Penguin, 1982), 35.

10 Maximillian Novak, *Economics and the Fiction of Daniel Defoe* (Berkeley: University of California Press, 1962), passim.

11 Sidney Godolphin, 1645–1712, was Lord Treasurer from 1702–10. He was created Earl of Godolphin in 1706.

12 Trade thus corresponds to the classic stereotype of the sinful woman who, because she recognizes no authority other than herself, brings about a great fall (i.e., Eve).

13 *A Review of the State of the British Nation*, 5.107 (1708), in *Defoe's Review: Reproduced from the Original Editions*, ed. Arthur Secord (New York: Columbia University Press, 1938). All subsequent references to this publication are abbreviated as *Review* and cited by volume and number parenthetically in the text (e.g., 5.107).

14 For a useful exploration of this trope, see Katherine Eisaman Maus, 'A Womb of His Own: Male Renaissance Poets in the Female Body,' in *Sexuality and Gender in Early Modern Europe: Institutions, Texts, Images*, ed. James Grantham Turner (Cambridge: Cambridge University Press, 1993), 266–88.

15 Anti-usury texts that associate the practice with female vice include Robert Bolton, *A Short and Private Discourse between M. Bolton and one M. S. concerning Usury. Published by E.B. by Mr. Bolton's Own Coppy* (London, 1637); Roger Fenton, *A Treatise of Usurie, Divided into Three Bookes* (London, 1612); and Miles Mosse, *The Arraignment and Conviction of vsvrie* (London, 1595).

16 Thomas Wilson, *A Discourse upon Usury*, ed. R.H. Tawney (New York: Augustus Kelley, 1963), 230.

17 Ibid., 366.

18 Ibid., 286–87.

19 *Politics* 1, §10, 1258b. The insertion in brackets is mine.

20 Cf. Credit 'has the effectual Power of Transmutation – For it can turn Paper into Money, and Money into Dross,' *Review* 5.107 (1708).

21 Terry Mulcaire, 'Public Credit; or, The Feminization of Virtue in the Market-

place.' *PMLA* 114 (1999), 1030. Implicit in Mulcaire's argument, and very much in line with my own thinking on this subject, is the claim that the 'financial revolution' enabled a transformation in the conceptualization of virtue – which could now be figured by a woman, and that this figure also enables a positively revalued symbol of the fecund potential of imagination.

22 See, for example, *Review* 3.5 (1706), and my discussion in 'The Mistress of the Marriage Market: Gender and Economy in Defoe's *Review*,' *ELH* 69 (2002), 359–83.

23 *Politics* 1, §7, 1256a; 6, 1254b; in *Complete Works of Aristotle*, 2:1993, 1990; cf. *Economics* I, §3; III, §1; 2:2131, 2147.

24 Cf. *Review* 3.6 (1706), in which Mr Review distinguishes 'Moderate Credit' from 'Exorbitant Credit.'

25 Cf. *Review* 3.5 (1706); 6.32 (1709).

26 On the development of the concept of 'genius,' see also Jonathan Bates, 'Shakespeare and Original Genius,' in *Genius: The History of an Idea*, ed. Penelope Murray (New York: Basil Blackwell, 1989), 76–97.

27 Thomas Hobbes, *Leviathan*, ed. J.C.A. Gaskin (London: Oxford University Press, 1996), 17.

28 See Dryden's famous formulation in 'An Account of the Ensuing Poem,' prefixed to *Annus Mirabilis* (1666).

29 *Leviathan*, 16.

30 Ibid., 16–17.

31 Ibid., 24.

32 Richard Kearney notes that rabbinical commentary interprets the imagination, the *yetzer hara*, or 'evil imagination,' as often identified with lustful impulses: 'bodily lust ... a symptom of the yetser or account of its origin in Eve's temptation and the subsequent fall into the historical order or sexual procreation.' *The Wake of Imagination: Toward a Postmodern Culture* (Minneapolis: University of Minneapolis Press, 1988), 44.

33 John Milton, *Paradise Lost*, ed. Scott Elledge (New York: W.W. Norton and Co., 1993), 5:100–13.

34 Thomas Rymer, *Preface* to the *Translation of Rapin's Reflections on Aristotle's Treatise of Poesie 1674*, in *Critical Essays of the Seventeenth Century*, ed. J.E. Springarn, 3 vols. (Oxford: Clarendon Press, 1908), 2:185.

35 Thomas Shadwell, in *Critical Essays*, 2:159.

36 The well-known story of the philosopher's seduction by a Phyllis, and love's triumph over reason, is thought to have begun with the fourteenth-century Henri d'Andeli's *Lai d'Aristote*. Jonathan Swift, 'A Digression concerning the Original, the Use and Improvement of *Madness* in a Commonwealth,' in *A Tale of A Tub, To Which is Added The Battle of the Books and the Mechanical Opera-*

*tion of the Spirit* ..., ed. A.C. Guthkelch and D. Nichol Smith (Oxford: Clarendon Press, 1958), 171.

37 John Locke, *An Essay concerning Human Understanding*, ed. Peter Nidditch (Oxford: Clarendon Press, 1975), 508, 161.

38 *The Pacificator* (London, 1700), 14.

39 See *The Generation of Animals* 1, §20, §21, and 4, §1–§2 in *Complete Works of Aristotle* 1: 1132, 1184, 1186.

40 Plato, *Timaeas* 71E–72A; Cicero, *De Oratore* 2.46.194 and *De Divinatore* 1.37.80; cited by Martin Kemp, 'The 'Super-Artist' as Genius,' in *Genius, The History of an Idea*, ed. Penelope Murray (New York: Basil Blackwell, 1989), 38.

41 On the politics of this ancient anxiety in Renaissance rhetoric, see Patricia Parker, *Literary Fat Ladies: Rhetoric, Gender, Property* (New York: Methuen, 1987). For an exploration of this theme in eighteenth-century poetry, see Marilyn Francus, 'The Monstrous Mother: Reproductive Anxiety in Swift and Pope,' *ELH* 61.4 (1994), 829–51.

42 Novak, *Economics and the Fiction of Daniel Defoe*, 15.

43 John D. Lyons, 'Descartes and Modern Imagination,' *Philosophy and Literature* 23.2 (1999), 302.

44 Jacques Derrida, *Writing and Difference* (Chicago: University of Chicago Press, 1978), 31–63.

45 I gratefully acknowledge and thank Martin Gierl, Sarah Kareem, Ann Kelly, Kathleen Lynch, Maximillian Novak, and Craig Yirish for their helpful comments on this paper.

chapter eight

# Living Forever in Early Modern Europe: Sir Francis Bacon and the Project for Immortality

DAVID BOYD HAYCOCK

The long eighteenth century seemed to represent a crisis for health and medicine in Britain. As Daniel Defoe observed in his *Tour* of 1724, the nation was clearly getting wealthier – but as the physicians saw, it was not getting any healthier. In fact, the opposite appeared to be true: if anything, the English were getting sicker. Dr William Stukeley was not alone in attributing the growth of nervous distempers to the rise in wealth and luxury: 'Our leaving the country for cities and great towns, coffeehouses and domestic track of business, our sedate life and excesses together, have prepar'd a plentiful harvest for these disorders,' he wrote in 1722. A decade later, his fellow physician George Cheyne famously declared that this was the 'English Malady.'[1] The 'English' had become 'Martyrs to ... Luxury and Wantonness'; those who transgressed the simple guidelines laid out in his 1725 *Essay of Health and Long Life* were 'guilty of a Degree of Self-Murder; and an habitual Perseverance therein' that was nothing less than 'direct Suicide.'[2] Given such profligacy, then, there was little wonder that, when looking back in time to the Ancient world of Adam and the Patriarchs, these doctors believed that men had lived for hundreds of years, seemingly free from illness. As the physician Edward Maynwaringe pointed out in 1670, 'In the Primitive Age of the World, mans life was accounted to be about 1000 Years: but after the Flood, the Life of Man was abreviated [*sic*] half.' It had continued to fall, such that by the time of Moses it was 'commonly not exceeding 120 Years.' And '[n]ow the Age of Man is reduced to half that: 60 or 70 years we count upon.'[3] What exactly had gone wrong? And what greater project could there be for the long eighteenth century than the recovery of life's full potential?[4]

It was Sir Francis Bacon who might be said to have launched just this project, this 'vast Undertaking, too big to be manag'd, and likely enough to come to nothing.'[5] In *The Advancement of Learning* (1605), he explained that he saw the duties of medicine as threefold: the conservation of health, the cure of disease, and the prolongation of life. The latter, he wrote, was 'the most noble of all' the objectives of physicians, but had been overlooked by them.[6] They had 'ignorantly ... mingled and confounded it as one and the same with the other two ... But to draw out the thread of Life, and to prorogue Death, for a season which silently steales upon us, by naturall resolution, and the Atrophie of Age; is an argument that no Physitian hath handled it according to the Merit of the subject.'[7] Though Bacon accepted that 'the retarding of Age, or the restoring of some degree of youth, does not easily purchase a beliefe,' he suggested that a man who had studied 'perfectly' the processes of the human body, and had investigated thoroughly the effects of diets, baths, ointments and 'proper Medicines,' might well be able to prolong life, or at least 'renew some degrees of youth, or vivacity.' Bacon thus expounded a complex scheme for prolonging youth and life. His methods included careful diet, exercise, dress, climate, and 'seasonable sleep.' Baths and unctions, regular purging, the letting of blood, and 'attenuating Diets, which restore the Flower of the Body' were all means that could be employed to reduce the effects of ageing.[8] But this was a life's work, and the alchemists' belief that immortality 'should be effected, by a few drops of some precious Liquor,' was foolish.[9]

In 1623 Bacon published *The Historie of Life and Death*, wherein he suggested that the youthful body's ability to repair itself 'might be eternall.'[10] Written (in Latin) over the winter of 1622, *The Historie* was to have been the last of six monthly essays that were to form the third part of his 'Great Instauration,' Bacon's radical and grand plan to overthrow traditional learning and refound scholarship on new, experimental, empirical, and essentially modern foundations. But Bacon decided to promote his thesis on the prolongation of life to second place, given 'the extreme profit and importance of the subject, wherein even the slightest loss of time should be accounted precious.'[11] There was to be no hanging around in the search for immortality.

But why did Bacon consider this project to be of such importance? In at least one respect, it seems a curious aspect of medicine on which to have focused such attention. As John McManners has meticulously shown, premature death was a prevalent condition in early modern Europe. There was, he estimates, only a one in fifty chance of a child born in France at the beginning of the eighteenth century reaching the age of fifty, and but a one in one hundred chance of its reaching seventy.[12] 'One-third of the human race perishes before reaching the age of twenty-eight months,' wrote the Comte de Buffon; 'half the human race perishes before the age of eight years.'[13] When Bacon was writing a century and a half earlier, these figures could only have been worse. To live long in early modern Europe, McManners observes, 'meant, most often, loneliness among a generation of strangers.'[14] Given such terrifying statistics of mortality, why did Bacon focus his attention on the prolongation of life? Why was this 'the most noble of all' objectives in medicine? Why not improvements in childbirth, or the fighting of disease, or pharmacoepia?

Bacon gave no direct reason for his concern with longevity in *The Advancement of Learning* or *The Historie of Life and Death.* But in the fragments known as Valerius Terminus, published around the turn of the seventeenth century, he had explained that 'the true ends of knowledge' were

> a restitution and reinvesting (in great part) of man to the sovereignty and power (for whensoever he shall be able to call the creatures by their true names he shall again command them) which he had in his first state of creation. And to speak plainly and clearly, it is a discovery of all operations and possibilities of operations from immortality (if it were possible) to the meanest mechanical practice.[15]

The quest for immortality thus fitted into Bacon's huge historical project for the re-establishment of learning as it had been in the earliest of times. For calling the creatures of the Earth by 'their true names' had been one of Adam's first acts in the Garden of Eden, and it was held that in Eden Man had lost the opportunity for immortality. In 1644 Richard Overton began his book *Man's Mortalitie* by explaining that God created Adam and Eve 'both innocent and free from sin, and so from Death and mortality: For the wages of Sin is Death, Rom. 5.12. I Cor. 15.56. Thus Man was gloriously immortall.'[16] For George Cheyne, mankind's shortened lifespan was likewise a punishment for original sin. Cheyne believed the globe we now inhabit was 'ruin'd and defac'd,' an imperfect version of the original perfection; life on this Earth, with this 'primitive Creation-Bod[y],' was a way of working off the 'Chains' of divine punishment and freeing our soul. Ultimately, however, both Earth and Body would be 'perfectly purified.' Then both the soul and the 'Creation-Body' would achieve immortality 'by the Divine Order and Appointment.'[17] In Cheyne's opinion, the decline in human lifespan resulted from God, after the Fall, giving Adam and Eve permission to eat meat. Thus, one part of God's 'Design in permitting animal Food, must of Necessity have been to shorten the Duration of human Life, to create, or at least to permit us by it, to multiply and enrage Diseases, Misery and Sufferings.'[18] It was clear to Cheyne (and he provided an argument for this opinion) that a diet that included meat was far more damaging to our bodies than one solely consisting of vegetables. Cheyne was, therefore, an early advocate of vegetarianism.[19]

Bacon's project, however, was far more sophisticated than Cheyne's. Recapturing immortality was not simply about extending life by avoiding meat; it was part and parcel of the 'Great Instauration,' the grand project to recover the lost knowledge of Adam. For it was firmly believed in Bacon's day that to know the 'true names' by which Adam had named the creatures of the Earth would be to truly understand and control them – for these original names had been lost in the confusion of languages following God's destruction of the Tower of Babel. From his feat of naming the beasts, Adam was credited with universal, encyclopaedic knowledge and a perfect memory. Indeed, according to the philosopher Joseph Glanvill, writing in 1661, Adam's knowledge had even been enhanced by the power of telescopic sight, so that he could see in all its finery 'the Coelestial magnificence' of the Heavens; and perhaps he also had microscopic sight, so that 'he saw the motion of the bloud and spirits through the transparent skin.'[20] The herbalist Nicholas Culpeper would be deeply moved by a vision he described in 1650, in which 'All the sick People in England presented themselves before me, and told me They had Herbes in their Gardens that might cure them, but knew not the vertues of them.'[21] This was the universal knowledge that Adam had at first possessed, and his descendents then had lost: the medicinal plants were out there in Nature to cure all the ills of Mankind, if only they could be discovered and identified. For there was seemingly nothing that Adam had not seen and understood.

This restitution of all knowledge partly explains the importance that Bacon laid on immortality. But long life was not simply the projected end of the restitution of learning, it was also one of its means. In the eighteenth century, the French writer the Marquis d'Argenson defined youth as that time 'when a man thinks himself immortal' and 'imagines projects of long duration.' D'Argenson believed that this age of optimism ended at around thirty-five.[22] But why should men age at all? If youthfulness could be maintained, would not that same confidence in pursuing 'projects of long duration' be retained? Indeed, in his *Historie of Life and Death*, Bacon argued that the youthful body's ability to repair itself 'might be eternall.'[23] What Bacon seemed to be suggesting was that given what a human being could achieve in one seemingly foreshortened and increasingly debilitated lifespan of seventy years, imagine what could be achieved in an almost perpetually youthful life of a thousand years? (It was commonly thought, for example, that it was the long lives of the antediluvians that had allowed them to perfect astronomy without the aid of the telescope.)

In explaining his concern with longevity, Bacon naturally cited the ancient dictum '*Life* is short, and *Art* long.' He then explained that 'therefore our labours intending to perfect *Arts*, should by the assistance of the Author of Truth and Life, consider by what meanes the Life of Man may be prolonged.' Longer life 'affoords [*sic*] longer opportunity of doing good Workes.'[24] He emphasized as much in the conclusion of his preface to *The Advancement of Learning*, where he declared 'that this Our Instauration is a matter infinite, and beyond the power and compasse of Mortality,' acknowledging that he was '*not unmindfull of Mortality, and Humane Condition, being it doth not promise that the* Deisgne *may be accomplisht within the Revolution of an Age only, but delivers it over to Posterity to Perfect.*'[25]

One of Bacon's major contributions to this project, therefore, would be the establishment of societies that would effectively perpetuate and improve down the ages the works of their members. The model for this would be laid down as Salomon's House in the posthumously published utopian tract *The New Atlantis*. This idea, that if death of the individual cannot be staved off, then immortality can be assured by the collective identity of the many pursuing the same philosophical and scientific goal, is certainly interesting. It is from this idea that we see the emergence of the first societies founded for the advancement of learning, and the establishment of what would become known as 'the Republic of Letters.' Of course, Bacon too had defined this idea in *The Advancement of Learning*, where he laid down his critical program for a 'Seate of the Muses.' There he advised King James that it was those who were 'fruitfull' in their offspring who had 'a fore-sight of their own immortality in their Descendants.' In this sense, he explained, Queen Elizabeth 'was a sojourner in the world, in respect of her unmarried life, rather than an inhabitant.' She enriched her own times, but it was through children that one extended one's life's work 'which succeeding Ages may cherish, and Eternity it selfe behold: Amongst which, if my affection to Learning doe not transport me, there is none more worthy, or more noble, *than the endowment of the world with sound and fruitfull Advancements of Learning*.'[26]

Thus, the decades after Bacon's death see the appearance in England of a number of societies founded for the collective advancement of knowledge. The Invisible College, the Oxford Philosophical Club, and then in 1660 the Royal Society of London were all at heart Baconian institutions. They would be followed in the early eighteenth century by such collective organizations as the Society of Antiquaries, the United Grand Lodge of Freemasons, and numerous provincial 'gentlemen's societies.' All these organizations shared a common interest in the advancement of learning and the improvement of society. I would also suggest that in many ways they encapsulated Bacon's project for immortality. If the man could not live forever, the legacy of his thought could, through generations of institutionalized philosophers pursuing the same goals. Bacon may not have had children to perpetuate his name, but his work was not without its clear offspring.

The institution was thus the individual collectivized and effectively immortalized. And as Thomas Sprat observed in his *History of the Royal-Society*, while one 'diligent Inquirer' might in his lifetime 'gather as much experience ... as a whole Academy,' one man alone, 'though he has the nimblest, and most universal observation, can never, in the compass of his life, lay up enough knowledge, to suffice all that shall come after him to rest upon, without the help of any new Inquiries.'[27] The Society was thus also the bridge that linked past research with future discoveries. Taking but one example to illustrate this point, the Society of Roman Knights was established in London in 1723 with the intention of improving knowledge of ancient Roman antiquities in modern Britain. As their principal founder, William Stukeley – a fellow of the Royal Society and secretary to the Society of Antiquaries – explained in his speech to the first meeting of its members: 'Time is but of two denominations, for the present is but a moment & scarce an entity or quantity, time past is the object of the enquirys of this Society & time future is the terminus.' Time, he explained, was but a 'mystical circle' of which the Society was at that moment the centre: 'here place its beginning, here find its medium & its termination.'[28] What Stukeley suggested was that the Society existed almost at all times: it was here in the short-lived, almost nonexistent present in which the inaugural meeting was taking place; it was also in the Roman past, with which it was concerned, and whose virtues and values it would rediscover and promulgate for the benefit of an increasingly imperial Britain; and also finally it existed in the future, when forthcoming members of the Society would both reap the rewards of their work and continue their mission.

In fact, the Society of Roman Knights lasted only a few years, but Stukeley's words speak eloquently for the founding ideals of all such societies of this period, many of which he was also a member. Stukeley echoes the belief held by numerous writers of this era: that it was history that brought with it true old age and wisdom, and the knowledge that could be won only by many lifetimes. Bacon had considered the establishment of libraries to be one of the four chief features of his 'Seates of the Muses,' and provision of a library and publication of fellows' communications formed two features of the Royal Society. Books, Bacon wrote, were '*as ships, passing through the vast sea of time*,' for they '*counite the remotest ages of Wits and Inventions in mutuall Trafique and Correspondency*.'[29] In books, history was to be found and knowledge transported. Thus, in 1618 the great bibliophile John Selden wrote, 'The many ages of Former Experience and Observation ... may so accumulat yeers to us as if we had livd even from the beginning of Time.'[30] And as Thomas Fuller declared in 1639, 'History maketh a young man to be old, without either wrinkles or grey hairs; priviledging him with the experience of age, without either the infirmities or inconveniencies thereof.'[31] This too is how David Hume saw it over a century later. He wrote in 1760 that the chief use of history was 'to discover the constant and universal principles of human nature, by shewing men in all varieties of circumstances and situations, and furnishing us with materials, from which we may form our observations, and become acquainted with the regular springs of human action and behaviour.' For Hume, history made up for the brevity of human life and the limits of knowledge, for without history 'we should be for ever children in understanding'; but history 'extends our experience to all past ages, and to the most distant nations.'[32]

The same case may be made for the most ambitious of all these seventeenth- and eighteenth-century societies, the Freemasons. Perhaps significantly, this society included a number of prominent Royal Society Fellows among its early members, including William Stukeley and the future president Martin Folkes. The Freemasons claimed a heritage that linked them directly with the Ancients. Defoe wrote that 'The Building of the Ark by Noah ... was the first Project I read of,'[33] and we find in James Anderson's Masonic Book of Constitutions published in 1723 that the Ark, along with the Tower of Babel and the Pyramids and Solomon's Temple, were all built 'according to the Rules of Masonry.'[34] Freemasonry thus appealed to a direct connection with the Ancient world, and it is significant to observe that in 1722, Robert Samber dedicated to the Freemasons of Great Britiain his translation of Harcouet's *Long Livers*, which boasted to contain 'A Curious History of Such Persons of both Sexes who have liv'd several AGES, and grown Young again: With the rare SECRET of Rejuvenescency of Arnoldus de Villa Nova, And a great many approv'd and invaluable Rules to prolong Life: As Also, How to prepare the Universal Medicine.'

But institutional longevity was only one facet of the Baconian project: physical longevity remained a very real goal. Perhaps surprisingly, though Edward Maynwaringe's '60 or 70 years' might be considered a fair maximum life expectancy for early modern England, the biblical lifespans of Abraham, Jacob, and Sarah were not thought wholly unattainable. A famous early modern guide book on how to live a long and healthy life was Luigi Cornaro's *Discorsi della vita sobria*, first published in 1558, when its author was ninety-one, and updated a number of times before his death in 1566. First translated into English in 1634, by 1727 it had reached a fourth edition under the title *Sure and Certain Methods of Attaining a Long and Healthful Life*.[35] One of the longest-lived Englishmen I have identified for the seventeenth century is the political philosopher Thomas Hobbes, who died in 1679 in his ninety-second year. Interestingly enough, Hobbes had for a brief period in the 1620s been Francis Bacon's amanuensis. John Aubrey records that Hobbes had a number of habits, such as singing (to exercise the lungs) and regular walks and occasional games of tennis (which he was still playing at the age of seventy-five), which 'he did believe would make him live two or three yeares the longer.'[36] Personal regimen seemed to be the key to long life for Hobbes. He had studied anatomy informally in Paris, and according to Aubrey 'was wont to say that he had rather have the advice, or take physique from an experienced old woman, that had been at many sick people's bedsides, then from the learnedest but unexperienced physitian.'[37]

Hobbes's ninety-two years, however, was as nothing compared with the two longest, unverifiable lifespans recorded in seventeenth-century England. Henry Jenkins, who died in 1670, was said to be 169, while the famous Thomas Parr, who died in London in 1635, was allegedly 152. The Earl of Arundel had discovered Parr, blind but living a healthy, humble life with his second wife in Shropshire. Parr had married again, it was said, at the age of 112: at a time when it was widely supposed that sexual interest ended at around 65, it was reported that Parr 'had had intercourse' with his new wife 'exactly as other husbands do, and had kept up the practice to within twelve years of his death.'[38] Arundel took Parr to London to meet the king, and showed him off for some weeks at the Queen's Head tavern in the Strand. Parr, however, soon took ill and died in the earl's house.[39] The renowned anatomist William Harvey, discoverer of the circulation of the blood, undertook Parr's autopsy on the king's command. He found no great signs of ageing in the old man's organs, however, and attributed Parr's death to a pulmonary consumption, or 'suffocation.' Having examined the stomach and intestines, Harvey deduced that 'by living frugally and roughly, and without cares, in humble circumstances, he in this way prolonged his life.' Indeed, he found that 'all the internal organs seemed so sound that had he changed nothing of the routine of his former way of living, in all probability he would have delayed his death a little longer.' Harvey blamed what actually appeared to be Parr's premature death on the smoky atmosphere of London compared to the fresh air of Shropshire, compounded by his sudden change to a diet richer and more varied than the simple one he was used to.[40]

In recognition of his longevity, Parr was buried in Westminster Abbey. No one seemed to doubt the authenticity of his great old age, and Harvey's autopsy report gave it an official imprimatur that even sceptically minded Victorians found hard to shift.[41] Of course, it was in fact hard to verify such lifespans in this era. Yet no questions of doubt appear to have been raised, and there were no requests made for documentary proof. In 1661 John Evelyn happily used Parr's seemingly untimely death as clear evidence for the harmfulness of London's polluted, smoky air.[42] Undoubtedly Parr and Jenkins were very old men, perhaps even nonagenarians, but in the absence of clear, accessible records, uncertainty encroached as great age was attained. This was as true for famous men as it was for the more humbly born. William Stukeley, for example, recorded that he smoked a pipe with Christopher Wren when Wren was a hundred years old; yet Wren was in fact ninety-two when he died. The Irish actor Charles Macklin, who died in London in 1797 in what was probably his hundredth year, was reported in his biography, published in 1799, to have 'attain[ed] to the great age of 107 years.'[43]

But it was more than documentary uncertainty that led to an acceptance of such longevity. Through the seventeenth and eighteenth centuries the authority of the biblical account was rarely questioned, and it was seemingly supported by numerous other sources. Bacon filled many pages of his *Historie of Life and Death* with examples from throughout history of men and women who had lived beyond their eighties to reach 100, 200, and even 300 and 500 years. Sir William Temple likewise noted many such examples in his *Essay Upon Health and Long Life* (1701). The native Brazilians, he observed, were said when that continent had first been discovered by Europeans 'to have lived two hundred, some three hundred Years.'[44] Temple had himself met a man and a woman in England aged 112, and a beggar at a Staffordshire inn who professed to be 124. The beggar told Temple that he ate milk, bread, and cheese, and meat only rarely, and drank mostly water. The innkeeper supported the beggar's claim, and Temple drew the same conclusion as Cornaro, Maynwaringe, Cheyne, and Harvey: that a simple diet, a life free from the stresses of great responsibility, and a sound environment were the mainstays of long life. There is no suggestion in Temple's work that long livers might be repositories of great wisdom, however. Indeed, Temple wondered whether long life was a blessing at all, advising 'that in Life as in Wine, He that will drink it good, must not draw it to the Dregs.'[45]

But there were those willing to ignore such sound advice. The most famous seventeenth-century individual known to have actively pursued the dream of antediluvian longevity was the French philosopher René Descartes. In his *Discourse of a Method* (1649), Descartes explained that

> if it be possible to finde any way of making men in the generall wiser, and more able then formerly they were, I beleeve it ought to be sought in Physick. True it is, that which is now in use contains but few things, whose benefits is very remarkable: But (without any designe of slighting of it) I assure my self, there is none, even of their own profession, but will consent, that whatsoever is known therein, is almost nothing in comparison of what remains to be known. And that we might be freed from very many diseases, as well of the body as of the mind, and even also perhaps from the weaknesses of old age, had we but knowledge enough of their Causes, and of all the Remedies wherewith Nature hath furnished us.[46]

Descartes then concluded his book by stating his resolution that he would 'employ the remainder' of his life 'in no other thing but the study to acquire some such knowledge of Nature as may furnish us with more certain rules in Physick then [*sic*] we hitherto have had.' Indeed, his 'in-clination' for the importance of this project drove him 'strongly from all other kind of designes.'[47] Descartes believed that the only things that could inhibit this 'enquiry of so necessary a Science,' were either 'the shortness of life' or 'the defect of experiment.'[48] Indeed, in a remark that alludes back to Salomon's House, Descartes added that for these very reasons he had published his Method, with the hope that by inviting 'all good Wits to endevour to advance farther in contributing every one, according to his inclination and power ... and so joyning the lives and labors of many in one, we might all together advance further then any particular Man could do.'[49] Public knowledge and institutional collaboration were the keys to philosophical and medical success.

These sentiments expressed in the *Discourse* were not idle daydreams. Descartes studied anatomy, advised friends on how to maintain their health, and kept a frugal diet. In a letter to Constantyn Huygens written in 1638, when he was forty-two, Descartes declared his hope that he might yet live 'more than a century' longer.[50] The English philosopher and alchemist Sir Kenelm Digby reported that Descartes had told him 'that he was very sure it was possible to lengthen out his [i.e., Man's] life to the period of the Patriarchs.'[51] Indeed, so confident was Descartes of his ability to realize his dream of longevity that his last patron, Queen Christina of Sweden, had the impression that he sought to live forever.[52] If the body was a mere machine, why should not the careful soul, through prudent diet, frequent exercise, and careful repair, be capable of extending its operation indefinitely? When Descartes died suddenly of pneumonia in February 1650, aged only fifty-three, his friend the Abbé Picot was stunned. He declared that if it hadn't been for that 'foreign and violent cause' that had 'deranged his "machine" in Sweden, [Descartes] would have lived for five hundred years.' An Antwerp newspaper was less impressed, however. It reported on 10 April 1650 'that in Sweden a fool has died who had claimed to be able to live as long as he liked.'[53]

William Temple's secretary, Jonathan Swift, was similarly unimpressed by such lofty ambitions. In the third part of his famous satirical travelogue *Gulliver's Travels* (1726), Gulliver visits the Grand Academy of Lagado in Laputa, Swift's parody of the ambitious new science of Francis Bacon and the experimenting natural philosophers of the Royal Society. Thereafter, Gulliver travels on to Luggnagg, where he is asked whether he has met any 'Struldbruggs or Immortals.'[54] It is explained to him that these immortals are occasionally born by chance and can be identified by a red circular spot on their foreheads. Gulliver cannot hide his excitement at hearing this news:

> I cryed out as in a Rapture; Happy Nation where every Child hath at least a Chance for being immortal! Happy People who enjoy so many living Examples of antient Virtue, and have Masters ready to instruct them in the Wisdom of all former Ages! But, happiest beyond all Comparison are those excellent Struldbruggs, who being born exempt from that universal Calamity of human Nature, have their Minds free and disengaged, without the Weight and Depression of Spirits caused by the continual Apprehension of Death.[55]

Gulliver is surprised, however, not to have met any of these immortals at the king's court, seeing as they would make 'such wise and able Counsellors.' If it had been his luck to have been born a Struldbrugg, he would have spent two centuries acquiring great wealth and would then have become a careful recorder of events and the changing customs of the times: 'By all which Acquirements, I should be a living Treasury of Knowledge and Wisdom, and certainly become the Oracle of the Nation.' Furthermore, he would create a 'lodge' of a dozen of his 'own immortal Brotherhood,' and together they would

> remark the several Gradations by which Corruption steals into the World, and oppose it in every Step, by giving perpetual Warning and Instruction to Mankind; which, added to the strong Influence of our own Example, would probably prevent that continual Degeneracy of human Nature, so justly complained of in all Ages.[56]

In addition, he would be entertained by the rise and fall of nations, and by discoveries in science such as the longitude, perpetual motion, universal medicine, 'and many other great Inventions brought to the utmost Perfection.'[57] This, of course, was Bacon's dream. But, as Gulliver soon discovered, dream and reality were much at odds. Struldbruggs soon wearied of immortality: by the age of eighty 'they had not only all the Follies and Infirmities of other old Men, but many more which arose from the dreadful Prospect of never dying.' They were 'opinionative, peevish, covetous, morose, vain, talkative,' and as repositories of history they were hopeless, having 'no Remembrance of any thing but what they learned and observed in their Youth and middle Age.' Persons seeking 'the Truth of Particulars of any Fact' were 'safer to depend on common Traditions than upon their best Recollections.'[58]

In fact, this was nothing less than what Harvey had related about 'Old' Parr. On dissection Parr's brain had been found to be 'sound, and quite firm and solid to the touch,' and in life though blind he had been able to 'hear very well and understand what he heard, answer questions readily, and react normally to situations.' But Parr's responses to interrogations about the events of his youth had been no more informative than those to be expected of a Struldbrugg. Harvey explained that Parr's 'power of memory ... had failed considerably so that he had no clear remembrance of his own actions as a young man, of the public events, famous kings and leaders, wars and civil disturbances in his early youth, of customs, men, prices of goods offered for sale or the other occurrences usually remembered by men. He remembered only his actions of most recent years.'[59] The experience of the aforementioned Irish actor Charles Macklin was similarly sorry. In 1788 he wrote to a friend explaining his failing health and his fear of onsetting senility: 'You are not eighty-eight. I never knew nor suspected the power of it before my illness ... for it is itself the Lord of all – Insanity I mean – that is left – & but a poor Butterfly called Life that flutters its breath as the other does its wings and then all is over.'[60] If this was life in old age, then what was it worth? As Macklin's biographer recalled of his friend's final years as he approached his centenary, 'Debility and decay appeared now to press hard upon this veteran of the stage; and it was pitiable to observe what havoc time had made upon his whole frame.'[61]

Despite these warnings, the idea of human longevity remained both a promising and rational project through the length of the long eighteenth century. By mid-century, it was possible for at least one gentleman, the Comte de St Germain, to make play with ideas of immortality. Of unknown origin and unknown parentage, but of great wealth, it was rumoured that he had been alive since the time of Christ. Casanova records how, at a dinner party, St Germain 'declared with impunity, with a casual air, that he was three hundred years old.' Casanova considered this 'very singular man' to be 'the most barefaced of all impostors,' but confessed, 'I found him astonishing in spite of myself, for he amazed me.'[62]

Karl Heinrich Baron von Gliechen, met the comte in Paris in 1759, and spent six months with him. Like a lawyer he collected evidence, and recorded the following:

> While not all the fables and anecdotes relating to the age of Saint-Germain merit the attention of serious people, it is true that the collection I have made of testimonies of persons of good faith, who have attested the long duration and almost incredible preservation of his person, has in it something of the marvellous. I have heard [the composer Jean-Phillippe] Rameau and an elderly relative of a French Ambassador to Venice affirm having known Saint-Germain there in 1710, looking like a man of fifty, and Monsieur Morin, who has since then been my secretary at the Embassy, for whose veracity I can speak, told me that he knew Saint-Germain in Holland in 1735, and was prodigiously astonished at finding him, now, not aged by so much as a year.

Such was the comte's reputation that in 1777, Count Philip Karl von Alvensleben, Prussian ambassador to Saxony, was asked by Frederick the Great to make secret investigations. Interviewing the comte, Alvensleben asked difficult and leading questions. As he explained afterwards to Frederick,

> I said that if the secrets he claimed could bring me to believe what people said of his great age, something he had just said, of the pleasure with which he had read Swift, made me doubt it, unless indeed he knew ancient times down to the last particulars. He replied that he knew them from the very circumstantial accounts of those who had lived in them.[63]

These remarks, as related in Jean Fuller's biography of the comte, are frustratingly vague. Fuller suggests that in alluding to the Struldbruggs, it 'sounds as though St-Germain had retorted that if von Alvensleben supposed longevity synonymous with felicity, he should read Swift.'[64] Or is the Comte confessing that he is playing a game, as surely he was – and admitting that he knows no more about the past than what anyone can learn by reading history? Thus, we are referred back to what I have already said on the subject of societies and institutions: that long life will only provide the knowledge that can also be wrought from history books. Inevitably, St Germain was rumoured to be both an alchemist and a Freemason, adepts thought to have mystical knowledge of great antiquity.

But what the story of St Germain really tells us is that the notions of longevity and prolonged youthfulness were not considered impossibilities. On the basis of the opinions and guidelines outlined by Bacon and the personal experiments of Descartes, as well as the many advice books that flourished through the long eighteenth century, it seemed that profound longevity was ultimately an achievable human reality. Thus, in 1780, Benjamin Franklin could write to his friend Joseph Priestly, 'The rapid progress true science now makes, occasions my regretting sometimes that I was born so soon. It is impossible to imagine the height to which may be carried, in a thousand years, the power of man over matter ... all diseases may by sure means be prevented or cured, not excepting even that of old age, and our lives lengthened at pleasure even beyond the antediluvian standard.'[65] Over a century and a half after Bacon's death, the project of prolonging human life was still capturing the imagination of Enlightenment philosophers. And as any casual perusal of the popular scientific press will reveal, it remains a compelling dream to this day.

## NOTES

The research and writing of this essay were supported by an Ahmanson-Getty

Fellowship at the Center for Seventeenth-Century Studies at UCLA. I am indebted for their assistance to the staff and librarians at the Clark Library, and for their guidance and support to Professor Maximillian Novak, Dr Martin Gierl, Dr Sarah Kareem, Dr Kimberly Latta, and Dr Alison O'Byrne. Since completion of this essay, many of its ideas have been developed into a full-length book, exploring projectors of immortality from the seventeenth to the twenty-first centuries. See David Boyd Haycock, *Mortal Coil: A Short History of Living Longer* (London and New Haven: Yale University Press, 2008).

1 George Cheyne, *The English Malady, Or, A Treatise of Nervous Diseases of all Kinds, As Spleen, Vapours, Lowness of Spirits, Hypochondriacal, and Hysterical Distempers, &c.* (London, 1733).

2 George Cheyne, *An Essay of Health and Long Life*, 2nd ed. (London: printed for George Strahan at the Golden Ball over-against the Royal Exchange in Cornhill; and J. Leake, bookseller at Bath, 1725), 4–5.

3 Edward Maynwaringe, *Vita Sana & Longa. The Preservation of Health, and Prolongation of Life. Proposed and Proved. In the due observance of Remarkable Precautions. And daily practicable Rules, Relating to Body and Mind, compendiously abstracted from the Institutions and Law of Nature* (London: printed by J. D., 1670).

4 This essay is indebted to, and inspired by, Gerald Gruman's 1966 essay 'A History of Ideas about the Prolongation of Life: The Evolution of Prolongevity Hypotheses to 1800,' in *Transactions of the American Philosophical Society* 56.9 (1966), 3–102. See also Charles Webster, *The Great Instauration: Science, Medicine and Reform, 1626–1660* (London: Gerald Duckworth & Co., 1975), esp. chap. 4, 'The Prolongation of Life'; and Marie Mulvey Roberts, '"A Physic against Death": Eternal Life and the Enlightenment – Gender and Gerontology,' in *Literature and Medicine during the Eighteenth Century*, ed. Marie Mulvey Roberts and Roy Porter (London: Routledge, 1994), 151–67.

5 Daniel Defoe, *An Essay upon Projects* (London, 1697), 13. Francis Bacon, like Robert Boyle after him, was prepared to accept the possibility of failure: what might be learnt in the attempt was enough to satisfy him, even if the attempt might ultimately fail. Bacon was certainly not the first to raise this possibility. Roger Bacon had studied the subject in the Middle Ages, and in 1683 his treatise was published as *The Cure of Old Age, and Preservation of Youth.*

6 Francis Bacon, *Of the Advancement and Proficience of Learning of the Partitions of Science*, trans. Gilbert Wats (Oxford, 1640), 199.

7 Ibid., 190–1.

8 Ibid., 201–3.

9 Ibid., 169–70. Though considered one of the fathers of rationalism, Francis

Bacon did not dismiss alchemy or magic out of hand: in *The Advancement of Learning* he ridiculed the idea that silver could be turned quickly into gold by 'a few graines of Elixir,' but did not reject the idea 'that Gold by an industrious and curious wit, may, at last, be produced' (ibid., 170). Alchemy and its seventeenth-century cognate, chemistry, would prove to be a very meaningful pursuit for a number of leading figures of the new science in England, in particular Samuel Hartlib, Robert Boyle, and Sir Isaac Newton. As well as his detailed researches into alchemy, Boyle was deeply interested in medicine. Indeed, the two subjects were intimately related. For as well as having the power to transmute base metals into silver and gold, the philosopher's stone (or elixir or 'medicine,' as it was also known) had the supposed power to cure all disease. In his 'Invitation to a Free and Generous Communication of Secrets and Receits in Physick,' published anonymously in 1655 in Hartlib's *Chymical, Medicinal, and Chyrurgical Addresses,* Boyle specifically mentioned the responsibility of chemists, if they knew it, to announce 'that great Elixer [*sic*] it self, they call the universal Antagonist of all diseases.' See Michael Hunter and Edward B. David, eds, *The Works of Robert Boyle: Publications to 1660,* 7 vols. (London: Pickering & Chatto, 1999), 1:8. The role of alchemists in the pursuit of long life in this period is hugely important, though their efforts were dismissed by Thomas Sprat, who recorded in his history of the Royal Society that the 'pretensions' of one group of 'chymists' was 'not onely to indow us, with all the benefits of this life, but with Immortality it self.' Thomas Sprat, *The History of the Royal-Society of London, for the Improving of Natural Knowledge,* 2nd ed. (London, 1702), 37. Unfortunately, this is too large a subject to be tackled here. To start, see Michael Hunter, *Robert Boyle (1627–1691): Scrupulosity and Science* (Woodbridge: Boydell Press, 2000) esp. chap. 8: 'Boyle vs the Galenists'; and William R. Newman, *Gehennical Fire: The Lives of George Starkey, an American Alchemist in the Scientific Revolution* (Cambridge, MA: Harvard University Press, 1994).

10 Francis Bacon, *The Historie of Life and Death, with Observations Naturall and Experimentall for the Prologning of Life* (London, 1638), 3–4. It may be noted that the question frequently arose whether death should be considered a disease or not. Aristotle and Galen both considered old age and death to be natural phenomena, and therefore not diseases. See Gruman, 'History of Ideas,' 15–17.

11 *The Works of Francis Bacon,* ed. by James Spedding, Robert Ellis, and Douglas Heath, 15 vols. (London: Longmans and Co., 1877), 5:215.

12 John McManners, *Death and the Enlightenment: Changing Attitudes to Death in Eighteenth-Century France* (Oxford: Oxford University Press, 1985), 5.

13 Ibid., 10.

14 Ibid., 73.

15 Bacon, *Works*, ed. Spedding, Ellis, and Heath (1876), 3: part 1, 222.

16 Richard Overton, *Man's Mortalitie: Or, A Treatise wherein 'tis proved, both Theologically and Philosophically, that Man (as a rationall Creature) is a Compound wholly mortall, contrary to that Common Distinction of Soule and Body: And that the Present Going of the Soule into Heaven or Hell is a meer Fiction* (Amsterdam, 1644), 1.

17 George Cheyne, *An Essay on Regimen. Together with Five Discourses, Medical, Moral, and Philosophical: Serving to Illustrate the Principles and Theory of Philosophical Medicin[e], and Point out Some of its Moral Consequences* (London and Bath, 1740), 40–3.

18 Ibid., 61.

19 See George S. Rousseau, 'Mysticism and Millenarianism: "Immortal Dr Cheyne,"' in *Millenarianism and Messianism in English Literature and Thought, 1650–1800*, ed. Richard Popkin (Leiden: Brill, 1988), 81–126.

20 Philip C. Almond, *Adam and Eve in Seventeenth-Century Thought* (Cambridge: Cambridge University Press, 1999), 42–3.

21 Nicholas Culpeper, *A Physical Directory* (London, 1650), quoted in Webster, *The Great Instauration*, 271.

22 René-Louis de Voyer de Paulmy, marquis d'Argenson, 'Pensées sur la vie et sur la mort,' in *Journal et Mémoires*, ed. E.J.B. Rathery, 9 vols. (1859–67), 5.214–15, quoted in McManners, *Death and the Enlightenment*, 78.

23 Bacon, *Historie*, 3–4.

24 Ibid., 1.

25 Bacon, *Advancement*, 19–20.

26 Ibid., 67–8.

27 Sprat *History*, 31.

28 William Stukeley, Bodleian Library, Oxford, MS Eng. misc. c. 401, fols. 22–3.

29 Bacon, *Advancement*, 64.

30 Selden, *Historie of Tithes* (1618), quoted in Achsah Guibbory, *The Map of Time: Seventeenth-Century English Literature and Ideas of Pattern in History* (Urbana: University of Illinois Press, 1986), 3.

31 Fuller, *Historie of the Holy Warre* (1639), 'the epistole Dedicatorie,' quoted in Guibbory, *Map of Time*, 2.

32 David Hume, *Essays and Treatises on Several Subjects* (London, 1760).

33 Defoe *Essay*, 13.

34 James Anderson, *The Constitutions of the Free-Masons. Containing the History, Charges, Regulations, &c. of that most Ancient and Right Worshipful Fraternity* (London, 1723), 10–13, 27–8.

35 Lewis Cornaro, *Sure and Certain Methods of Attaining a Long and Healthful Life: With Means of Correcting a Bad Constitution, &c.* 4th ed. (London, 1727).

36 John Aubrey, *Brief Lives, Chiefly of Contemporaries, Set Down by John Aubrey,*

*Between the Years 1669 and 1696*, ed. Andrew Clark, 2 vols. (Oxford: Clarendon Press, 1898), 1:347, 351.

37 Ibid., 350.

38 Geoffrey Keynes, *The Life of William Harvey* (Oxford: Clarendon Press, 1966), 223. The autopsy report was printed in John Betts's *De ortu et natura sanguinis* (1669), 319–25; an abstract of the report appeared in the *Philosophical Transactions* (1668), 3:886–8. On sex and age, see McManners, *Death and the Enlightenment*, 80–1.

39 Keynes, *Life of Harvey*, 221.

40 Ibid., 224.

41 See William J. Thoms, *Human Longevity: Its Facts and Fictions* (London: J. Murray, 1873).

42 John Evelyn, *Fumifugium: Or the Inconvenience of the Aer and Smoake of London Dissipated* (London, 1661) 21.

43 James Kirkman, *Memoirs of the Life of Charles Macklin, Esq.*, 2 vols. (London, 1799), 2:438.

44 William Temple, *Miscellanea. The Third Part* (London, 1701), 112.

45 Ibid., 195.

46 René Descartes, *A Discourse of a Method, For the Well-Guiding of Reason, And the Discovery of Truth in the Sciences* (London, 1649), 101–2. For a full investigation of what he calls 'Descartes's medical project,' see Steven Shapin, 'Descartes the Doctor: Rationalism and Its Therapies,' *British Journal of the History of Science* 33 (2000), 131–54.

47 Descartes, *Discourse*, 127.

48 Ibid., 102.

49 Ibid., 103.

50 Quoted in Gruman, 'History of Ideas,' 78.

51 Quoted ibid., 79.

52 Ibid.

53 *Extra ordinarisse Posttijdinghe*, 10 April 1650, quoted in G.A. Lindeboom, *Descartes and Medicine* (Amsterdam: Rodopi, 1979), 94.

54 Jonathan Swift, *Gulliver's Travels*, ed. John Hayward (London: Nonesuch, 1934) 202.

55 Ibid., 203.

56 Ibid., 205.

57 Ibid., 206.

58 Ibid., 207–8.

59 Keynes, *Life of Harvey*, 224.

60 William Appleton, *Charles Macklin: An Actor's Life* (London: Oxford University Press, 1961), 226.

61 Kirkman, *Memoirs of Macklin*, 2:424–5.

62 Quoted in J. Rives Childs, *Casanova: A New Perspective* (New York: Paragon House, 1988), 85.

63 Quoted in Jean Overton Fuller, *The Comte de Saint-Germain: Last Scion of the House of Rákóczy* (London: East-West, 1988), 231. Alvensleben reported that he thought the comte to be about seventy years old – an accurate estimate, as historians believe he was born about 1706.

64 Ibid., 231.

65 Quoted in Gruman, 'History of Ideas.'

chapter nine

# Johnson before Boswell in Eighteenth-Century France: Notes towards the Impossible Project of Reclaiming a Man of Letters

HOWARD D. WEINBROT

Recorded eighteenth-century French translations of Johnson's *Life of Savage*, one; portions of the *Rambler*, five; of the *Idler*, one; of the *Adventurer*, two; of *Rasselas*, ten; of the Preface to *Shakespeare*, one; of the *Journey to the Western Islands*, two.[1] Studies of these twenty-one works, their implications, their many possible cousins, or their possible influence – none. Samuel Johnson as perceived in eighteenth-century France and francophone culture remains terra incognita.

The dissuasive force is not a mythical cartographic monster, but the real Dragon Preconception. Since Johnson was unknown or irrelevant in eighteenth-century France, we do not investigate whether Johnson was known or relevant in eighteenth-century France. The distinguished French anglicist Michel Baridon puts it this way while debunking the notion of an Age of Johnson: 'Seen from the other side of the Channel, Johnson's intellectual stature has never assumed the size described by Anglo-Saxon literary historians.' That aberration is 'bien anglais,' and seems to Frenchmen 'very typical of their neighbours' endearing taste for oddities.' It is among those British things 'to be wondered at, humored and never understood, like cricket or savouries at the end of an Oxford dinner.' Johnson's importance thus 'has been exaggerated by an amiable perversion of national feeling.'[2]

The analogy of Johnson and bad British cuisine surely is *bien français.* More important, it imposes a lamentable twentieth-century French neglect of Johnson upon laudable eighteenth-century French attraction to Johnson. If we consider the breadth of Johnson's achievement, it would be strange for him not to be valued in an often shared French literary culture. Periodical essays, lexicography, biography, canon formation, critical theory and practice, travel narratives, moral tales, voyages, and, increasingly if grudgingly, the reading and performing of Shakespeare were familiar parts of cross-channel intellectual life. Exploration of Johnson as perceived in France thus has much to teach us: about the shape of Johnson's career and authority; about his role for continental Europe's most powerful arbiter of taste; about cross-cultural modes of adaptation; and about new documents for the bibliography of Johnson's works.

Moreover, Johnson long has been paired with Boswell, who so memorably recorded Johnson's heard brilliance and viewed oddities. Whether then or now in the anglophone world, literary, verbal, and physical Johnson are part of his biographical portrait.[3] What would happen if another culture knew Johnson only as a man of letters, if it knew neither his golden speech nor tarnished body? I hope, at least in part, to determine how Johnson the man of letters was received in francophone Europe. I hope thereafter to suggest how the biographical impulse changed these literary perceptions.[4]

The evidence for Johnson's pre-Boswellian literary presence in eighteenth-century France is so substantial that I can only focus on representative responses to his major works. There is more to be said, found, and done – including study of Johnson as translated.[5] So far as I now can tell, though, his French reputation from about 1750 to 1825 had five sometimes overlapping phases that are roughly similar to his reputation in Britain. The first is response to Johnson the moralist, to Mr Rambler.

## I. Johnson the Man of Letters and Moral Writer

Matthew Maty's February 1751 (4:235) *Journal britannique* announces publication of the *Rambler*, to which it soon returns. In April, Maty uses twenty-five pages to discuss individual *Rambler*s (4:363–87). In July and August of 1752, he adds thirty-two pages regarding its complete reprint. Along the way, Maty considers or mentions some ninety-two different *Rambler*s and provides numerous extended and translated extracts (8:243–74). Pierre Clément's *Nouvelles littéraires de France et d'Angleterre* briefly twice mentions the *Rambler* in October 1751 and again in October 1752.[6] Thereafter, the *Journal étranger*'s first volume and second month in 1754 analyses the first three *Rambler*s (May, 1:183–92). In June it adds the next three, and thus provides nineteen pages of extracts and intelligent evaluation of a periodical elevated by association with the respected *Spectator* (1:227–35). Between 1754 and 1756, Urbain Roger's *Le Traducteur* in Copenhagen translates at least twenty four *Rambler*s and six *Adventurer*s.[7]

When these journals comment, they generally agree that Johnson's style is demonstrably un-French, scholastic, and more elevated than entertaining or lively. It lacks fashionable lightness, and could have 'un peu plus de vivacité, de variété & d'enjouement' (*JB*, 4:371). There also is broad agreement regarding Johnson's admirable breadth of topics (*JB*, 4:364, 381; 8:259–60), sustained elegance and quality of mind, confidence in the reader, and what Maty calls 'un grand fond d'humanité' in the sometimes misanthropic author (8 [1752]: 254). The *Journal étranger* is especially keen on those numbers concerning morality, criticism, and philosophy (1 [May 1754]: 183).

That *Journal* also exemplifies early French engagement with Johnson's intellectual positions, what he called 'exchange of mind.'[8] It sometimes laments Johnson's puzzling remarks (1 [May 1754]: 187), but its mixed tone changes when it responds to his already important fourth *Rambler* on the moral basis of the novel. The 'raisonnable' exclusion of machines and romance fiction raises some problems, for the novel borrows these devices from successful epic practice. More important, Johnson rightly urges that the novel should inspire 'l'amour de la vertu & l'horreur du vice. J'approuve celle ci de toute mon ame, & j'exhorte les Romanciers de mon siecle à ne s'en jamais écarter.' The *Journal* then praises Johnson's rejection of the overlapping of vice and virtue in a character: 'L'auteur a bien raison de combattre ce systéme' (1 [June 1754]: 228–9, 230). Johnson soon has select essays filched without acknowledgment by, among others, Arnaud Berquin and Lewis Sebastien Mercier, and with some acknowledgment in the *Journal helvétique*, the *Censeur universel anglois*, and the *Bibliothèque universelle des romans.*[9] Extended versions of the *Rambler* appeared as *Morceaux choisis du Rambler, ou du Rodeur* in Paris in 1785, and as *Le Rodeur* in Maestricht in 1786. The Paris translator A.M.H. Boulard reminds us that Johnson was an 'écrivain très distingué,' a judgment probably shared by Aubin Louis Millin, who soon borrowed four of Boulard's translated *Rambler*s for his own *Mélanges de littérature étrangère* (1785–6). In 1826, C.G. Lambert, baron de Chamerolles, used five volumes for a complete translation of the *Rambler* and its author's 'immortel écrits.'[10]

Reviewers further spread Johnson's fame. *L'Année littéraire* warmly seconded Boulard's judgment and thanked him for rendering 'un vrai service aux lettres en nous donnant la traduction d'un ouvrage dont la morale est aussi saine que le goût' (6 [1785]: 232). The *Journal encyclopédique* was kinder still: 'On trouve dans les morceaux ici traduits un observateur profond, & un écrivain agréable.'[11] Johnson's reputation as

Mr Rambler was so substantial that when the *Dictionary* appeared in 1755, the *Bibliothèque des sciences et des beaux arts* called him 'le savant & ingénieux Mr. *Samuel Johnson,* qui dans l'incomparable feuille périodique intitulée le *Rambler,* apprenoit à ses compatriotes à penser avec justesse sur les matières les plus intéressantes' (2:482; see also 36 [1771]: 348).

Such relative popularity extended to the moralism of *Rasselas* (1759) as well. In 1760 Octavie Guichard Belot provides the first French translation of *Rasselas.* Her title page identifies Johnson as '*Auteur du Rambler.*' Madame Belot admires this 'roman philosophique' because, like Voltaire, Johnson knows that 'le bonheur est une chimere' (iij–iv).[12] As with Johnson's other major early works, French knowledge of *Rasselas* was spread through reviews. Fréron's *L'Année littéraire* finds that it sometimes languishes; but he admires Imlac's 'très belle dissertation sur la poësie que je conseille à nos jeunes Poëtes de lire,' but whose excellent rules nonetheless are unreachable (3 [1760]: 150–1).[13] Fréron contrasts *Rasselas* with *Candide,* the work of his enemy Voltaire, and distinguishes between kinds of response to human depravity. Voltaire leads to despair with the world and within ourselves. Johnson makes us the object of our own compassion and seeks to make us better (165–6). When in 1798 the Comte de Fouchecour adds his own translation and commentary, he acknowledges 'la célébrité de l'immortel auteur de Rasselas' whose beauties place it 'à la tête des romans anglois.' By the end of the eighteenth century, *Rasselas* would be used to teach English to French readers, as early in the nineteenth century that 'joli petit roman ... par le célebre Samuel Johnson' would entertain them.[14]

## II. Dictionary Johnson

Much of the distinction of the 'savant et ingénieux' M. Rodeur nonetheless stemmed from his achievement as the recording voice of the English language. He indeed soon becomes Dictionary Johnson in France as in England. In 1747, for example, the *Bibliothèque raisonée des ouvrages des savans de l'Europe* praised Chesterfield's role as patron and encourager of 'l'Auteur qui se nomme *Johnson.*' In his *Plan* of the *Dictionary,* Johnson used a pure and elegant style to write brilliantly about language. If the completed work is as good, the English finally will have their standard lexicon. In 1751, both Maty and Clément describe Johnson as an admired author whose dictionary is expected, desired, and encouraged.[15] The *Journal des sçavans* for September–October lists the *Dictionary* among new English publications. By its January–March 1756 number, the books had arrived 'à la Veuve Cavelier & fils, Libraires, rue S. Jacques.' The same issue cited the *Dictionary*'s full title, its hefty price of 150 livres, and its role for the English language as that which 'le Dictionnaire de l'Académie Française est pour notre langue.'[16]

That refrain would be familiar among French commentators, but the primacy of Johnson's lexicon was not instantaneous. It had to leap the same hurdle in France that it did in Britain – namely, replacing the competition. Until well into mid-century, the best known English dictionary was Thomas Dyche's *The New General English Dictionary*, probably in the revised version by William Pardon in 1740 and thereafter. A new edition of the Abbé François Prévost d'Exiles's *Manuel lexique ou dictionnaire portatif* appeared in Paris in 1755; Esprit Pezenos issued his *Nouveau dictionnaire universel des arts et sciences, français, latin, et anglois* in Avignon in 1756 and in Amsterdam in 1758. Each makes overt its debt to Dyche, who is at once a source and an argument on authority. Johnson's appearance on the French lexical scene, then, both parallels the Académie's efforts and replaces Dyche's as the norm of British lexicography. As we will see, it was at first a rocky and then a definitive incursion. By midsummer of 1755, indeed, the *Dictionary* begins a second phase of Johnson's literary reputation in France and evokes major and contradictory reviews in Maty's *Journal britannique* and the Abbé Prévost's *Journal étranger*.

Maty's long, occasionally admiring but generally hostile, discussion appeared in the *Journal britannique* for July and August of 1755. The fiction that he will proceed 'avec réserve mais avec impartialité' (17:221) soon evaporates. Maty's first loyalty was to his patron Chesterfield, as indeed he thinks Johnson's should have been. Maty's review thus only is sandwiched by high praise. Johnson's earlier work, he says, shows him as a 'Philosophe profond, Littérateur solide, Écrivain harmonieux.' His dictionary will immortalize him as the dictionaries of the French and Italian academies immortalized them – except that Johnson wrote his alone, and thus is 'en quelque sorte une Académie pour son Isle' (17:219). The review ends with Maty's praise of Johnson's erudition, and his confidence that Johnson offers England's 'première idée d'un veritable Dictionaire' (17:244). That rich if thinly sliced bread, however, scarcely holds the rotten meat within. Maty scolds Johnson for not reprinting the fine 1747 *Plan* dedicated to Chesterfield, for adding the presumably redundant Preface, and thus for hiding an obligation to his aristocratic Maecenas. Maty will not seek the reasons for such conduct, but will consider whether the *Dictionary* is worth its price and whether it fulfils the *Plan*'s promise (17:220–1).

In addition to other matters – like the too symmetrical, rhythmic, obscure, and self-pitying Preface (17:222–3) – Maty finds six serious faults in the *Dictionary*: (1) Its principle of selection for words is too broad; (2) it is too impartial in recording politics and religion; (3) it is too reverent towards rough seventeenth-century English prose at the expense of superior and elegant modern prose; (4) it lacks dates for illustrative quotations; (5) its orthography, pronunciation, etymology, and accents are helpful, but neither its definitions nor illustrations adequately explain a word's different shades of meaning; and (6) its often useful illustrations should have been chronologically ordered.

Maty largely is negative for reasons extrinsic to the *Dictionary*'s achievement but intrinsic to his alliance with Chesterfield. Johnson had every reason to say of Maty: 'the little black dog! I'd throw him into the Thames.'[17] Johnson had no reason to throw the Abbé Prévost into the Seine. His *Journal étranger* twice exposes French readers to the *Dictionary* and perhaps also negatively alludes to Maty's points.

The July 1755 *Journal étranger* includes the *Dictionary* in its section on 'Grammaire,' but considers far more and refutes Maty in all ways. Prévost, for example, thinks that the Preface expresses 'beaucoup de sçavoir' (132). He acknowledges that Johnson is solicitous 'en faveur des Etrangers' and believes that the 'solidité de son jugement' (141) makes the *Dictionary* useful for those who want to learn English (148). Prévost thinks that Johnson has revived several obsolete words that contribute 'à l'beauté du langage' (140) and that by looking back to the relatively near Renaissance, Johnson 'a mieux aimé remonter à la source de la vraie diction' (147–8). Prévost also admires the research and work behind the illustrative quotations that 'doivent être agréables à ceux qui ont du goût pour la Littérature Angloise' (148). Unlike Maty again, Prévost admires Johnson's definitions and thinks 'qu'il est parvenu à connoître par quels degrés un mot a pris differentes significations' (147). Both men, however, agree that the remarkable *Dictionary* is a surprising achievement for 'un seul homme' (144).

The *Journal étranger* was so impressed with the *Dictionary* that in December of 1756 it returns to it in the section on 'Angleterre' and bluntly begins: 'Cet Ouvrage est un des plus importans de la Langue Angloise; & c'est un des plus grands qui ait jamais été composé par un seul homme, dans aucune Langue; L'Auteur est M. *Samuel Johnson*' (111). Better still, the Preface is a model for 'ceux d'entre nous qui seroient tentés de rendre le même service à la Langue Françoise.' To facilitate that effort, Prévost translates the entire Preface 'avec quelqu' exactitude' (112).[18] He thus overcomes Maty's grousing and further enhances Johnson's reputation among French readers.

For example, in 1765 Jean-Baptiste-Antoine Suard's *Gazette littéraire de l'Europe* elevates the *Dictionary* to a wrongly perceived unique role: Johnson gave the world 'le premier Dictionnaire & le seul qui existe,' presumably of the English language. In 1786 *Le Censeur universel anglois* reports that Johnson, the 'grand génie du siècle,' refused to gratify the proud and ungenerous Chesterfield by dedicating the *Dictionary* to him.[19] Comparable friendly judgments of the *Dictionary* and its maker begin the nineteenth century. In 1804 *La Decade philosophique, littéraire et politique* reports that the English regard 'le célèbre' Johnson as '*le colosse de la littérature.*' His works include 'un Dictionnaire de leur langue que beaucoup de gens regardent comme l'ouvrage le plus parfait qu' existe en ce genre' (40:222–3). In that same year, the *Revue philosophique, littéraire et politique* describes the *Dictionary* as 'si supérieur à notre Dictionnaire de l'Académie française' (42 [1804]: 40).[20]

## III. Shakespeare

By 1765, however, Johnson was not always thought so colossal in France. He then encountered an audience beginning to acknowledge Shakespeare's virtues but still impressed by his vices. Many such readers probably judged Shakespeare through Voltaire's dark spectacles. His *Letters concerning the English Nation* (1733) savaged Shakespeare as the tasteless unorthodox genius who both created and destroyed the English theatre. Addison's regular *Cato* should have been England's proper dramatic model.[21]

Johnson's *Shakespeare* thus initiates a third phase of his reception in eighteenth-century France. It indeed becomes contentious when the Preface engages French criticism in general and Voltaire in particular.[22] France's great polemic man of letters even uses the term 'un crime de haute trahison' when in 1776 he angrily responds to the growing popularity of Shakespeare in France.[23] Adaptations by Jean-François Ducis and the royally patronized translation by Pierre le Tourneur (1776–82) opened French gates to English barbarians.

Jean-Baptiste-Antoine Suard tries at least partially to protect France in general and Voltaire in particular from the English invader. Late in 1765, Suard's *Gazette littéraire* introduces its discussion with a harsh judgment of Shakespeare's form, moral structure, language, editors, and Johnson himself. His edition and its Preface have not been as successful as one might expect, and he said much good but also much bad about his author: 'Les éloges paroîtront vraisemblament aux Etrangers un peu exagérés, & les critiques ont révoltés la plûpart des Anglois' (7:171–2). Suard nonetheless prints extracts from the Preface that extend from the beginning to the end of the critical sections and include softened versions of Johnson's skirmish with Voltaire.

The translation thus properly includes Voltaire among those who blame Shakespeare for indecorously showing a drunken king (7:176) and allows him to be called one of the 'petits esprits' (7:177). That, though, is as far as Suard will go, and he even modifies certain insults. Here is Johnson's doublet regarding Dennis, Voltaire, and others who judge Shakespeare with uncongenial artificial rules: 'These are the petty cavils of petty minds.'[24] For Suard, that is 'des chicanes de petits esprits' – a translation that eliminates half of the diminutives. More significantly, Suard eliminates Johnson's blunt attack upon Voltaire's insistence on the unities and his likening of Voltaire to Rymer and Dennis. Appropriate violations, Johnson said, 'become the comprehensive genius of Shakespeare, and such censures are suitable to the minute and slender criticism of Voltaire.' Johnson then quotes Lucan's *Pharsalia,* in which the petty laws of conquered lands gladly surrender to triumphant Caesar (YE 7:79–80). Suard deletes both the insult and the quotation.

Suard soon reprinted his translation in *Variétés littéraires* (1768–9). That text also is the basis for the *Journal helvétique*'s July 1769 partial reprint of the translated Preface, so Voltaire was triply leagued among the 'petits esprits' (4:71).[25] He retaliated in *Questions sur l'encyclopédie*'s discussion of the English theatre. He there characterizes Johnson as a drunken practical joker who talked nonsense about Shakespeare.[26]

By the third quarter of the eighteenth century, however, Voltaire's judgments were often questioned and Shakespeare's genius often exalted. Pierre le Tourneur's preface to his own Shakespeare translation scolds Voltaire without naming him. No doubt to the noise of Voltaire spinning in his grave, it also regularly cites Johnson's edition. As le Tourneur says in 1776 regarding his text, 'le plus communément nous suivons celle de Johnson.' Such an accolade drew more attention to Johnson's Preface. The *Journal anglais* often praises and cites le Tourneur's edition and at one point confronts the unnamed Johnson on the unities (1776, 3:199–200).[27] By the early nineteenth century, Johnson's Preface appears on both sides of the Shakespeare debate. In 1813, the *Mercure étranger* claims him as one of the sane English critics who recognize Shakespeare's faults (1:206). A decade later, Stendhal's *Racine et Shake-*

*speare* (1823, 1825) uses Johnson's 'célèbre préface' to argue against the classical unities and in favour of romanticism, whose patrimony he traces to Johnson.[28]

The fourth phase of French response provides further light on Johnson, who already was an English basso profundo in the French literary chorus.

## IV. Biographer, Traveller, and Critic

The January 1765 *British Magazine or Monthly Repository* (6:1–66) reprinted large portions of Johnson's *Life of Savage* (1744). By May, this in turn was further abridged in Pierre Rousseau's *Journal encyclopédique* (1 [1765]: 81–95). When Johnson's *Shakespeare* appeared in October of that year, French readers thus had been alerted to his biographical as well as critical acumen. One of those readers was the Shakespearean Pierre le Tourneur, who in 1771 published *Histoires de Richard Savage et de J. Thompson, Traduites de l'anglois* (Paris). Given Savage's obscurity, le Tourneur fears that in France his *Life* will not repeat its British success; but he knows that the biography 'présente quelquefois au Sage des réflexions aussi intéressantes & non moins utiles' (ij).

Perhaps the most interesting aspect of le Tourneur's *avertissement*, however, is its discussion of Savage as a mixed character appropriate both for biography and for tragedy (ij–iv). Le Tourneur seems to have learned from and approved Johnson's Preface to *Shakespeare* and its theory of complex human rather than monolithic theatrical character. Perhaps as well such an attractive, if bizarre, a biographical character is the reason why in 1771 le Tourneur's translation was praised and abstracted both in the *Bibliothèque des sciences et des beaux arts* (36 [1771]: 348–67) and in *L'Année littéraire*. It was further abridged in 1774 in the *Journal des sçavans*.[29] In the *Bibliothèque* we read that *Savage* is by 'l'ingénieux auteur du *Rambler*,' and that the elegant translation indeed is likely to fare as well in France as the original did in Britain with its 'succès le plus brillant & le plus soutenu' (2 [1771]: 337). That success is confirmed in 1776 when the *Journal anglais* prints another unattributed and abbreviated version of *Savage*, whom it regards as one of the victimized heroes of literary history (2:449; text 2:449–73). That judgment may also have been behind another reprinting in 1806.[30]

Johnson's later works would be equally sustaining. France's long political concern with Scotland blends with increasing European Celtomania and affection for Macpherson's Ossian poems. In 1775 the *Journal ency-*

*clopédique* thus reviews Johnson's *Journey to the Western Islands* (1775) with special emphasis on English colonial and punitive response to Scotland. The *Journal* does not fully understand Johnson's complex attitude towards Anglo–Scots relations, but it nonetheless recognizes that whatever the Highlanders' barbarism, their civilized occupiers are worse: 'nous croyons voir dans son récit une horde d'Indiens détruite par une horde plus sauvage & plus nombreuse' (6.1 [1775], 251).[31]

Johnson's *Journey* also was recognized as a valuable guide to the still romantic Scottish Islands. In 1785 an abridged *Journey* appears in Pierre-Henri Mallet's *Nouveau recueil de voyages au nord de l'Europe et de l'Asie* (Geneva and Paris). The *Journey* is both among '*les plus estimées*' of such narratives and by an author the English regard as 'entre leurs meilleurs Ecrivains & leurs plus profonds moralistes.' Mallet himself knows that the *Journey* embodies '[l]'esprit philosophique' (2:1).[32] Thereafter, it guides B. Faujas-Saint-Fond's *Voyages en Angleterre, en Ecosse et aux îles Hébrides* (1797), again is separately issued by Henri-Noël-François Huchet, comte de la Bédoyère, in 1804, and is duly praised in *La Decade philosophique* (40 [1804], 222–3). The reviewer shares Huchet's belief that among modern visits, the *Journey* 'mérite le premier rang par l'étendue et la justesse de ses observations.'[33]

The major publishing event of Johnson's final years of course was the *Lives of the Poets* (1779–81), and we again see active interest, translation, and hopes for further translation. In 1779 the *Journal encyclopédique* reviews the first two of the four volumes. It pays special attention to Johnson on Milton and concludes that his equitable remarks are 'aussi justes, aussi vraies que délicates & neuves' (6.1 [1779], 277–8). It returns to the *Lives* and concludes that the two other volumes of the 'belle collection ... ne méritent pas moins d'éloges par la saine critique qui y regne' (6.3 [1779], 458). By the later eighteenth century, Johnson's relevant lives are cited in numerous comments on English poets.

In 1785, for example, the *Journal des sçavans* reviews the first volume of *Mélanges de littérature étrangère*, abstracts key parts of 'Cowley,' urges that it merits being read, and notes that the 'savant' Johnson himself was 'aussi distingué par son érudition que par son esprit & son goût' (771). In the same year the *Censeur universel anglois* urges that a full and good French version of the *Lives* would contribute 'également à la réputation et à la fortune' of the translator. To help that process, the *Censeur* itself provides 'Gray' in 1785, and 'Prior' and 'Otway' in 1786. Much of *Pope* appears in Aubin Louis Millin's *Magazin encyclopédique* in 1796 as borrowed from Boulard's translation. At the end of the long extract Boulard adds a note advising readers where to find the French versions of 'Cowley,' *Savage*, and 'Gray,' and concludes with a puff for the still unpublished complete French *Lives* by 'Le citoyen Sinson' (4:124n).[34] No wonder that Johnson's Necrologie in the *Nouvelles de la république des lettres* observes that the *Lives* enjoyed 'le plus grand succès' and that Johnson was 'un Ecrivain supérieur & digne de l'admiration des étrangers' (29 [1785], 228). No wonder also that the *Censeur universel*'s laudatory epitaph on Johnson praised 'l'universalité de sa science' and 'son talent critique & biographique' (1:83). That admiration no doubt was enhanced by the publication of Boulard's version of the life of 'Milton' in 1797, of an English abridgment of the *Lives* sold in Paris in 1805, of Boulard's Milton and Addison in 1805, of 'Butler' twice in 1813, and of the first volume of the larger *Vies des poètes anglais* in 1823.

## V. Hennet's *Poétique anglaise*

By the mid-1780s, then, Johnson in France was a literary icon whom his contemporaries engaged and respected for his manifold contributions to European intellectual life. Such respect was made clear in a work that clarifies Johnson's fifth phase as a man of letters in France. Albin-Joseph Ulpien Hennet's three-volume *La Poétique anglaise* in 1806 characterizes Johnson as 'l'oracle littéraire de son pays' (2:438), and one who helps him to understand the strange British poetry and people. When he began to read that poetry he thought himself 'transporté dans un monde nouveau. La poésie anglaise, en effet, diffère entièrement de celle des autres peuples' (1:12). The English are 'plus hardie, plus fière, plus libre' (1:13) than the French or other nations. He nonetheless attempts to make those alien works understood even to those who do not know the language (1:15). Hennet's second volume includes sometimes perfunctory and sometimes extensive and perceptive discussions of 122 English major and minor poets. Johnson clearly is Hennet's guide to the poets included in his *Lives of the Poets*, which probably were open before him as he wrote.[35] Though they sometimes were partial and judged from the heart, they have 'une narration rapide, attachante, développée avec ordre et clarté, une critique sage et judiceuse sur leurs ouvrages' (2:456). Johnson the wise critic had several functions for Hennet.

The first was as collaborator, guide, and argument on authority. Hennet's pages are fairly punctuated with remarks like 'ici un passage de Johnson' (1:118), 'observe Johnson' (1:128), 'dit Johnson' (1:156, 248, 2:21, and many others), 'Johnson l'accuse' (1:248), 'Johnson s'élève' (2:68), 'Johnson appelle' (2:148), 'Johnson ajoute' (2:248), and 'Johnson a retrouvé' (2:297). Johnson also appears in extended quotations designed to support Hennet's points, to clarify aspects of the poet's achievement, to provide context, and along the way to show Johnson's own acuity (e.g., 2:243). At one place, he quotes a long passage in which Johnson praises Pope's achievement in the *Iliad* translation (2:302) that proves so much superior to a French translation that Hennet exclaims, 'en anglais, quel rhythme enchaneur! quelle harmonie imitative' (1:103).

Johnson's overt appearance in the *Poétique anglaise*, however, is dwarfed by Hennet's unacknowledged but recognizable use of the *Lives of the Poets.* This is clear in several of his discussions, including those of Addison, Milton, and Pope. The most striking example is Hennet's thirty-four-page section on Savage that epitomizes Johnson's longer biography added to the *Lives.* He eliminates some of Johnson's moral instruction as well as his comments on Savage's poetry and includes only one 'dit Johnson' to suggest proper authorship (2:361). Hennet is not plagiarizing so much as popularizing – on behalf of English poetry and of Johnson as biographer. Hennet also uses Johnson to encourage debate and alternative viewpoints.

Johnson's strong opinions set Hennet thinking about significant aspects of English versification and critical evaluation. He rejects Johnson's severity towards Collins's *Persian Eclogues* (1:256), and he thinks that 'Johnson, si judicieux d'ailleurs, me paraît, en général, trop partisan de la rime' (2:189). He later characterizes the difference of opinion regarding Johnson and Mason on Gray's poetry and concludes: 'C'est au lecteur à decider entre le critique et le panégyriste' (2:419). Hennet also extracts Johnson's 'Parallèle de Dryden et de Pope' and its final bias towards Dryden. In so doing, Hennet invokes history and European judgment, which demonstrates that Pope regularly is reprinted and translated as Dryden was not. 'Pour moi,' he concludes, 'je suis tenté de dire que Dryden est un grand poëte; mais que Pope est le dieu même de la poésie' (2:319).

By early in the nineteenth century, the *Censeur universel anglois* had published Johnson's 'The Winter's Tale' in English and in French (1785, 1:82), the English *Lives of the Poets* (1805) sold in France reminded readers that Johnson was a poet (2), and Jean-Baptiste-Nelson Cottreau had published his translation of *London* (1808). Hennet thus was part of a small but genuine movement to acknowledge the poetic part of Johnson's achievement, as he does at the end of his second volume. Hennet candidly believes that Johnson holds a higher rank 'parmi les littérateurs, que parmi les poëtes' (2:450), but he knows enough of Johnson's poetry to emphasize its distinction and its high place in the British canon (2:126). He is among the 'redoutables adversaires' to the very best, like Shakespeare, Milton, and Dryden (2:184). Goldsmith and Johnson are 'deux poëtes célèbres' who have simplicity and elegance (2:194). Hennet praises individual poems as well. The 'Drury Lane Prologue' is 'parfaitement écrit ... On le lira avec plaisir dans ses oeuvres' (1:320). *Irene* was 'écrite d'un style nerveux, brillant de sentiment et de poésie' (2:455), and the satire *London* is 'peut être le plus parfait' of its kind (1:205). While praising that poem, Hennet recognizes a Johnsonian trait especially congenial to a survivor of France's bloody revolution.

Hennet ascribes the origin of *La Poétique anglaise* to the therapeutic and liberating functions of art during the Terror. On those desolate evenings he would distract himself by translating Pope and Thomson. When enclosed owing to danger in the streets, his free ideas could take him to happier lands and 'le charme de la poésie soutenait mon âme abattue' (1:7). Such fear of violence also affected Hennet's judgment of Pope's and of Churchill's personal and violent satires. Who, he says of Churchill, 'en sortant d'un revolution aussi terrible que celle de la France, [pouvait] aimer des poëmes dictés par le plus violente démogogie?' (2:87). In contrast, though Johnson's *London* was an Opposition party piece, its attacks were general not brutally personal. As a satire, *London*'s style

> est noble et vigoreux, sa versification est sage et généreuse; point d'injures grossières à de plats auteurs, point de personnalités jalouses contre des rivaux ou des ennemis: son pinceau fidèle, quoiqu'un peu rembruni, trace hardiment une peinture générale des moeurs et des caractères: jamais la satire n'eut un but plus noble, une exécution plus brillante. (1:205–6)

Such judgments clearly were neither parochial nor unexpected. Hennet's volumes in 1806 culminate almost sixty years of French judgment and broad if not universal enjoyment of Johnson as a critic, moralist, biographer, travel writer, lexicographer, and poet whose humanity permeates his work. Hennet's Johnson is 'l'écrivain le plus instruit, le plus fécond qu'ait produit l'Angleterre.' Hence, 'tout ce qu'il a fait méritait d'être conservé' (2:451). Such admiration reflects the many years in which Johnson's periodical essays were regularly translated and discussed, the *Dictionary* became the envy of French linguistic theorists, *Ras-*

*selas* even gained Voltaire's approbation, *Shakespeare* sparked admiration and debate, the *Journey to the Western Islands* became a controversial but *philosophe* document, and the *Lives of the Poets* and its criticism became guides to British letters.

We note that Hennet's role in this sequence lacks three major words or concepts: 'Boswell,' 'conversation,' and 'idiosyncrasy.' Hennet acknowledges but dismisses Johnson's quirks, and mentions only Hawkins's unsatisfactory *Life* of Johnson (2:450–1, 456–7). Others' responses, though, already had begun to change after Johnson's death, when the outpouring of biographies blends with, begins to supplement, and then replaces knowledge of Johnson's work. For example, the compiler of the *Detail authentiques des malheurs et de la fuite de prince Charles Edouard dans les Hébrides* (Paris, 1786) translates the relevant section from Boswell's *Journal of a Tour* and records 'beaucoup de Pensées recueillies des conversations du Docteur Johnson' (on pp. 30–8). He also translates 'plusiers autres que j'ai choisies dans ses différens Ouvrages' (1), many of which are from *Rambler*s, and all of which assume an epigrammatic voice of wisdom (30–8).

At about the same time, though, French response to Boswell's *Journal of a Tour* to the Hebrides with Johnson and biographies of him begin to turn the literary man into the the talker and personality. Amplifying his powerful voice diminishes the power of his written works and thus begins the double tradition of Johnson studies. Knowledge of the later *Life of Johnson* then becomes even more dominant. The earlier modest imbalance changes dramatically upon the canonization of Boswell's *Life of Johnson* and its interpreter Macaulay. They often become nineteenth-century France's portals to Johnson, and make clear that his reclamation as a man of letters is scarcely more than a project.

## VI. Johnson in France after Boswell and Macaulay

In 1785 the *Censeur universel* used six articles to reprint portions of Thomas Tyers's biographical sketch of Johnson. Shortly thereafter, the *Journal encyclopédique* and their five double-column pages abstract Mrs Thrale's *Anecdotes ... of Johnson* (2 [1786], 429–33). Its review notes that the talents and moral character of the justly celebrated writer are the principal subjects of these interesting and instructive anecdotes (5.1 [1786]: 231). Unfortunately, aspects of the instruction include Johnson's rudeness in conversation, severe manners, and probable exclusion from polite company if not for his purity of morals and useful precepts. The

*Journal encyclopédique* concludes: 'Il faut convenir avec Mistriss P., que tout ce qu'il disoit étoit dur, & tout ce qu'il faisoit estimable' (235).

That judgment was similar to the one Guillaume Dubois de Rochefort reached in the *Journal des sçavans* for August of 1786. He warmly acknowledges Johnson's achievement as author of the *Rambler, Lives of the Poets,* and, among other works, the 'fameux Dictionnaire qui porte son nom.' That very name excites curiosity, and few men of letters would not be interested in his life and thought. Mrs Piozzi includes too many trivial details, but she also makes plain that Johnson was singular, morose, bizarre and severe (536–7). Though the *Journal* indeed acknowledges some of Johnson's intellectual and literary achievement, it is 'le caractere impérieuse & dur que M. Johnson montroit dans la societé' that dominates the review (541).[36]

This personalizing process already had advanced substantially with two antagonistic reviews of Boswell's *Journal of a Tour* that set much of the tone for later discussion of Johnson. In March 1786 the *Censeur universel anglois* severely scolds Boswell for revealing Johnson's many prejudices, follies, caprices, weaknesses, animosity, superstition, intolerance, and peurile pronouncements. Boswell has prostituted his homage to Johnson's follies (1:268, 267). In April the *Journal encyclopédique*'s reviewer cannot easily distinguish the diverse sensations that reading Boswell's book induced. It grants that there are occasional judicious remarks and clever conversations, but in general it finds successive severe sarcasms, childish vanity, and 'des choses si plates, que plus d'une fois on jette le livre de dépit ... Une chose qui étonne encore plus, c'est l'attention des personnages pour Johnson. Ils lui témoignent un respect qui va presque jusqu'à l'adoration' (3.1 [1786], 223).[37] We then see a translated portrait of Boswell's 'aristarque anglois' that emphasizes his character, politics, manners, religion, and conversation and only briefly touches on some of the literary works that previously evoked such respect (224–5). By August of 1788, the review of *Two Dialogues containing a Comparative View ... of Chesterfield and ... Johnson* observed that the dragon of virtue Johnson has been widely written upon. 'Malheureusement, aucun de ses biographes n'est parvenu à le faire aimer, à justifier la dureté de son humeur.' Here indeed is 'le philosophe sauvage' (6.1 [1788], 56, 59, but with mangled printing). Intensifying Anglo-French warfare limited subsequent reviewing, for I have not yet found reviews of Boswell's 1791 *Life of Johnson.*[38] When these surface, it seems reasonable to assume that they too will emphasize Johnson's character, rudeness, talk, and public incivility, as in fact he later was characterized. Anecdote after anecdote regularly appear regarding verbal exchanges, his hostility to the Scots, and his harsh and bizarre character.

Hermile Reynald, for example, published his doctoral dissertation on Johnson in 1856. He discusses and admires Johnson's work, but he is especially taken by Johnson the odd man and brilliant talker, as shown by the 'fameux Boswell.' Indeed, Johnson's conversation, has 'plus de force peut-être et plus d'éclat' than his books; his talk thus more ably disciplines in order to ameliorate both letters and manners. Reynald later adds that, like Macaulay, he too wonders how someone so indifferent to the form of government could be so angry against the partisans of 'liberté.'[39]

The union of Boswell and Macaulay as movers of judgment is even clearer in Hippolyte Taine's influential *Histoire de la littérature anglaise* in 1863. His book orients itself around national differences and national race. Taine admired English politics, economics, and education, but like many of his predecessors, he regarded the English as splendidly vital but violent, rude aliens, who should be approached with caution.[40] At one point Taine tries an anthropological experiment: what if one of the brutes found himself in a civilized country with civilized manners? What if we transported Johnson, 'this ruler of mind, ... into France, among the pretty drawing-rooms, full of elegant philosophers and epicurean manners; the violence of the contrast will mark better than all argument, the bent and predilections of the English mind.'[41]

We can predict the consequences when Taine describes the blunt, brown-suited, oddly shaped, tick-besieged, facially scarred, half-blind, politically rabid, dogmatic, religious zealot inflicted upon Taine's Gallic epicureans: 'Frenchmen of the present time, the admirers of the *Contrat Social*, soon feel, on reading or hearing all this, that they are no longer in France' (2:187). Upon further exposure, 'the astonishment of a Frenchman redoubles ... we yawn' (2:189). Johnson's periodic essays are blandly wholesome 'national food. It is because they are insipid and dull for us that they suit the taste of an Englishman' (2:190). Taine knows why Johnson behaved as he did: his 'eyes were English, and the senses are barbarous. Let us leave our repugnance behind us, and look at things as Englishmen do' (2:191). Accordingly, 'Frenchmen will say that such lessons are good for barbarians,' but will barely like such lay preaching by Johnson and others. 'I reply that moralists are useful, and that these have changed a state of barbarism into one of civilisation' (2:192).

Taine in the nineteenth century anticipates Professor Baridon in the twentieth century: for each, Johnson is strange English food for the strange English nation. Each commentator includes a covert theory of causation overt in Reynald: influence by Boswell and Macaulay and their emphasis on conversation and peculiar personality rather than on literary achievement. Emphasis on such texts changes to an appoximate balance between talk and text, and then to the idiosyncratic aristarque. After about 1786, for many across the channel Johnson becomes a learned version of the English eccentric, a quirky and diminished John Bull rather than the widely admired extraordinary man of letters able to instruct the Académie Française in the art of lexicography. Post-Boswellian commentators reject and redefine the earlier French history of literary response. Johnson the author then had been called England's most instructive and fecund writer, all of whose works deserved preservation. Once the verbal and visual replace the intellectual Johnson, once those parts replace or distort the whole, he is merely the repugnant if well-meaning alien barbarian, respect for whom is an 'amiable perversion of national feeling.' In such a case, the project of recovering Johnson as a man of letters indeed becomes impossible. *Tant pis à nous.*[42]

## NOTES

A shorter version of this essay was published in Howard Weinbrot, *Aspects of Samuel Johnson* (Newark: University of Delaware Press, 2005).

1 I draw this partial list from David Fleeman's indispensable *Bibliography of the Works of Samuel Johnson* (Oxford: Clarendon Press, 2000). The terminal date for the count is 1800, though I go somewhat further in my own discussion. This essay will have added several items to Fleeman's lists. Francophone culture includes relevant publication and commentary in the low countries and often culturally French Germany and Scandinavia.

2 Michel Baridon, 'On the Relation of Ideology to Form in Johnson's Style,' in *Fresh Reflections on Samuel Johnson: Essays in Criticism,* ed. Prem Nath (Troy, NY: Whitston Publishing Co., 1987), 85. See also René Wellek, *A History of Modern Criticism: 1750–1950. The Later Eighteenth Century* (New Haven: Yale University Press, 1955), 84: Johnson 'has been dismissed – especially on the Continent – as a "British superstition."' Miriam Bridenne more recently asserts that without Boswell's *Life of Johnson,* 'le nom de Johnson n'aurait peut-être pas atteint les rives de la posterité.' *Samuel Johnson. Le Paresseux. Traduit de l'anglais par M. Varney* (Paris: Éditions ALLIA, 2000), 119.

3 For aspects of physical and parodic Johnson, see Morris Brownell, *Samuel*

*Johnson and the Arts* (Oxford: Clarendon Press, 1989), 91–104, and Philip Smallwood, 'The Johnsonian Monster and the *Lives of the Poets*: James Gillray, Critical History, and the Eighteenth-Century Satirical Cartoon,' *British Journal for Eighteenth-Century Studies* 25 (2002), 217–45. The world aware of Johnson as talker also knew that Johnson himself distinguished between the dialectics of speech and of writing. As Boswell properly said in the *Journal of a Tour to the Hebrides* (1785), though Johnson 'owned he sometimes talked for victory; he was too conscientious to make errour permanent and pernicious, by deliberately writing it.' See *Boswell's Life of Johnson together with Boswell's Journal of a Tour to the Hebrides,* ed. George Birkbeck Hill, rev. L.F. Powell, 6 vols. (Oxford: Clarendon Press, 1934–50), 5:17. Subsequent references to the *Life* are from this edition.

4 Johnson's own symbolic and real visits to France generally reinforced his image as a man of letters. The symbolic visit was on 12 June 1755 when Ruvigny de Cosnes, the British chargé des affaires in Paris, presented a copy of the *Dictionary* to the Académie Française. (I am indebted to Dr James Caudle for information regarding de Cosne.) It gratefully acknowledged receipt, promised its own pending dictionary in return, and recognized the international importance of Johnson's national achievement. Members then in attendance included Marivaux and the anglophile Shakespeare-loving Comte de Bissy, who later gave Sterne's Yorick a passport for travel in France. Johnson's two ample folio volumes probably attracted the attention of academicians still revising their own lexicon, with which Johnson's was to compare favourably. About 1769 Louise Flint in France translated Johnson's post-annotation judgments on Shakespeare's plays. These are likely to have circulated among small groups, but they were not published and apparently are lost: see also n. 26 below. For the Académie, see Institut de France, *Les Registres de l'Académie Françoise 1672–1793,* ed. Charles Camille Doucet (Paris, 1895), 3:67. See Johnson's French letter to Miss Flint, dated 31 March 1769 in *The Letters of Samuel Johnson,* ed. Bruce Redford, 5 vols. (Princeton: Princeton University Press, 1992), 1:321–2, and 1:321 n. 1; referred to hereafter as Johnson, *Letters.*

Johnson's real and generally less pleasant visit to France was in October and November of 1775. He there met Elie-Catherine Fréron, to whom he denigrated Voltaire and whose wish to translate Johnson's works was sadly denied by his death in 1776. In Paris on 25 September 1775 Johnson met the Abbé Roffette who leapt out of his chair to embrace Johnson for his 'sublime' Latin celebration of *Paradise Lost.* Johnson obviously spoke in France, but he normally spoke Latin with other men of letters and thus reinforced his image as a learned visitor – as also was made plain by most of Johnson's pre-

Boswellian Gallic readers. For the visit to Fréron, see *The French Journals of Mrs. Thrale and Doctor Johnson*, ed. Moses Tyson and Henry Guppy (Manchester: Manchester University Press, 1932), 173, and Boswell, *Life*, 2:406, where Johnson thus labels Voltaire: 'Vir est acerrimi ingenii et paucarum literarum.' Mrs Thrale offers two sometimes differing accounts of the meeting: *French Journals*, 85–6, and her *Anecdotes of the Late Samuel Johnson* (1786), as recorded in *French Journals*, 85–6 n. 1.

5 See, for example, the translation by Antoine-Marie Henri Boulard in *Morceaux choisis du Rambler, ou du Rodeur; ouvrage dans le genre du Spectateur, traduit de l'anglois de Johnson* (Paris, 1785; but see n. 10 below). Boulard significantly rearranges Johnson's order of publication. He follows Johnson's meaning with reasonable, but not consistent, fidelity but often changes his style along the following lines: fewer passive-voice sentences; simpler diction and sentence structure, especially for Johnson's periodic sentences and doublets; changing of words to French concepts when the English word does not have a French equivalent; and eradication of many colloquialisms that seemed gross to French taste. We badly need a full study of the Johnson translations in France.

6 Pierre Clément's *Nouvelles littéraires* were reprinted as *Les Cinq années littéraires ... 1748–52*, 2 vols. (Berlin, 1756); see Lettre 86, dated Londres, 15 October 1751, 2:157. For association with the *Spectator*, see 2:157. These volumes were further reprinted as *Lettres critiques sur divers sujets de littérature* (Amsterdam, 1767). See also Aubin Louis Millin in *Mélanges de littérature étrangère* (Paris, 1785–6), 1:211, and the *Revue philosophique, littéraire et politique* 42 (1804), 40.

7 For these, see Urbain Roger, *La Traducteur. Ou Traduction de divers feuilles choisies tirées des papiers periodiques anglais.* I have been able to see only volumes 2 (1754–5) and 4 (1756) of these, only available at the Bibliothèque Nationale de France. I assume that volumes 1 and 3 have several other *Rambler*s and perhaps *Idler*s. The *Rambler*s translated are numbers 8, 28, 29, 44, 33, 34, 45, 42, 46, 40, 54, 41, 43, 47, 49, 50, 57, 71, 74, 75, 176, 77, 58, 55.

8 Johnson complained that Goldsmith 'never exchanged mind with you': Boswell, *Life*, 3:37. I have considered some aspects of Johnson and such exchanges in 'Johnson and the Arts of Narration: *The Life of Savage, The Vanity of Human Wishes*, and *Rasselas*,' in *Samuel Johnson: Commemorative Lectures delivered at Pembroke College, Oxford*, collected by Magdi Wahba (Beirut: Librarie du Liban, the Egyptian International Publishing Co.–Longman, 1986), 13–38.

9 We recall Arthur Murphy's fortunate fall. One day when in the country with Samuel Foote, he found it necessary to return to London to find a subject for his next *Gray's Inn Journal.* Foote suggests that he merely translate a nice little essay in a handy French magazine that, alas, turns out to be a French version

of Johnson's *Rambler* no. 190 (1752). Later discovery of the error allows red-faced Murphy to apologize to Johnson and become his life-long friend and admirer: Boswell, *Life*, 1:356 and n. 2. I have not yet been able to identify the 'French magazine.' The retranslated *Gray's Inn Journal* was no. 38, 2nd ser., 15 June 1754. It is omitted in contemporary reprints. For Berquin, see *Choix de tableaux tirés de diverses galeries*, with the *Idler* no. 21 (1759), as praised in *L'Année littéraire* 2 (1775), 74–80. Mercier elaborates on *Rambler* nos. 204, 205 (1752) in *Éloges et discours philosophiques* (Amsterdam, 1776), 165–79, and again in *Fictions morales* (Paris, 1792), 1–28, with slight changes, as 'Ou est le bonheur?' I am indebted for Mercier to Paul K. Alkon, who alerted me to Riikka Forsström, *Possible Worlds: The Idea of Happiness in the Utopian Vision of Louis Sébastien Mercier*, Bibliotheca Historica 75 (Helsinki: Suomalaisen Kirjallisuuden Seura, 2002), 138 n. 236. The *Bibliothèque universelle des romans* also reprints these as 'Histoire du roi Seged' with the *Rambler* but not Johnson acknowledged: 2 (1777), 26–32 and 1 (1783), 181–98. The *Journal helvétique* reprints the *Rambler* no. 91 (1751) as 'L'Esprit et La Science, Allegoire Angloise tirée Du *Rembler*' (*sic*), 168–70, also without mentioning Johnson. The *Censeur universel* prints Boulard's version of this as 'Histoire allégorique de Patronage' in its review of the *Morceaux choisis du Rambler* (3:291–3).

10 The *Morceaux chosis* with which I have worked is dated 1789, but it is reviewed by the *Journal encyclopédique* in 1785, the *Journal des sçavans* in 1786, and the *Censeur universel* in 1786 (3:291–3), and obviously is borrowed and acknowledged by Millin in 1785–6. The *Censeur* also notes publication of the 1786 Maestricht text (3:78), and observes that 'cet Auteur justement célèbre' is more difficult to translate even than most English poets: '& son *Rambler* est de tous ses écrits, celui où il a le plus travaillé à caracteriser sa diction par l'énergie des termes q'il a souvent crées, & la hardiesse des inversions que l'Anglois autorisé' (3:73). I have not seen the 1785 edition, but assume that it is the same as the 1789. See *Melanges* 2:143–95. Millin follows Boulard in calling the *Ramblers* 'Discours': Millin's 'Discours I' also is Boulard's 'Discours I' and Johnson's *Rambler* no. 22; Millin's 'Discours II' is Boulard's 'Discours VIII' and *Rambler* no. 19; Millin's 'Discours III' is Boulard's 'Discours XXI' and *Rambler* no. 82; Millin's 'Discours IV' is Boulard's 'Discours XVII' and *Rambler* no. 24. All of Johnson's essays were from 1750. The *Censeur* observes that 'Tous les Journaux ont déja fait l'éloge de discernement qui l'a guidé dans ce Recueil, de l'exactitude de sa traduction & de la sagesse, ainsi que de la pureté de son style' (3:291).

11 [Boulard], *Morceaux choisis du Rambler*, [v]. See also notes 6 and 10 above. Boulard is most interested in the moral and domestic rather than the literary

numbers, and is especially attracted to the English genre of the periodical essay. He translates the *Rambler* in his own rather than in Johnson's chronological order; but he gives Johnson's numbers in the left margin of his 'Table des discours.' His copy text is '*la huietième Edition du* Rambler, *faite à Londres en* 1771,' 500. *L'Année littéraire* reviews the *Morceaux choisis* in 6 (1785), 217–32, 6:232 quoted, as does Pierre Rousseau in the *Journal encyclopédique* 8 (Nov. 1785), 70–4, 71 quoted. See 7:74, in which the editors cite enough to show 'avec quel agrément l'auteur anglois traitoit les sujets les plus sérieux, & avec quelle pureté son traducteur écrit dans sa langue.'

The fame carried over to the *Idler* and was enhanced by the later *Dictionary*. The *Gazette littéraire* 8 (Dec.–March [1766]), 285–9 translates no. 37 (1758) as 'L'Or et le Fer.' It is '*Discours Moral traduit de l'Oisif* [The Idler], *Ouvrage anglois dans le goût du* Spectateur, *& composé par M.* Johnson, *Auteur du* Dictionnaire de la Langue angloise, *du* Rambler *& de plusieurs autres ouvrages estimés.*' The *Idler* was translated in two volumes by Jean-Baptiste Varney as *Le Paresseux, par le Docteur Johnson* (Paris, 1790). Given the fame and distinction of its author and the multiple editions in English, Varney says, he cannot understand why it has so long been neglected in France. Johnson's own name recommends it 'si puissamment' (1:vij).

12 Even Voltaire was either polite or persuaded by Madame Belot's version, to whom on 16 May 1760 he responded with thanks for her gift: *Rasselas* 'm'a paru d'une philosophie aimable, et très bien écrit': *The Complete Works of Voltaire, Correspondence and Related Documents*, ed. Theodore Besterman *et al.*, 141 vols. (Banbury, Oxfordshire: The Voltaire Foundation, 1970–2006), 21:309, D8913. Johnson knew and was pleased by such rapid translation of his philosophic tale. In 1766 he presented a copy of the French *Rasselas* to Mrs Thrale, which she in turn presented to William Conway in 1820. Her marked copy includes marginal annotation at the end of chapter 46, when Imlac reminds the astronomer of his human insignificance before divine vastness: 'Souvenez-vous surtout que vous êtes seulement un atome dans la masse de l'humanité, & que vous n'avez ni telle vertu, ni telle vice qui puisse vous attiroir exclusivement des faveurs ou des afflictions si surnaturelles.' A marginal forearm and hand with a long index finger points to these words. The wrist area also is pierced by two intersecting but almost parallel lines, as if also to suggest marking an important passage. See Yale University Beinecke shelfmark, 1979.55, *Histoire de Rasselas, prince d'Abissinie. Par M. Jhonnson* [*sic*], *Auteur du Rambler, & traduite de l'anglois par Madame B***** (Amsterdam, 1760), 214; there are markings on p. 19 as well. Handwritten notes on the verso of a front marbled page indicate the respective gifts to Mrs Thrale and to William Conway. 'N.F.M.' later purchased the book 'at the sale of Mr. Con-

way's effects, 24 May 1828.' I do not know whether the annotation is Mrs Thrale's or another reader's. Johnson's gift to Mrs Thrale, however, suggests that he wanted her to know about his relative French fame.

Johnson was pleased by all the several translations of *Rasselas*. On 4 March 1773 he tells William White: 'The little Book has been well received, and is translated into Italian, French, German, and Dutch. It has now one honour more by an American edition.' See Johnson, *Letters*, 2:13.

13 Fréron, *L'Année littéraire ... par M. Fréron, des Académies d'Angers, de Montaubon & de Nancy* (Amsterdam, 1760); various longueurs 3:160, 163, 164; on Imlac's dissertation, 3:150–1. As with several other French journals, the volume number represents the volume for the year only, not for the entire *Année littéraire*.

14 *Rasselas. Prince d'Abissinie. Roman. Traduit de l'anglais de Dr. Johnson, par le Comte de Fouchecour. Enrichie de taille douce* (London, 1798), [iii]. Jean-François-Louis-Marie-Marguerile de Salivet de Fouchécourt (*sic*) was married in England on 26 December 1793 to Charlotte-Agathe Grant de Vaux. As an émigré in England to save his neck, he gratefully observes: 'Heureux si mes efforts me méritent l'approbation de mes lecteurs, et si la nation loyale et hospitalière à la quelle je fais hommage de ma traduction l'accueille avec indulgence' [iii]. By 'roman' Fouchecour probably means *conte philosophique*. *Rasselas*'s sense of the futility of search for a perfect world must have resonated among the French Revolution's enemies.

In 1787 the May *Journal des sçavans* optimistically reports that English now is so widespread in France 'qu'elle est devenue aujourd'hui une partie de l'éducation de la jeunesse' (312). This was included in a review of a prospectus of a series of English books that included 'Johnson's Lives of the English Poets' (313). For Johnson as an agent in that language-learning, see, for example, *La Vraie maniere d'apprendre une langue quelconque, vivante ou morte, par le moyen de la langue françoise; suite de la grammaire Angloise, ou Traducteur littérale d'un ouvrage anglois, intitulé The Prince of Abissinia, Histoire de Rasselas, Prince d'Abyssinie* (Paris, 1787) – with facing French and English translations. There were comparable pedagogic versions. In 1818 'Johnson est reconnu pour l'autuer le plus classique de l'Angleterre.' *Rasselas* is written 'avec autant de force que de correction et d'élégance': *Nouveau cours de langue anglaise, contenant l'histoire de Rasselas, du docteur Johnson, et le poëme du Village Abandonnée, de Goldsmith* (Paris, 1818), 1:iij. For 'joli petit roman' see *La Vallon fortuné, ou Rasselas et Dinarbas*, trans. [ ] McCarthy (Paris, 1817), [3]. The 1819 bilingual edition of *Rasselas* referred to 'Le Docteur Samuel Johnson qui fut à juste titre considéré comme le plus brillant ornament du 18e siècle': *Rasselas, Prince of Abyssinia; A tale by Dr. Johnson. With the Life of the Author. Both in English and French* (Paris, 1819), [2–3]. The honorific 'Dr. Johnson' in such titles suggests further knowledge of the English literary scene. Diderot was not among *Rasselas*'s friends. He says of Madame Belot's translation and its prefatory linking of *Rasselas* and Voltaire's *Candide*: 'tout le monde a trouvé le Candide français très-gai et très original, et le prétendu Candide anglais, ennuyeux et détestable.' See *Correspondance littéraire, philosophique et critique par Grimm, Diderot, Raynal, Meister,* ed. Maurice Tourneux, 16 vols. (Paris, 1877–82), 4:231, Diderot to Grimm, 15 April 1760. 'Tout le monde' was larger than Diderot thought.

15 *Bibliothèque raisonnée* in article XIV, 'Nouvelles Literaires de Londres,' in its July-August-September number: 39 (1747), 233–4. I am indebted to James Sledd and Gwin J. Kolb for bringing this to my attention: *Dr. Johnson's Dictionary: Essays in the Biography of a Book* (Chicago: University of Chicago Press, 1955), 82–3, 98–9, and 219 n. 132. Given the warm praise of Chesterfield, one may reasonably suspect that Maty in London wrote or encouraged the *Bibliothèque*'s praise of Johnson's *Plan.* The *Bibliothèque* says that 'pour être bon Critique il faut être bon Philosophe' (39:233–4). In the *Journal britannique* Maty himself praises what he expects will be rare in dictionaries, Johnson's philosophical and critical approach (4:236). Given Maty's consequent stern review of the *Dictionary*, one must again suspect that puffing Johnson was surrogate stroking of Chesterfield. For Clément, see *Les Cinq années littéraires* 2:157 n. (i). I have not yet found the relevant original issue of October 1751 to determine whether the note was consistent with Maty's 1751 notice, or whether it was added for the 1755 edition, when publication of Johnson's *Dictionary* seemed imminent and the booksellers renewed their advertisements.

16 *Le Journal des sçavans,* Sept.–Oct. 1755, Octobre in 'Nouvelles litteraires. Angleterre. De Londres,' 2081; and listed in the annual 'Bibliographie,' 2682; arrived in the book shop, Jan.–Mars, 1756, 'Mars. Nouvelles litteraires. De Paris,' 571.

17 Here is the fuller context for Johnson's later outburst against Maty. At one point in the *Journal britannique*'s discussion of the *Rambler* Maty makes a dangerously dim remark regarding Johnson's prose. He crudely associated Johnson with Colley Cibber: 'Jamais C-bb-r ne sera sage, / Ni J-hns-n ne sera plaisant' (4:370). Johnson apparently knew this remark, which would have increased his anger regarding the *Dictionary* review and its defence of Chesterfield. When Johnson awaited publication of the *Dictionary*, he wanted to begin a periodical that would 'give my countrymen a view of what is doing in literature upon the continent.' Dr William Adams informs Johnson that Maty had just terminated his *Journal britannique* and might serve as Johnson's assistant. That is the time of Johnson's outburst. See Boswell, *Life*, 1:284. Maty ends his *Journal britannique* in December 1755. In the *Life* Adams, or Boswell,

mislabels it the *Bibliothèque britannique.* See also the Hill-Powell note, 1:284 n. 3. Boswell gives the impression that this happened in March 1755, before the *Dictionary* appeared on 15 April. I suspect that he is mistaken and that Johnson's remark was in December and a response to the *Journal britannique*'s unpleasant review. Maty of course knew the *Dictionary*'s proud claim to be free of patronage. I can only speculate that he also knew Johnson's already familiar harsh letter rebuking Chesterfield and sought to rebuke Johnson in turn.

18 Maty gave several extracts, as did Prévost in the first *Journal étranger* discussion of the *Dictionary*. The translation in December 1756 is complete. Boswell thought that Diderot translated this, or perhaps another, version, but does not cite his evidence and I can find none. See Boswell to Temple, 6 November 1775, in the Yale Edition of the Private Papers of James Boswell, Research Edition, vol. 6, *The Correspondence of James Boswell and William Johnson Temple, 1756–1795,* ed. Thomas Crawford, 13 vols. (Edinburgh and New Haven: Edinburgh University Press and Yale University Press, 1997), 403 and 404 n. 5, in which Crawford and C.B. Tinker also 'have been unable to trace this translation.'

19 *Gazette littéraire* 7 (Sept.–Nov. 1765), 171. See below for further information regarding the *Gazette* on Johnson's Preface. Thereafter, the *Journal géneral de la littérature étrangere* notes the appearance of John Ebers's five-volume *Englisch deutsches, und deutsch englisches woertebuch* (Leipzig, 1799): Ebers 'a pris pour guide le grand Dictionnaire *de* Johnson, édition de 1785': 4.1 (1801), 27. This *Journal* also announces Herbert Croft's attempted revision and correction of Johnson's *Dictionary*: see 4.3 (1801), 137–8. The *Journal des sçavans* notes the appearance of the revised fourth edition (1773) of the *Dictionary*: July 1773, 'Nouvelles littéraires. Angleterre de Londres,' 502. The *Censeur universel* for 21 October 1786 reported that the imprudent Chesterfield lacked the generosity of soul to be a proper literary patron. Johnson 'dédaigna de satisfaire sa vanité,' and wrote the memorable letter: it 'vivra éternellement dans la mémoire de ceux qui ont eu le bonheur de l'entendre' (3:382). The article from which this remark comes, like many others in the *Censeur universel,* may translate an English source.

20 See also Hennet's *La Poètique anglaise* (Paris, 1806), discussed below. Hennet calls the *Dictionary* 'ce grand ouvrage, le meilleur qui ait jamais existé dans ce genre, aussi vaste dans sa conception, aussi riche dans son exécution que le dictionnaire de l'académie française est faible et imparfait' (2:454).

21 See Letter 18, 'On Tragedy,' in Voltaire's *Letters concerning the English Nation.* Voltaire's attitude towards the English is summed up in this remark in a different context: 'It is inadvertently affirm'd in the Christian Countries of *Europe,* that the *English* are Fools and Madmen' (73). In literature, such men

are best suited for the wildness of the ode and the terror of tragedy. I have further considered aspects of relevant problems in 'Enlightenment Canon Wars: Anglo-French Views of Literary Greatness,' *ELH* 60 (1993), 79–100, and 'Censoring Johnson in France: Johnson and Suard on Voltaire: A New Document,' *Review of English Studies*, n.s. 45 (1994), 230–3.

French attitudes towards Shakespeare remained at the least uncertain for much of the eighteenth century. In 1765, for example, the Parisian *Gazette littéraire de l'Europe* says that British enthusiasm for Shakespeare is 'une chose incompréhensible pour les Etrangers' (5:174). That incomprehension was encouraged by regular bowdlerizing of Shakespeare for performance or translation. Their response to Shakespeare in English performance in English theatres often was consternation at the drama, audience, and actors. See Simon Nicolas Henri, *Annales politiques, civiles, et littéraires du dix-huitieme siecle: ouvrage périodique, par M. Linguet* (Paris, 1777), 1:171–80, a Frenchman's visit to Covent Garden with his sister, where they sit in the upper gallery, are puzzled and intimidated by the rowdy behaviour, and clearly dislike the brutality in *Othello*. See also 2:51–2, in which Pope's and Hume's praise of Shakespeare seems like nonsense, and 3:264 (also in 1:180), where the ironically labelled 'divin *Shakespear*' is played in a vulgar, rowdy, plebeian theatre peculiar to English taste. Linguet even writes a decidedly mixed estimate of David Garrick and is puzzled by the virtual state funeral more appropriate for a national hero: 5 (1779), 229–56. The French visitor, above, could not understand English. Nor did Shakespeare's nominal French translator (1783), Jean-François Ducis. See Joseph H. McMahon, 'Ducis: Unkindest Cutter?' *Yale French Studies* 33 (1964), 14–25. One can see Ducis's work in *Oeuvres de J. F. Ducis* (Paris, 1813). For other discussion of Shakespeare in France, see J.J. Jusserand, *Shakespeare en France sous l'ancien régime* (Paris 1880); Thomas R. Lounsbury, *Shakespeare and Voltaire* (New York: Charles Scribner's Sons, 1902); C.M. Haines, *Shakespeare in France: Criticism Voltaire to Victor Hugo*, Shakespeare Survey (London: The Shakespeare Association, 1925); Theodore Besterman, *Voltaire on Shakespeare: Studies on Voltaire and the Eighteenth Century*, vol. 54 (Geneva: Institute et Musée Voltaire, 1967).

22 For some aspects of this conflict, see Studies on Voltaire and the Eighteenth Century, vol. 179, *Voltaire and the English* (Oxford: The Voltaire Foundation, At the Taylor Institution, 1979), and especially David Williams, 'Voltaire's War with England: The Appeal to Europe 1760–1764,' 79–100. Josephine Grieder has studied larger aspects of anglomania and anglophobia, in Histoire des idées et critique littéraire, no. 230, *Anglomania in France, 1740–1789: Fact, Fiction, and Political Discourse* (Geneva: Librairie Droz, 1985).

23 'Lettre de M. de Voltaire à l'Académie Française,' 25 August 1776, as in *Oeu-*

*vres complètes de Voltaire, Mélanges,* 52 vols. (Paris 1877–85), 9:352. Such Shakespearean corruption threatens 'immoler la France' (358) and 'humilier sa patrie' (359). Grace à dieu, French letters survived the sauvage, bas, absurde, barbarian onslaught.

24 The Yale Edition of the Works of Samuel Johnson, vol. 7, *Johnson on Shakespeare,* ed. Arthur Sherbo, intro. Bertrand H. Bronson (New Haven: Yale University Press, 1968), 66. Subsequent citations from this edition are cited in the text as YE with the volume and page.

25 See '*Observations sur Shakespeare, tirées de la Préface que M. S. Johnson a mise à la tête d'une nouvelle édition des oeuvres de ce Poëte*' (4:65–94). There are several nuances, corrections, grammatical changes, and printer's variations (like et for &) in the *Variétés littéraires*'s version; but the extracts and translations essentially are the same as those in the *Gazette littéraire.* See the *Journal helvétique* for July 1769, 3–29, for its observations on Shakespeare 'tirées de la Préface' of Johnson: 9 for Voltaire's remarks as among 'des chicanes de petits esprits.' Some scholars see more similarity between Johnson and Voltaire than is usually thought. See Donald Greene, 'Voltaire and Johnson,' in *En-lightenment Studies in Honour of Lester G. Crocker,* ed. Virgil Topazio (Oxford: Voltaire Foundation, 1979), 111–31, and Mark J. Temmer, *Samuel Johnson and Three Infidels: Rousseau, Voltaire, Diderot* (Athens: University of Georgia Press, 1988).

26 Voltaire, *Questions sur l'encyclopédie,* 2nd ed., 9 vols. ([London], 1771), 2:171. Voltaire petulantly refers to Suard's term 'petits esprits,' which is scarcely an inevitable abbreviation of Johnson's 'petty cavils of petty minds' (YE 7:66): 'Je ne veux point soupçonner le sieur *Johnson* d'être un mauvais plaisant, & d'aimer trop le vin.' Voltaire subtly enlarges his revenge in 1776, when he tells the Académie Française that Rymer correctly criticized irregular Shakespeare. See 'Lettre de M. Voltaire à l'Académie Française,' 25 August 1776, in *Oeuvres complètes de Voltaire,* 30:363. See also Johnson's letter to young Louise Flint, *Letters,* 1:321, who would become the unfortunate wife of the counter-revolutionary student of language, Antoine Rivarol. He abandoned her shortly after their marriage and left her and their son in France during the Revolution. Madame Rivarol supported herself in part with translations from English into French. These included Edmund Burke, *Appel des Whigs modernes aux Whigs anciens* (1791) and Robert Dodsley, *Encyclopédie morale, contenant les devoirs de l'homme en société, ou économie de la vie civile* (1803; 2nd ed., 1821). She began this process of translation as the companion to Frances Reynolds in Paris, from whence she sent Johnson her versions of his individual comments on Shakespeare's plays. Johnson replied to her in French on 31 March 1769 and mentioned 'des traductions les plus belles' (*Letters,* 1:322).

27 The *Journal anglais* no. 13, for 15 April 1776, included a long synopsis of

*Othello* (2:257–81), as from le Tourneur's *Shakespeare.* It later offers synopses of *The Tempest,* no. 15, May 1776 (2:385–404), *Julius Caesar,* no. 20, 30 July 1776 (3:193–203), and *Henry IV,* January 1778 (3:437–43). The discussion of *Julius Caesar* attacks Shakespeare's violation of the unities and asks about the bases for the unnamed Johnson and others' support for their own systems: 'sont-elles solides?' (3:199). Long synopses suggest French lack of familiarity with the plays. The *Journal des sçavans* notes the 1773 appearance of the Johnson-Stevens Shakespeare, on which le Tourneur's French versions are based: 'Nouvelles littéraires. Angleterre de Londres,' January 1774, 51.

28 For le Tourneur, see *Shakespeare traduit de l'anglois, dédié au roi* (Paris, 1776) 1:4 n. 1. Le Tourneur's *Shakespeare* also soon was joined by major continental commentaries and translations, including those by Johann Joachim Eschenburg (1771, 1778, 1787) for whom Johnson, often through the Stevens-Johnson revision, also is a regular source of notes and guidance. For Stendhal: *Stendhal. Oeuvres complètes. Racine et Shakespeare,* new ed., ed. Pierre Martino and Victor Del Litto (Geneva and Paris: Cercle du Bibliophile and Librairie Ancienne Honoré Champion, 1970), 59, 61, and the notes on p. 337, where we learn that in 1819 Stendhal called Johnson's Preface 'Le pére du romanticisme.' Such interest in Shakespeare and English literature was joined by comparable interest in German literature. For many, these nations' letters were thought more 'romantic,' liberating, and appropriate for modern France than its more 'classical' literary ancestors.

29 The *Journal des sçavans*'s belated October 1774 notice of le Tourneur's *Savage et ... Thompson* (653–66) does not acknowledge Johnson's authorship. The review probably was designed to help publicize le Tourneur's pending Shakespeare translation, whose Prospectus it reprints in June 1775 (429–32).

30 Diderot was not persuaded and was ambivalent regarding *Savage*: 'Cet ouvrage eût été délicieux, et d'une finesse à comparer aux *Mémoires du Comte de Grammont,* si l'auteur anglais se fût proposé de faire la satire de son héros; mais malheureusement il est de bonne foi': *Denis Diderot: Oeuvres complètes,* ed. Jules Assézat and Maurice Tourneaux, 20 vols. (Paris, 1875–7), 9:451.

31 The reviewer quotes Johnson at length on the lack of a written Erse language, and therefore the impossibility of Ossianic Erse manuscripts. It is hard to tell whether its concluding remark is ridicule or reverence: 'Ainsi cette langue, qui a fait du bruit dans le monde littéraire, est reléguée par l'auteur dans le pays des chimeres, plus vaste encore que l'on ne pense. La voilà rangée parmi ces problèmes d'antiquités dont tant de sçavans ont cherché la solution, qui ont enfanté une multitude de volumes, & pas un rayon de lumiere' (256). The *Journey*'s discussion of Ossian remained known and contested. See *Bibliothèque britannique ... littérature et sciences et arts* (48 [1811], 111). This is part of

its review of Henry Mackenzie's 1805 *Report of the Committee of the Highland Society of Scotland* concerning the authenticity of the Ossian poems.

32 The *Journal des sçavans* reviews the *Nouveau recueil* in its March 1786 number and briefly discusses the 'Voyage en Ecosse ... par le Docteur Johnson' (131–6, 132 quoted).

33 For Huchet, see *Voyages dans les Hébrides, ou iles occidentales d'Écosse, par le D^r^. Johnson. Traduite de l'anglais* (Paris, An. XII), [i]. Huchet goes on to say that 'rien n'a échappé à ... son attention' regarding the Hebrideans' government, prejudices, superstitions, and the like. Johnson also offers 'réflxions ingénieuses et profondes' in powerful style: 'Son style extraordinairement travaillé, est plein de nerf et ne manque pas d'une certaine grâce; mais l'art s'y fait trop sentir' (i). Huchet praises his valued friend the Abbé Ricard for urging him to do the translation. Johnson's *Western Islands* clearly were familiar to the French literati.

34 See *Magazin encyclopédique* 4 (1796), 94–124 for its printing of Boulard's '*Jugement de Samuel Jonhson* [*sic*] *sur les oeuvrages de Pope*' (4:94). Millin translated 'Cowley' in the first volume of his *Mélanges de littérature étrangère*, le Tourneur's translation of *Savage* and 'Gray' appears in Couret-de-Velleneuve's *Matinées d'été*, and *L'Esprit des journaux*, January 1780, reviewed the first four volumes (124n). The *Bibliothèque Britannique ... littérature et sciences et arts* offers 'Remarques' on the 'Life of Milton': 32 (1806), 314–37, and further 'Remarques' on Johnson's view of English versification: 470–93. Johnson, 'ce grand critique' (473), clearly remained an engaging and fertilizing force in French criticism as in French knowledge of English (489, n. 1). Boulard's 'Citoyen' Sinson apparently is the M. Sinson who translated Shaftesbury's *Les Soliloques, ou entretiens avec soi-même* (London and Paris, 1771). I have not attempted to determine whether the putative manuscript has survived in private or in pubic hands.

35 Hennet must also have been using one of the other compendia of English poets, probably Chalmers or Bell. See Thomas F. Bonnell, 'John Bell's *Poets of Great Britain*: The "Little Trifling Edition" Revisited,' *Modern Philology* 85 (1987), 128–52, and Bonnell, 'Bookselling and Canon-Making: The Trade Rivalry over the English Poets, 1776–1783,' *Studies in Eighteenth-Century Culture* 19 (1989), 53–69.

36 De Rochfort's review was thoughtful, but his familiarity with Johnson's canon was general rather than specific. Discussion of Johnson's poetry, for example, takes only one paragraph and is limited to a few 'des vers latins' (542). He seems to think that the Preface to Shakespeare is part of the *Lives of the Poets*, though this may only be careless writing (543). He also thinks that Johnson's and Britain's denigration of Corneille by means of Shakespeare is strange and misguided (543).

37 Contrast the tone and substance of these reviews with the warmly positive review of Boulard's *Morceaux choisis du Rambler* (81 [1785], 70–4) cited in n. 11 above. Perhaps thanks to Boulard and the Maestricht *Le Rodeur*, the *Journal encyclopédique* also prints what it calls 'De la retraite,' a translation of Johnson's *Rambler* no. 6 (1750).

38 For example, in 1797 Boulard, who had read in a German journal that Hayley was going to publish a new life of Milton, includes this remark in his translation of Johnson's *Vie de Milton et jugement sur ses écrits*: 'Je suis fâché que la guerre ne m'ait pas permis de me le procurer' (140). The *Journal des sçavans* ceased publication for almost five years, from December 1792, when it might have reviewed Boswell's *Life of Johnson.*

39 Reynald, *Samuel Johnson: Étude sur sa vie et sur ses principaux ouvrages. Thèse de doctorat* (Paris, 1856), ix (Boswell), x (conversation), 197–8 (Macaulay and *liberté*). Reynald often quotes Johnson's apparent harsh language towards others: 'Ces violences nous étonnent' (199). Johnson makes limited appearances in other French nineteenth-century literary history. In Philarète Chasles he is part of the army of English eccentrics and a pedantic bear: *Le Dix-huitième siècle en Angleterre* (Paris, 1846), 2:28. At one point Abel-François Villemain lists Johnson as one of Macpherson's stern enemies: *Cours de littérature Française: Table de la littérature au xviii$^{e}$ siècle*, new edition (Paris, 1868), 3:7–8. In other places Johnson appears briefly as a commentator, as in Villemain's *Vies des principaux poètes anglais* (Münster, 1870?), 56 (vs Voltaire on Shakespeare), 115 (on Milton), 133 (on Young), 156–7 (on Pope's Homer), 169–70 (vs Macpherson).

40 Parochial views of course are not limited to French literary historians. For many on the other side of the channel, the worst of sins was not being British. Johnson himself was scarcely immune to that infection. For a full study of Taine's relations with England, see Bibliothèque de la revue de Littérature Comparée, vol. 6, F.C. Roe, *Taine et l'Anglettere* (Paris: Édouard Champion, 1923). Taine thought that Carlyle overrated Samuel Johnson (68); Taine himself admired Macaulay (see 76–8), who almost certainly influenced his view of Johnson.

41 Taine, *History of English Literature*, trans. H. van Laun, 3 vols. (New York, 1871), 2:185. Subsequent citations are given parenthetically in the text. By then Mrs Piozzi's *Anecdotes* long were eclipsed by Boswell's mingled fame and notoriety. See, for example, *Biographie universelle ancienne et moderne* (Paris, 1843–65). After noting the errors of Hawkins and Murphy, we read that the best life of Johnson is that of the Scottish Boswell, 'son admirateur et son ami: ... et, malgré son extrême prolixité, elle a été rémprimée un très grande nombre de fois ... On recherche encore les *Anecdotes sur le docteur Johnson*, par Madame Piozzi' (21:108). Her work may have moved on slightly away from

the nineteenth-century back benches when included as part of George Birkbeck Hill's *Johnsonian Miscellanies* in 1897.

42 I end this essay with omnibus thanks and a plea. My friend and colleague Eric Rothstein read an earlier version and made characteristically helpful comments. Lance Wilcox invited me to present that version at the Midwest American Society for Eighteenth-Century Studies in October of 2003. Max Novak similarly invited a later version for his 'Projects' series as Clark Professor in 2003–4. The essay keeps changing as I discover more and more evidence at great repositories of knowledge. In America I worked at my home institution, the University of Wisconsin, Madison, as well as Harvard University's Houghton and Widener libraries, the Newberry Library in Chicago, Princeton University's Firestone Library, Yale University's Beinecke and Sterling libraries. In Europe, the British Library was invaluable, after which the Bibliothèque Nationale in Paris and, briefly, the Bibliothèque St Geneviève supplied valuable materials not elsewhere available. Brian Williams helped me to prepare this version for the press. The continuing support of the William Freeman Vilas Trust of the University of Wisconsin has made such work possible.

My plea is for further research in the areas of Anglo–French relations in the eighteenth century. At least two generations of largely French or French-trained comparatists made remarkable contributions to our understanding of the various contacts between Europe's then two greatest cultural powers. In the process of turning over about a thousand French literary volumes, I discovered numerous references to British authors and texts, including new translations. It is time to rewrite earlier narratives of Anglo–French relations. This essay may be a start, and perhaps a guide to future colleagues who wish to work on Johnson as translated in France. There surely are more texts to be discovered, and more insights to be gained about cultural transmission and cross-cultural literary judgment.

chapter ten

# Art from Nowhere: The Academy in Utopia

ALBERT BOIME

In this paper I will explore some links between utopian thought and academic institutions in the hopes that we can shed light on the conventional stereotype that academies are unyielding bulwarks of conservatism and that utopias are unalloyed dreams of benevolent wisdom. In the course of my investigation, I was surprised to discover that academic and utopian ideals have more in common that might at first be supposed, sharing similar ideological and cultural aims in both developing models for social change and reform and, at the same time, maintaining and reinforcing certain cultural norms common to their respective historical points of origin.

First, let us define the academy: with one or two exceptions I use this term to designate a private or public institution with a twofold mission of instruction and the conferring of honorary distinction.[1] The academy aims at training and indoctrinating neophytes in some specialized branch of higher learning or creative practice and second, it establishes an exclusive body of membership that bestows prestige on its founders and participating professionals. By so doing the academy gains legitimacy as a model for public standards in cultural production. One way it achieves civic influence is by organizing public exhibitions of works meeting its standards while discouraging alternative exhibition sites, and assuming some of the functions of a museum, including forming a permanent collection that may also double for study purposes. The membership is called upon to organize the pedagogical program on behalf of its private or public and national sponsorship. I would argue that social status and elevated political standing lie at the heart of the founding of academies. Often, however, the participating practitioners and their acolytes function as mere cogs in the wheel of fortune that benefits the prince or patron who sponsors the institution. In this context, it may be worthwhile remembering the etymological origin of the term 'academy.' Derived from the Greek *Akademeia*, the garden or olive grove named after Akademus, a local hero of the Trojan wars who owned the property just outside Athens. Hence, the erudite philosopher Plato's fixing his name to that of the patron's served the purpose of glorifying both their mutual public images.

Hundreds of formal academies were established throughout the world from the mid-sixteenth century through the nineteenth century.[2] Supported by court, state, and private organizations of aristocrats and bourgeoisie, these academies concentrated on the study of language, history, science, and the visual arts, and often engaged these studies with explicitly nationalistic goals. Now, at first glance, the notion of an academy in utopia may seem somewhat oxymoronic if not downright moronic, and with good reason: in the modern age, the academy has been identified with impossible rigidity and resistance to independent thought and innovation – characteristics hostile to the free and easy lifestyle imagined by authors of utopian institutions. Yet academies of art are on the rise again in a postmodern age, academies devoted to the revival of traditional figurative forms in which life drawing has once again become the centrepiece of the curriculum. It seems that society needs some bastion of academic authority to legitimize art forms in a period when so much visual culture seems to verge on what is perceived as 'decadence.' This is understandable in hierarchical societies, but even the utopian enterprises were typically organized around centralized institutions for the gathering and dissemination of culture. Even if these utopian projections were more democratically run and publicly oriented than their predecessors, the fact remains that the new world could not be ordered without them. It is precisely this phenomenon that I wish to explore here, the role of the Academy – traditionally an authoritarian voice – in the projected utopian schemes. Perhaps such an exploration may help us better appreciate the need for such institutions even in so-called democratic societies.

Like the Academy, the notion of utopia has been in and out of fashion at various times during its near five-hundred-year history, but it would seem that it too, as a distinct narrative form enlisted in the service of broad social critique and reform, has shown surprising resilience and durability. Utopian ideals go to the heart of the notions of projectors and projecting. Whether cast in the far-flung future or in a geographically remote region, the utopian model typically performs a schizophrenic act in drawing a contrast between conditions in the present and some ideal universe. Inevitably, however, the utopian model falls back on extrapolating from the present the possibilities for that future transformation.[3] It is perhaps just the necessity for guarding a semblance of reality and practicality that relates the model to the academy.

We may begin with the academy's own formulation of its mission. The very first article of the statutes of the fine arts Accademia di San Luca in Rome published in 1796 established the institutional aims in two italicized terms – *conservare* and *propagare*: keep the faith and pass it on.[4] Over two centuries later, the 1919 annual report of the American Academy in Rome announced:

> The object of the American Academy in Rome is not to afford opportunities for a few individuals to perfect themselves for the practice of their chosen professions. The ideal is to create an atmosphere in which a limited number of carefully selected artists and scholars may develop that synthesis of intellectual culture which will make them worthy to preserve and continue the great traditions of the past, in order that the standard of art and literature may be handed on from year to year, constantly strengthened and improved.[5]

The rhetoric has not essentially altered except for a slight shift in emphasis, one crucial for advancing the possibility of a utopian role for an academic culture. To conserve the traditions of the past is the academy's retrogressive feature, but clearly the academicians believed that these traditions were somehow connected with the improvement and perfection of both art and society. Diderot and d'Alembert's *Encyclopédie* articles on the academies distinguish between the university and the academy by downplaying the pedagogical function of the latter and setting forth as its central aim the 'securing of perfection' of its discipline.[6] Yet, in many ways, academies constituted programmatic attempts to provide an alternative forum for the acquisition of knowledge. Universities with outdated curricula based on Aristotelian doctrine overtly served the political and social ends of local principalities and governments. In contrast, academies might be seen as idyllic refuges independent of politics, despite the fact that the personal proclivities of academic sponsors sustained their social and political function. In the section of the *Encyclopédie* on the French Academy at Rome, it was claimed that the institution had been the primary instrument in bringing about 'the perfection of art in France.' The notion that art could attain to perfection is, of course, a part of a larger Enlightenment discourse that predicted the progress of the human mind to a peak of 'absolute perfection' – a notion taken up and developed by D'Alembert's disciple Condorcet in *L'esquisse d'un tableau des progrès de l'esprit humain* (1795). The academy thus performed the role of a kind of 'think tank' for satisfying the cultural needs of the state. This meant that academicians and their acolytes took, as their role models, the canonical geniuses of the past whose careers unfolded at the highest plane of fine-art culture. Academic regulations invariably insisted that their members and disciples exercise the highest morality and display refined dispositions – the social complement of the declared aesthetic principles.

Academies may best be thought of as research centres attempting to centralize and monopolize early modern cultural production. At the most concrete and literal level, they provided a space or venue within which both theoretical and practical issues could be discussed, investigated, and debated – sometimes in relation to potential legislation. By centring control and production in a particular space, the academy could be seen in relationship to other venues of cultural centralization often found in utopian writing: the museum, the gallery, the cabinet, and the salon.[7] The subject set in 1779 for the Grand prix of the French Academy of Architecture was a museum bringing together the products of both science and art. The academy's program specifically requested

> a museum to house the productions of the sciences, the liberal arts, and natural history. The repository of the sciences will embrace a library, a cabinet of medals, several spaces for geography and prints; that of the arts will embrace galleries for painting, sculpture, and architecture; that of natural history will include rooms for the repository of anatomy, injection, taxidermy, plants, and shells.[8]

It would seem that Étienne-Louis Boullée's plan of 1783 had been inspired by this academic competition. He proposed a dramatic monumental scheme, probably with the knowledge that it had a very slight chance of being built (figure 10.1).[9] Movement between these kind of spaces could be very fluid: Francis Bacon gave prime space to a gallery of statuary honouring important inventors and scientists in *New Atlantis* (1627), itself the model for the Royal Society.[10] The comment on Boullée's drawing reads, "In the centre of the museum is a Temple of Fame destined to contain the statues of great men." His plan for a museum uses the term in its original sense, as a precinct devoted to the practice of the arts, rather than as a mere repository of art.

Figure 10.1 Étienne-Louis Boullée, *Museum, Interior View*, ink and wash drawing with watercolour, ca. 1783. Bibliothèque Nationale, Paris.

Thus, there are two aspects of the academy that can be related to utopian ideals, its construct of the cultural perfection of society and the individual and its role as a centre of instruction and experimentation in vivid, foursquare architectural settings that constitute such a fundamental component of utopian writing. Almost all utopian texts feature such institutions of knowledge at the core of their imagined communities. At the same time, academies actively served as training grounds for visionary plans that could be seen as a counterpart of that utopian world.

Take life drawing, for example, the keystone of the academic curriculum and rooted in an idea of the Beautiful derived from classical norms. It was held that the nude symbolized the human species in a state of nature; that in its association with antiquity, it conveyed the clearest image of artistic nobility and grandeur. It was perceived as the paradigm for the study of nature generally; the forms of the body, with their contracted and expanded volumes, concavities and convexities, yielded the greatest variety of examples for mastery of the physical world. Finally, there was the profound identification with the forms generic to the species; the fascination with the live model reflects, in fact, the commitment to some construct of perfection of the human body and its unique existential status. Again, this concept proved to be oppressive to the human spirit, as seen in the contrast of the norm with actual live models whose forms had to be 'corrected' by the academician. When this construct could be rationalized in the name of religious worship, the human body could be considered 'la forma divina.'[11] In Johann Valentin Andreae's *Christianapolis* (1619) there is an Anatomy Factory, for 'there is nothing so close to a miracle as the workings of the bodies of living creatures, and above all the body of man, which they call a miniature model and epitome of the world.'[12] For Leonardo da Vinci, life drawing revealed the mysteries of the universe and improved the morals of the practitioner, an attitude carried over in a statement by the American academician Kenyon Cox, who commented early in the twentieth century: 'Life drawing sharpens the senses, broadens the powers, and stimulates the intelligence, making the student a finer and in every way a more efficient being.'[13]

One other innovation of the early-nineteenth-century French Academy was the introduction of a Prix de Rome competition and practice in landscape painting. Although held only every four years instead of annually, like the Prix de Rome for history painting, it stressed the heroic-classical form in which celebrated personages and events were represented but subordinated to their environment. The landscape tradition in France, typified in the work of Claude and Poussin, consistently invoked the topography of Italy, which the young painters mined for benchmarks of nascent Western civilization as well as luxuriant gardens and coastal scenes warmed by an eternal sunshine. The heroic-classical modality did not reveal nature in its everyday guise, but nature as it ought to be, freed from contingency and contradiction and submitted to the artful arrangement of the classically trained artist. The pioneering theoretician of this category of painting, Pierre-Henri de Valenciennes (1750–1819), was a political moderate who repudiated revolution, offering science as the universal panacea for the human condition. He believed that the human mind was capable of rationally ordering society and rectifying defective social and cultural institutions. To point the way to this ideal in imagery he proposed the model of the historical landscape – his metaphorical equivalent of Enlightenment utopia. The landscapist's goal was to take his close study of an imperfect reality and transform it into an imagined ideality, exploiting the paradoxical vision of antiquity as the perfect setting (figure 10.2).[14]

What unites both the academy and utopians is precisely the Enlightenment assumption of the perfectibility of humanity. Although the entrenched biases of the academy often proved to be an insuperable obstacle to the realization or even practical application of this idea, in principle the academy believed it was providing visual role models for society's uplift. Its hated buzz-word was the 'ugly,' and academicans would have agreed with H.G. Wells's assessment in his *A Modern Utopia* that 'ugliness is the measure of imperfection.'[15] Scientists shared this social construct, as in the case of the British geologist Gideon Algernon Mantell, who was disgusted by the sight of the hordes attending the Great Exhibition of 1851, and contrasted them with the sculpture on view: 'Vulgar, ignorant, country people: many dirty women with their infants were sitting on the seats giving suck with their breasts uncovered, beneath the lovely female figures of the sculptor. Oh! how I wished I had the power to petrify the living, and animate the marble: perhaps a time will come when this fantasy will be realized, and the human breed be succeeded by finer forms and lovelier features, than the world now dreams of.'[16] We shall see later how the academy's refusal of the notion of the contingent, and its vision of the Beautiful, biased and time-bound, overlapped with the utopian ideal of perfection that ultimately proved so detrimental to women, working classes, and non-European racial groups.

In addition to the idea of the academy as a forum for individual fulfilment, academicians perceived their productions as models of enlighten-

Figure 10.2 Pierre-Henri Valenciennes, *The Ancient City of Agrigentum: Composed Landscape*, Salon of 1787. Musée du Louvre, Paris.

ment and harmony for the public. Especially in the realm of architecture, students were asked to visualize physical embodiments of institutions devoted to the quest for universal knowledge. In this sense they paralleled the utopian writers who supplied the details, right down to the bricks and mortar, of their schemes of social reconstruction. As already pointed out, the models for the visions anticipating future academies and museums were classical antiquity or some idealized vision of a past epoch like the Renaissance – perfection of the present signifying back to the future – that needed to be justified by accepted standards of excellence in the past. These structures were vast and symmetrical in plan in complementary harmony with cosmic regularity, thus metaphorically linking microcosm to macrocosm through geometrical ordering and bilateral symmetry. One of the most ingenious pictorial solutions to this ideal is Raphael's fresco of the *School of Athens* in the Vatican, depicting a model 'think-tank' of the great minds of history – a kind of superlative faculty whose members probably would be denied tenure if alive today and caught up in the budget crunch (figure 10.3). Referencing Plato's Academy, Raphael's painting synthesizes the pedagogic and honorary functions, where the savants actively engage in philosophical discourse, teaching, talking, pondering – totally absorbed in the problems at hand.[17] In the centre, Plato points upward to indicate the source of Platonic ideas, while Aristotle, standing next to him, gestures downward to suggest his preoccupation with terrestrial phenomena. We see the geographer Ptolemy with Euclid demonstrating a geometrical theorem. Significantly, the relationship of the figures to the architecture is crucial, a stately vaulted plaza that seems to exist independently, which neither constricts nor overwhelms but metaphorically suggests the loftiness of this learned society's enterprise.[18]

Above all, utopian fiction consistently focused on the possibilities of the ideal city. Starting with Thomas More, and his almost foursquare city on the hill, Amaurot, utopias have been consistently visualized as symmetrical urban spaces (figure 10.4). Mumford's trenchant reading of these spaces as ultimately authoritarian has a compelling logic, though utopian authors pretended to re-enact in the fictive romances an earthly harmony reflective of cosmic order.[19] Notions of the good life were inevitably bound up with the urban environment and the conveniences it provided. In this utopian view, the city also represented the creation of human intelligence as opposed to the rustic or primitive paradise, an artifice that depended upon rational ingenuity and creative inspiration tweaked to perfection through a process of education. The ideal city is

Figure 10.3 Raphael, *School of Athens*, fresco, 1512. Vatican, Rome.

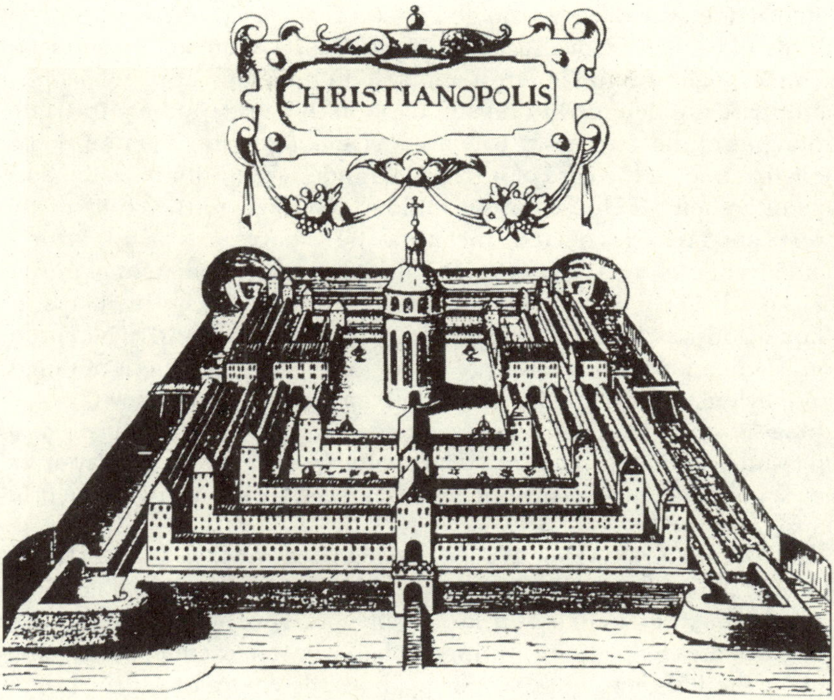


Figure 10.4 *View of Christianapolis*, from Johann Valentin Andreae, *Christianapolis* (1999), 144.

rationally planned and systematically organized, with its specialized activities unfolding within a strongly managed framework. It embraces all the arts and sciences now brought to perfection, a material expression of encyclopedic knowledge archived for the benefit of its inhabitants. In the management and dissemination of this information, the academy – or some analogous institution that embraces a museum-like component – has a key role to play.

Here the academy demanded from its architects and architectural protégés the visualization of the organization and management of the ideal city, and although rarely spelled out in detail in the texts, in any projection this implied task of urban planning – especially from the eighteenth century on – would have been unthinkable without this institutional contribution. Much of this ordered environment is physically shaped and symbolically arranged in accordance with the core concept of the individual utopia. Again, in More's utopia there are 'magnificent temples, built with great effort and able to hold a great many people,' and these are designed to remain dark on the inside to help focus the thoughts of the devotees.[20] Bacon's *New Atlantis* on the island of Bensalem featured, at the core of its religious and social institutions, the Society of Salomon's House, an institution 'dedicated to the study of the works and creatures of God,' including the sciences, arts, manufactures, and inventions of all the world. This institutional embodiment of universal knowledge is a trope of all utopias. One of the wise members, or 'father' of the House, stated the aim even more precisely: 'The end of our foundation is the knowledge of causes, and secret motions of things; and the enlarging of the bounds of human empire, to the effecting of all things possible.' Amid the astonishing parade of experimental labs dealing with light, colour, and sound are extensive galleries for display of the latest inventions, including one in which there are statues of all the principal inventors.[21]

In Tommaso Campanella's *La Città del Sole*, or *The City of the Sun* (1623), the polis is situated on a hill and is divided into seven large circuits, named after the seven known planets. Four gates face the four points of the compass and are approachable by four avenues. At the summit of the hill there is a spacious plain at 'the center of which rises an enormous temple of astonishing design.'[22] The temple is perfectly circular and has no enclosing walls but rests on well-proportioned columns. Its large dome has a cupola at its centre with an aperture directly above the single altar in the centre of the temple. On the altar rests a huge celestial globe on which the heavens are described, and beside it is a terrestrial globe. On the vault of the dome appear all the larger stars with their names and the influences each of them have upon the earth. A metaphysician runs the government and there are three collateral princes, including Wisdom, who has charge of all the sciences and the liberal and mechanical arts. He has commissioned imagery of all the sciences for the walls and the ravelins, both inside and out, a veritable encyclopedia of every known object in the universe. Andreae's *Christianapolis* (1619) also features institutions of knowledge as fundamental to their imagined communities. The entire community is square in plan, with each of its sides seven hundred feet long, accurately aligned with the four quarters of the earth. At the physical centre of Christianapolis is a magnificent circular temple – both opulent and artistic – but at the heart of Christianapolis is the four-storey-high College, square in shape and enclosed by four towers. The governors of the community, those in charge of Religion, Justice, and Learning, have their seat there – 'human perfection' gathered into one place.[23]

Also in Christianapolis is found a Museum of Natural History, where the whole of natural history is depicted upon the walls with tremendous skill – the phenomena of the heavens, the face of the earth in different climatic zones, the races of humankind, the forms of growing things, and every variety of stone and gem. There is also an Astronomical Museum that is as remarkable for its representations of the heavens as the Natural History Museum is for its depictions of the earth. Another centre of information, the Pharmacy, is a compendium of the natural sciences all carefully classified. Opposite the Pharmacy is the School of the Visual Arts, whose resources are widely exploited by the community. Architecture and the arts of fortification and encampment are taught there, and it provides the whole of Christianapolis with pictures relating to the movements of the earth as well as paintings used as teaching aids in the education of the young. In addition, there are pictures and statues of famous individuals strategically located to serve as role models. Finally, Andreae notes the influence of the 'beauty of shapes' on the inner soul in inculcating 'the elegance of Christian life.'[24]

Although in *Erewhon* Butler is at pains to heap ridicule on art schools and the universities that house them, still the protagonist discovers, in the heart of the country, the enchanting metropolis with its 'great towers and fortifications, and lofty buildings that looked like palaces,' and on the grounds of the terraced garden landscape of one venerable *palazzo* he espied 'statues of the most exquisite workmanship.'[25] In Bellamy's *Looking Backward*, when Julian West gazes for the first time on the Boston of the year 2000 he is shocked to see

> [M]iles of broad streets, shaded by trees and lined with fine buildings, for The most part not in continuous blocks but set in larger or smaller inclosures, stretched in every direction. Every quarter contained large open squares filled with trees, among which statues glistened and fountains flashed in the late afternoon sun. Public buildings of a colossal size and an architectural grandeur unparalleled in my day raised their stately piles on every side. Surely I had never seen this city nor one comparable to it before.[26]

Here architecture is central to the utopian project, and it is a conspicuous historical fact that architects and urban planners, especially in the modern epoch, are the ones who project plans for the 'cities of tomorrow.' Since architecture was inevitably taught in academies, these institutions can be credited with providing the support mechanism for developing the ideal city.

Not all the academies devoted to the utopian enterprise were enclosed in physical structures, however; the famous early-eighteenth-century Roman academy, Accademia degli Arcadi, made a garden its major meeting place.[27] Like the classical landscapists, the membership looked to a sylvan past for their model of happiness, a return to a pastoral ideal and the 'simplicity of nature' – a goal shared by certain utopian projects that operated in tension with a squalid urban environment and sought a salubrious alternative for its inhabitants. Subsidized by John V of Portugal, the Arcadians purchased land on the slopes of Rome's Janiculum Hill and commissioned the architect Antonio Canevari to design the physical setting for their formal meetings and festivities. Although dominated by the literati, artists, philosophers, and professionals of all types participated in the ceremonies. In a setting that combined elements of a suburban villa, garden, and rustic countryside, they dressed as shepherds and shepherdesses and adopted rustic names to live out their fantasy of harmonizing with nature. Although the Arcadians sponsored public lectures, they did not conduct classes for neophytes and thus lacked an important function of traditional academies. Nevertheless, their membership rolls reached nine thousand or so in the eighteenth century, many of whom belonged to local institutions like the French Academy at Rome and the Accademia di S. Luca, as well as to foreign-based organizations. The membership clearly had the potential to exert an enormous influence on the younger generation.[28]

The stated purpose of the academy was the diffusing of good taste, *buon gusto*, throughout the peninsula, a goal its members thought achievable by their pastoral retreat and example. Good taste then meant more than a narrow association with canned tuna, but rather the community saw themselves as missionaries for a new order capable of promoting a positive Italian cultural identity and providing society with a key to enable it to distinguish the true and the beautiful from the false and the deviant. Of course, the membership enjoyed the distinct advantage of operating within a safe haven – evidently their activities did not threaten the status quo even though they subscribed to Cartesian principles and the scientific method. At a time when the Office of the Inquisition branded as heretics those mathematicians and natural scientists who espoused a scientific methodology, it would seem that poets and artists escaped censure by displacing their rationality to dreams of a golden age.

A more serious attempt at overhauling society was the program of the Accademia dei Lincei (named for the lynx with its piercing sight) in the early seventeenth century, which could boast the all-around talents Galileo and Giovanni Battista Della Porta as its members.[29] The group construed itself as a 'philosophical militia' that would radically reform and reorganize the production of knowledge. The members have been compared to the elders of Salomon's House described by Bacon in the *New Atlantis.* They conceived plans for training cadres of youth in the disciplines of natural history and astronomy organized around seminar discussions. They even proposed an intense publishing program to spread the results of their investigations to the wider public, and had plans for an international network of satellite institutions. They were determined to rid themselves of envy, pride, and other common vices and promote only virtuous conduct and disinterested search for knowledge – hallmarks of the utopian institutions. In this way the number of learned individuals would multiply, 'and human perfection will result from the natural desire to know.'[30]

As we have seen in the case of Boullée's visionary museum, members of fine arts and architectural academies also indulged in utopian projections by demanding of their young disciples their visionary blueprints for a future society. We see a number of such projects in France in the second half of the eighteenth century, at the height of the influence of Enlightenment philosophy and reformist ambitions. For Boullée, a professor of the Académie d'Architecture who lived through the revolutionary period, his profession was a marriage of art and science. Like the Arcadians, Boullée believed that nature held the key to inspirational inventions in architecture. He established an environment of gardenlike harmony that resembled the Elysian Fields beloved of the antique poets and realized in the present through the ingenuity of the architect.[31] He too believed that central to the architectural ideal was good taste, what he termed with a capital letter, 'Goût.' The essence of *Goût* is that it provides delicacy, subtlety, and elegance rather than strength and ostentation.[32] His ideal shape was the sphere, the shape of the globe and the planets, which in its symmetry and regularity would project 'the image of perfection.' Indeed, for Boullée symmetry appeals to the eye just "because it is the image of order and perfection."[33] His conception was grounded in Burke's notion of the Sublime, one component of which is Grandeur, an experience related to the human desire 'to embrace the universe' in the imagination.[34] The aim of the architect was to choose a site where he can make a museum that gathers unto itself 'all the scattered beauties of nature and where the spectator finds all that is useful to life and thus all that serve to prolong life.' Taking a page from Condorcet, Boullée insisted that everything in nature marches towards 'the goal of perfection.'[35]

Boullée shared with Condorcet the notion of perfectibility of the species through education and was committed to a planned environment that would promote this outcome. One of Boullée's favourite projects was the plan for the National Library (originally the Bibliothèque Royale), devoted to housing the products of those who made their age illustrious and in turn provide inspiration for the future. The masterpieces of great authors inspire others to want to follow in their footsteps and give rise to lofty thoughts for the improvement of society. Boullée confessed that his plan was heavily influenced by Raphael's composition of the *School of Athens* and tried to realize it in three dimensions on a scale vast enough to achieve a sense of exaltation (figure 10.5).[36] He opens his 'Memorandum' accompanying his sketches with the statement that the national building most crucial to the Nation is the one that 'houses all acquired knowledge.' An enlightened ruler will always favour institutions that contribute to the progress of the arts and sciences, and Boullée settled on the idea of an immense basilica lit from above that would house not only the literary heritage of the past but also the work to come inspired by it.[37] Sheltered beneath one huge vault – symbol of the cosmos – it would constitute a 'vast amphitheatre' embracing the sum of all human learning.

Boullée believed that the Academy of Architecture, like the Academy of Painting, had the obligation to demand of its disciples projects that not only would reveal the talents of the proposed members but that also would contribute to the heritage of the academy. It should create the context for this possibility by drawing up a master plan for Paris, a plan that would include all the components that add to the smooth functioning and beauty of a major metropolitan centre. The sites considered appropriate for buildings would be indicated, and suitable programs for the individual structures would be given in advance. In turn, so as to encourage their best efforts, the individual disciple would contribute their own memorandum with their designs. The academy would then present the best designs to the government and keep them on file as models to be continually consulted. The public would inevitably be drawn into this process and acquire a knowledge of architecture that would gradually lead to a general improvement in the quality of the urban environment. This would be accomplished through public

Figure 10.5 Étienne-Louis Boullée, *Plan of Interior View of Bibliothèque Nationale*, ink and wash drawing with watercolour, ca. 1780. Bibliothèque Nationale, Paris.

exhibitions that would allow for lively open debate before any official decision.

Gradually, the academy would establish communication with the provincial architects, providing them with the latest innovations and ideas and encouraging them to follow its example. Thus, the Academy of Paris would affect the functioning and beauty of the whole of France. Civil engineers throughout France would construct maps showing plans for roads, canals, bridges, and other projects belonging to the realm of civil architecture. Boullée encouraged the incorporation of civil engineers in the midst of the august body in order 'that their projects offer the most complete museum of everything related to the field of architecture.' Whenever a member of this body presented his candidature for admission to the academy, he would submit a project in this field not as a simple exercise but as a practical contribution to the improvement of the urban or rural environment.[38]

Boullée's younger contemporary Claude-Nicolas Ledoux published his utopian treatise on architecture, *L'Architecture considérée sous le rapport de l'art, des moeurs et de la législation,* a few years later in 1804.[39] Ledoux had been heavily invested in town planning and landscape design for private patrons and their estates, but his utopianism grew in response to revolutionary pressures and he is often credited with anticipating the communal architecture of utopian socialism. Akin to Boullée, Ledoux received his greatest support in the development of his public architecture from the Académie Royale d'Architecture, to which he was elected in 1773. The stimulus to his new developments was sparked by programs formulated within the Académie during the mid-1780s. In a response to Rousseau's notion of a primitive utopia uncorrupted by artificially induced desires, he called for a return to the salubrious environment of the countryside. In the post-revolutionary period, when Ledoux was concerned with salvaging his ruined reputation for his previous aristocratic associations, his preoccupations turned to a social world united by brotherhood and communal values – concerns that show the regenerative influence of Freemasonry on his thought. His inventions in the countryside would be instruments of reform and happiness for the population of Franche-Comté, where he had previously worked on the Saline de Chaux, the saltworks in the town of Chaux designed for the farmers-general who collected the salt tax.

The idea of a space capable of uplifting social mores by means of its particular shape and symbolic representation was deeply embedded in Enlightenment ideals. Boullée and Ledoux, both Freemasons, were heavily invested in the notion that architecture could ameliorate the human environment by transforming the individual's physical and, ultimately, mental horizon, thereby decisively contributing to the regeneration of society.[40] This is evident in several of Ledoux's utopian designs for Chaux (figure 10.6). But he took the idea one step further in his commitment to customized spaces that promoted social exchange and fraternal gatherings. Here, he took as his models the salons and Masonic lodges that involved members of all the estates – thus advancing the idea that venues of association and fraternity could encourage social democratization and progress in the arts. The newly revitalized Masonic movement was crucial here, following the consolidation of French Freemasonry under the aegis of the Grand Orient in 1773. Masonry was an institution ostensibly dedicated to the improvement of society, a community dedicated to worldwide brotherhood under the aegis of the Great Architect of the Universe. Since architecture and building are key components of the Masonic symbolic repertoire, it is no wonder that utopianist architects enthusiastically embraced its community. Finally, Freemasonic goals meshed with Enlightenment ideals in the revolutionary and immediate post-revolutionary periods, thus allowing its membership a more open stance and a little more breathing space in the public sphere.[41]

I would like to conclude my paper with a brief discussion of Louis-Sébastien Mercier's popular utopian novel *L'an deux mille quatre cent quarante*, first published in 1771, which projected futuristic glimpses of the ideal new city.[42] This is the work not of an architect but of a civil engineer belonging to the Académie des Sciences.[43] Mercier was himself a Freemason caught up in the wave of contemporary architectural utopianism. The protagonist of the novel finds himself in a gleaming urban environment where every corner offers the sight of a beautiful fountain issuing forth pure and transparent water and every house is equipped with the necessities and luxuries that make everyday life pleasurable. Suddenly he comes upon a vast temple of knowledge, the Cabinet du Roi, whose pediment carries the inscription *Abrégé de l'Univers* – Microcosm of the Universe. Once on the inside, he discovers 'four wings of vast extension' surmounted by an immense dome, and on display he encounters every phenomenon and specimen of nature and human invention. Students and the general public can be found studying in an unprecedented non-competitive environment in which learning is an expression of freedom. Here the encyclopedic institution functions less as the province of an exclusive institutional body of Great Men than as one devoted to maximizing the diffusion of knowledge.[44] Similarly, his visit to the Académie de Peinture – again marked by its immense proportions and scale –

Figure 10.6 Claude-Nicolas Ledoux, *Gateway to the Park of Bourreville*, engraving, ca. 1789.

reveals magnificent objects in good taste, each of which is the equivalent 'of a moral and instructive book.'[45] There are no more bloody battles or shameful orgies of the gods, but images of uplift instilling feelings of greatness and virtue – generosity, courage, and a mistrust of idleness. The academy's competitions are now open to foreigners, and the petty jealousies and rivalries that marked the old institution have totally disappeared. The work clearly displays a healthy independence, and cold copying is no longer practised because there is no need to flatter a tyrannical master. There are also a series of lesser academies teaching every conceivable medium where the students work under the watchful eyes of the public.[46]

Until now, we have been mainly stressing the ideals of the utopian project, but the reality could be far different. As in several utopian works, Mercier has problems with Jews; in a bizarre section of his book, his so-called enlightened project comes to an abrupt halt as he presents a compendium of every anti-Jewish stereotype imaginable.[47] It is evident that, for a number of the utopians, the paradise they envisioned would be troubled by the presence of this 'stiff-necked' people. This is not altogether surprising given the Christian contexts of the early utopias, and the later visions of a lily-white society of Nordic-looking individuals. Bacon, ever the empiricist, deals with the issue ambivalently; Jewish families reside in Bensalem, but unlike their brethren elsewhere, they do not hate the name of Christ nor do they harbour secret rancour against their host population. On the contrary, they acknowledge the Saviour and allow him 'many high attributes,' and they 'love the nation of Bensalem extremely.'[48] Bacon practically denominates these local Jewish folk as converts, but nevertheless allows them their 'Jewishness' in the end. Andreae's Lutheran-inspired *Christianapolis*, of course, speaks for itself. Hence the utopians, whatever their supposed nationality, were, in reality, Western European Christians or (occasionally) Deists, right down to William Morris, with his homogeneous, handsome, and 'regular of feature' types. No room for Jewish angst or Jewish noses in utopia. It would appear that difference had to be proscribed in the ideal society and that prejudice got in the way of a truly inclusive society; like the ideal community projected by T.S. Eliot, who advocated homogeneity and, even more important, upheld the idea of unity of religious background, where 'reasons of race and religion combine to make any large number of free-thinking Jews undesirable.'[49]

Finally, I want to return to the bodily self-image that formed the core of the academic curriculum and was a central tenet in many of the utopian texts. In William Morris's *News from Nowhere*, beauty is taken as the overriding aim of society, affecting the production of every object from building practice to the clothes worn by the people. Morris draws a connection between the handsome physical appearance of the people in his utopian wonderland and the cardinal standard of beauty that permeates the whole of society.[50] By invoking the notion of the Beautiful as the inevitable and preferred outcome of his utopian fiction and its physical imprint on humanity, he inadvertently reaffirmed the long-standing ideological program of the Royal Academy and its sister academies everywhere to refuse the contingent and brute realities of everyday life and to project images of perfection (actually sameness) and natural harmony as models of social uplift.

Of course, 'beauty' is a variable social construct or cultural norm that evolves in tandem with other social and cultural factors, and at no time was this more obvious than in the imperialist Victorian context of Morris's novel. Feminist studies have demonstrated that the concept is particularly oppressive of women, pressuring them into physical conformity with a stereotype that was reinforced during the Victorian era with medical and biological arguments that rationalized traditional sex roles as rooted in female anatomy and physiology, while ignoring their damaging effects on women's self-image. Even more telling of Morris's era, categories of the Beautiful and the Ugly were often deployed against different racial groups – including Jews and Africans – as a means of justifying both domestic repression and colonialist aggression.

Here would be grounds for anti-utopian sceptics whose pessimistic view of human nature leads them to conclude that no matter how right the environment or favourable the circumstances, human creatures cannot avoid succumbing to the sins of pride and prejudice. This exclusionary tendency may also throw light on the limitations of such exclusive bodies of membership as academies, whose priestly governors also rule in Utopia. In the end, academies and utopias strike me as more alike than different, both embodying the cultural biases and elitism of the period in which they were conceived.

## NOTES

I wish to thank Maximillian Novak for his invitation to participate in 'The Age of Projects' conference that inspired this paper, and to Myra Boime, Vernon Hyde Minor, and Stephen Eisenman for their suggestions and support. Andrew

McClellan's article on museums as utopian spaces (see n. 7 below) was also a source of inspiration.

1 A. Boime, 'The Cultural Politics of the Art Academy,' *The Eighteenth Century: Theory and Interpretation* 35 (1994), 203–22. For a recent systematic investigation of the academic idea see C. Goldstein, *Teaching Art: Academies and Schools from Vasari to Albers* (Cambridge: Cambridge University Press, 1996), esp. chap. 1–5.

2 See N. Pevsner, *Academies of Art Past and Present* (Cambridge: Cambridge University Press, 1940), 7–8.

3 See F.E. Manuel and F.P. Manuel, *Utopian Thought in the Western World* (Cambridge, MA: Harvard University Press, 1980), 1–29; K. Kumar, *Utopianism* (Buckingham: Open University Press, 1991), 1–19.

4 *Statuti dell'insigne accademia del disegno di Roma, detta di San Luca evangelista* (Rome, 1796), 1.

5 Quoted in Lucia and Alan Valentine, *The American Academy in Rome* (Charlottesville: University of Virginia Press, 1973), 81.

6 D. Diderot and J. Le Rond D'Alembert, *Encyclopédie*, 3rd ed., 36 vols. (Geneva and Neufchâtel, 1778), 1: 226, 238.

7 See Andrew McClellan, 'From Boullée to Bilbao: The Museum as Utopian Space,' in *Art History and Its Institutions: Foundations of a Discipline*, ed. Elizabeth Mansfield (London: Routledge, 2002), 46–64.

8 Jean-Marie Pérouse de Montclos, *Étienne-Louis Boullée* (Paris: Flammarion, 1994), 252, no. 55.

9 Jean-Marie Pérouse de Montclos, *Étienne-Louis Boullée (1728–1799): Theoretician of Revolutionary Architecture*, trans. James Emmons (New York: Braziller, 1974), 24.

10 F. Bacon, *Essays and New Atlantis* (New York: Clasedo Club, 1942), 300–1.

11 Albert Boime, 'Curriculum vitae: The Course of Life in the Nineteenth Century,' in *Strictly Academic: Life Drawing in the Nineteenth Century* (Binghamton, NY: University Art Gallery, 1974), 5–15.

12 J.V. Andreae, *Christianapolis* (1619), trans. Edward H. Thompson (Dordrecht: Kluwer, 1999), 211.

13 K. Cox, 'The Museum and the Teaching of Art in the Public Schools,' *Scribner's Magazine* 52 (1912), 127.

14 For a discussion of these developments see A. Boime, *Art in an Age of Counterrevolution* (Chicago: University of Chicago Press, 2004), 449–56.

15 H.G. Wells, *A Modern Utopia* (London: Chapman and Hall, 1905), 110–11.

16 Eliot C. Curwen, ed., *The Journal of Gideon Mantell, Surgeon and Geologist* (London: Oxford University Press, 1940), 273–4.

17 R.E. Lieberman, 'The Architectural Background,' in M. Hall, ed., *Raphael's 'School of Athens'* (Cambridge: Cambridge University Press, 1997), 75.

18 This idea is inspired by McClellan, 'From Boullée to Bilbao,' 48.

19 L. Mumford, 'Utopia, the City and the Machine,' in F.E. Manuel, ed., *Utopias and Utopian Thought* (Boston: Beacon Press, 1967), 3–24.

20 T. More, *Utopia*, trans. and ed. H.V.S. Ogden (New York: Appleton-Century-Crafts, 1949), 77.

21 Bacon, *Essays and New Atlantis*, 271–2, 288.

22 T. Campanella, *The City of the Sun: A Poetical Dialogue*, trans. D.J. Donno (Berkeley: University of California Press, 1981), 29, 31.

23 Andreae, *Christianopolis*, 186–7.

24 Ibid., 210–15.

25 Samuel Butler, *Erewhon or Over the Range* (New York: New American Library, 1961), 76.

26 Edward Bellamy, *Looking Backward 2000–1887* (New York: Modern Library, 1917), 27.

27 See the major discussions by V.H. Minor, 'What Is "buon gusto"? The Arcadian View,' *Antologia di Belle Arti: Studi sul Settecento II*, n.s. 59–62 (2000), 70–82; and 'Ideology and Interpretation in Rome's Parrhasian Grove: The Arcadian Garden and Taste,' *Memoirs of the American Academy in Rome* 46 (2001), 183–228.

28 L. Barroero and S. Susinno, 'Arcadian Rome, Universal Capital of the Arts,' in *Art in Rome in the Eighteenth Century*, ed. Edgar Bowron and Joseph Rishel, Philadelphia Museum of Art (London: Merrell, 2000), 47–75.

29 See D. Freedberg, *The Eye of the Lynx: Galileo, His Friends, and the Beginnings of Modern Natural History* (Chicago: University of Chicago Press, 2002).

30 R.E. Bindman, 'The Accademia dei Lincei: Pedagogy and the Natural Sciences in Counter-Reformation Rome,' PhD dissertation, University of California, Los Angeles, 2000, 120.

31 See Boullée's essay 'Architecture, Essai sur l'Art' and its translation in Helen Rosenau, *Boullée and Visionary Architecture* (London: Academy Editions, 1976), 88. Both versions may be found on 81–143.

32 Ibid., 89, 124 for English and French versions.

33 Ibid., 86.

34 Ibid., 86–87.

35 Ibid., 9.

36 Ibid., 104.

37 Ibid.; Pérouse de Montclos, *Etienne-Louis Boullée (1728–1799)*, 37.

38 Rosenau, *Boullée and Visionary Architecture*, 110–11.

39 See A. Vidler, *Claude-Nicolas Ledoux: Architecture and Social Reform at the End of*

*the Ancien Régime* (Cambridge, MA: Harvard University Press, 1990), 375–87. My discussion of Ledoux rests in large measure on the exemplary research of this book.

40 For Ledoux's affiliation, see ibid., esp. 336–60; for Boullée's Freemasonic sympathies see Pérouse de Montclos, *Étienne-Louis Boullée,* 122–3.

41 A. Boime, 'Les thèmes du Serment: David et la franc-maçonnerie,' in *David contre David,* Actes du colloque organisé au Musée du Louvre par le service culturel du 6 au 10 décembre 1989, ed. Regis Michel, 2 vols. (Paris: La Documentation francaise, 1993), 1:261–91.

42 I am using the 1786 edition: L. Lemercier, *L'an deux mille quatre cent quarante,* 3 vols. (Paris, 1786).

43 Ibid., 1:52.

44 Ibid., 2:25–58.

45 Ibid., 59–67.

46 Ibid., 64–5.

47 Ibid., 3:167–73.

48 Bacon, *Essays and New Atlantis,* 279–80.

49 T.S. Eliot, *After Strange Gods: A Primer of Modern Heresy* (London: Faber and Faber, 1934), 63.

50 William Morris, *News from Nowhere or an Epoch of Rest* (Oxford: Oxford University Press, 2003), 13, 54, 119, 139, 172, 181.

# PART III

# ENVISIONING THE FUTURE

chapter eleven

# Composing Westminster Bridge: Public Improvement and National Identity in Eighteenth-Century London

ALISON F. O'BYRNE

> Publick Buildings, commodious Streets, well paved, clean, and lighted, good Roads, navigable Rivers, and useful Bridges, have always appeared of the greatest Service to a State, and Evidences of a good Police, as well as of the Wealth, Industry, and Prosperity of a People.[1]

One of the most frequently articulated complaints about London in the eighteenth century was its lack of magnificence. As London, Westminster, and Southwark grew into one large metropolitan environment, politicians, statesmen, and 'public-spirited' gentlemen aired their concerns and offered their ideas about the shape the modern city should take. The above observation, from a pamphlet outlining the steps taken towards the building of Westminster Bridge in the two years following the act of parliament for its construction, argues that the appearance of a city, especially a capital city, should articulate a country's political, commercial, and national identity. The author describes a city opened up for viewing, implicitly by foreign visitors, in the form of public buildings, commodious, safe, and navigable streets and rivers, and, the author explicitly mentions, bridges that assist in creating both an open metropolis and an easily passable river. A city at once elegant and convenient is a service to both the state and its people, displaying the nation's good 'Police,' meant here in the broader eighteenth-century sense of public order and regulation, 'Wealth, Industry, and Prosperity.'

This chapter examines the building and representation of Westminster Bridge in the eighteenth century. I am self-consciously invoking the title of Wordsworth's sonnet 'Composed upon Westminster Bridge, 3 September 1802' as a means of exploring the ways in which the bridge was composed legally, architecturally, and, ultimately, via paintings, engravings, and various descriptions, aesthetically in the public imagination. One of the most important issues in this 'composition' was the abandoning of the original plans for a timber bridge and the building of a stone bridge, the façade of which was made entirely of Portland stone, one of the most expensive building materials in the period. Although proposals for a bridge at Westminster were rejected in 1722, a parliamentary act was passed in 1736 that allowed for the bridge's completion and opening to the public in 1750. From the first proposals for a new bridge, through its development and completion, the bridge was discussed, debated, and represented under the rubrics of a number of central eighteenth-century issues and concerns about taste, national identity, public magnificence, and public versus private development. It was the impetus behind a series of systematic improvements to the appearance of London and Westminster that drew on theories of the circulation of the blood to examine ideas about the health, circulation, and openness of cities.[2] The project's success offered a starting point for a discourse of public works and improvements that was especially prominent in the two decades following the bridge's opening.[3]

Westminster Bridge was, to borrow Miles Ogborn's term, one of eighteenth-century London's 'spaces of modernity.' In Ogborn's account, 'spaces of modernity' are sites that not only are, in themselves, new, novel, and therefore modern, but that also emerge out of 'modern' processes. Study of these spaces, Ogborn argues, reveals larger cultural shifts in the period that we associate with modernity.[4] In the case of Westminster Bridge, two processes of modernity are especially important. First, the process of scientific inquiry played itself out in all aspects of the bridge building: the various plans and inquiries into how and where a bridge might be built; proposals and designs for various types of bridges sold as pamphlets; and, finally, the interest in the bridge works in progress, including the various new machines and devices designed specifically for its building. Second, and more important, Westminster Bridge emerges out of, and is shaped by, the development of what Habermas has termed the 'bourgeois public sphere,' which envisages itself as actively engaging with government.[5] Ogborn has examined the role of the emergent public sphere as the context for the Westminster Paving Acts of 1762, exploring the role of this public sphere in debates about the order and cleanliness of the 'urban built environment' and the ways in which this becomes a matter of 'political authority, polite sociability, self-control and commodity exchange.'[6] The growing self-consciousness about the appearance of the streets of London and, as home to national government, Westminster in particular, that Ogborn traces through an examination of political philosophy, is also a result, I would argue, of two developments in the period. First, it is necessary to place this concern about the appearance of the metropolis in the context of the development of tourism in its modern sense, complete with guidebooks (distinct from pocket histories), pocket-sized maps, and souvenir items catering to native as well as foreign visitors in London. Second, the building of Westminster Bridge had gone a long way towards forming this public of which Ogborn speaks, a public that is interested in airing its concerns and adding its ideas and opinions to debates about the shape the metropolis should take.

As a public work in a period that, according to the architectural historian John Summerson, witnessed very little public building, Westminster Bridge, unlike Blackfriars Bridge, built in the 1760s, was the subject of an incredible amount of print in the form of debates, proposals, surveys, and views.[7] Despite the contention and concerns surrounding the bridge – and the sinking of its central pier shortly before the bridge was due to open – it became immediately celebrated in histories, guidebooks, and prints with accolades announcing it as a bridge worthy of the metropolis of Great Britain. During its building and after its completion, Westminster Bridge functioned in eighteenth-century histories, topographies, guidebooks, engravings, and paintings in much the same way as, according to Karen Newman, in her *Cultural Capitals*, the Pont Neuf functioned in seventeenth-century descriptions and representations of Paris. It was, to paraphrase Newman, the privileged sign of eighteenth-century London's modernity: a feat of engineering and an elegant design that, unlike its medieval neighbour London Bridge, was devoid of shops and houses, leaving broad open passages for pedestrians and wheeled traffic alike. But while, as Newman has demonstrated, representations of the Pont Neuf helped to create an ideology of the urban that celebrated the mixing of classes, of commercial and leisure activities, of crowds engaged in economic exchange and conspicuous consumption, representations of Westminster Bridge strove to articulate a different kind of urban ideology. Although the bridge was open to all, crowds are noticeably absent in representations of it. Instead, the bridge is described and represented as a space of polite entertainment. Visual representations often show small groups of men and women situated at a distance from one another engaged in conversation, while guidebooks and topographies after the bridge's opening describe it as a place of entertainment on a Sunday, full of men and women promenading on top and pleasure boats passing under its arches. Accounts of the bridge after its opening reveal the gap between its envisaged use, imagined as rendering commerce and everyday life more convenient, and its appropriation as an emblem of London's politeness and modernity.

The bridge's geographical positioning, linking Lambeth with the centre of national government at Westminster, provided a new, scenic entrance into London for visitors from the Continent who marvelled at its baroque design and the view of London and Westminster it provided. In his *Picturesque Tour through the Cities of London and Westminster* of 1792, Thomas Malton begins with a description of the approach to London that a majority of visitors from the Continent would have followed and the first views of the metropolis the visitor would experience, illustrated here (figure 11.1) with a detail from John Rocque's *Plan of the Cities of London and Westminster* (1744–6). Once the bridge was opened, and before the area south of the river was developed towards the end of the eighteenth century, the visitor to London, passing through the 'spacious area of St. George's Fields,' would enjoy a 'view [that] comprehended almost the whole extent of the Cities of London and Westminster ... [and] that altogether filled the mind with expectations of grandeur suitable to the Capital of the British Empire.' Continuing onwards, Westminster Bridge would rise 'boldly before the spectator, enriched and enlivened by a multitude of objects in motion, and the buildings on the Middlesex shore.' As the traveller passed across the bridge, he or she would be offered a panoramic view of London, the view to the north, towards St Paul's and the city, providing, in Malton's words, 'a prospect, perhaps, not to be equalled by any view of the kind in Europe.'[8] Malton's description draws on an aesthetics of grandeur and spaciousness that were at the heart of representations of the bridge, and, by extension, London, throughout the century.

The bridge was planned and developed in a period when there were growing concerns about the appearance of London – especially to foreign visitors. Discussions about improving the city's appearance date back to its rebuilding after the Great Fire, when architects and public figures, including Christopher Wren and John Evelyn, offered proposals that would replace the destroyed medieval city with an Enlightenment plan featuring grand avenues intersected by radial streets. These improved plans for the city, designed to showcase a range of public buildings and ornaments, including, in Evelyn's plans, libraries and fountains, were never put into place. The sense of urgency to get the city

Figure 11.1 Detail, John Rocque, *A Plan of the Cities of London and Westminster and Borough of Southwark* (1744–6), showing Westminster Bridge, several of the new roads that were part of the bridge works, including Bridge Street, Parliament Street, and The New Road, and the as yet undeveloped St George's Fields. William Andrews Clark Memorial Library.

up and running, the cost of totally transforming its appearance, and the debates about property boundaries resulted in a rebuilding scheme which merely improved upon the pre- fire layout, widening and straightening streets and requiring the use of brick or stone.

In topographies of London and descriptions of its buildings throughout the eighteenth century, the rebuilding was seen as a 'missed opportunity,' with architects and self-confessed men of taste left to lament London's lack of magnificence. The author of *A Critical Review of the Publick Buildings, Statues, and Ornaments* (1734), after outlining the benefits of 'noble and elegant buildings,' including the 'fame' of a nation and its people and the draw for foreign visitors, announces that ''Tis high time therefore for us to look about us too, and endeavour to vie with our neighbours in politeness, as well as power and empire ... No nation can reproach us for want of expence in our publick buildings, but all nations may for want of elegance and discernment in the execution.'[9] Britain's military strength, its 'power and empire,' may be unquestionable, but, the author urges, it is now time to begin to temper this reputation with 'politeness' and to demonstrate a balance between power and refinement via elegant and discerning public buildings. It was not only military power that elegant public buildings were meant to temper and refine: in 1766 the architect John Gwynn described the importance of public works that lend magnificence to a capital city as 'a national concern ... worthy of a commercial people.'[10] In his *London and Westminster Improved* (1766), Gwynn summed up the arguments in favour of a systematic plan to halt the expansion of London through private development and begin to transform it through a series of public works.

While the building of Westminster Bridge, and the transformation of its surrounding area, had gone some way towards effecting this plan, Gwynn, writing sixteen years after the opening of the bridge, was quick to remark on

> what might have been done, and how little has been done, when so fine an opportunity presented itself; certainly the building of the new bridge, and the powers with which the commissioners were vested, demanded much more, and had a general plan of improvements been duly considered, it is as certain that a very different use would have been made of so desirable a field for the exertion of taste, elegance and magnificence.[11]

Attention to public magnificence, Gwynn argues, would 'stimulate the powers of invention, and ... create employment for great numbers of artists.' It would improve Britain's reputation as a nation of taste and refinement by rendering noblemen accustomed to 'grand and elegant objects' before embarking on the Grand Tour, and would stimulate the national economy by drawing foreign visitors 'of distinction and taste.' Works of public magnificence in London would produce 'a refinement of taste, which in a nobleman produces true magnificence and elegance, [and] will in a mechanic produce at least cleanliness and decorum.'[12] According to Gwynn, public works were not merely ornamental to a city, but also improved the lives of its inhabitants.

In the half-century leading up to the building of Westminster Bridge, London was expanding rapidly. According to architectural historian Elizabeth McKellar, London was transformed between 1660 and 1720 into 'a modern metropolis of brick and stone which broke its traditional bounds and spilled out in all directions.'[13] The 'rage for building' was transforming the West End, where private estates were being developed into fashionable squares surrounded by town houses. Daniel Defoe registered this transformation in his *Tour thro' the Whole Island of Great Britain* (1724-5), referring to London 'in the modern acceptation ... tak[ing] in all that vast mass of buildings, reaching from Black-Wall in the east, to Tot-hill Fields in the west' plus Lambeth and Southwark, the whole, he calculates, totalling over thirty-six miles in circumference.[14] London 'in the modern acceptation,' expanding through private development, was in need of structural urbanization, improvements to help the modern metropolitan environment function, especially to assist the flow of traffic. As Nicholas Hawksmoor explained, in favour of the building of a second bridge over the Thames: 'Necessity and Convenience calls for another Bridge to expedite and forward the transacting of Business, and make it easy for Passengers, whether Citizens, Gentlemen, or others in the Country.'[15] Westminster Bridge was the first effort to address the needs of the modern metropolis, triggering, as we shall see, a number of systematic improvements to help transform and modernize what was in almost every other way already a modern city.[16]

Although suggestions for a bridge at Westminster date back to the Elizabethan era, the first serious proposal for a bridge in Westminster in the eighteenth century appeared in 1722. William Pulteney chaired a committee of enquiry into the building of a bridge, which found that 'the City of Westminster, and Parts adjacent are, of late Years, very much increased; and that such a Bridge as is desired, is become absolutely necessary; and will not only be a great Advantage to the said City, and parts adjacent, but also to the neighbouring Counties.' The committee also surveyed the area and found a location that would not 'in the least prejudice the Navigation' of the river, and which would make a very firm foundation.[17] 'A Bill for Building a Bridge cross the River Thames' was presented to parliament in 1722, and the architect Colin Campbell, in his *Vitruvius Britannicus; or the British Architect* ([1722?]) presented a plan for the bridge. Both were rejected.

The reasons for these rejections were captured in two publications of that year: *Reasons against Building a Bridge from Lambeth to Westminster* and *The Westminster Bubble*. Despite the fact that one of the arguments put forth for the need of a bridge was the growth of Westminster, which in turn increased the size of the whole of London, the objections to the bridge revealed the still highly divisive lines between Westminster, home of the Court, and the City, with its ancient privileges and freedoms. Both publications make use throughout of the terms 'project,' 'scheme,' and 'bubble,' terms associated with underhanded or deceptive financial dealings and drawing on the public's fears of financial ruin in the wake of the recent South Sea Bubble. The author of *Reasons against Building a Bridge* argues that a second bridge would entirely shift the balance of commerce in London. Westminster, already home to many of the City's merchants and tradesmen who left after the fire and never returned, would soon, he fears, become home to their businesses, as well as the markets and shops where they purchase their provisions, thus destabilizing the division between the centres of court and commerce. Drawing heavily on ideas of 'publick Benefit' and 'the publick Good' in a period when the meaning of these terms was still heavily contested, the author warns that the proposed bridge is 'at the best a precarious Project, and carries no certainty with it.'[18] He argues that the bridge would spell financial ruin for the City and destroy the lives of its watermen and the poor, a change that, by lessening the number of young men groomed for a career in the navy, would in turn destroy the nation.[19] *The Westminster Bubble* also presents the debate as one between the Court and City, suggesting in its very title the dubiousness of the proposal. Plans for a bridge at Westminster seemed to have been laid firmly aside when Daniel Defoe lamented, in his *Tour*, that such a project would have to await a future generation.[20]

A decade later, the movement for a bridge at Westminster was revived when what are variously described in eighteenth-century histories of the bridge as 'several publick-spirited Gentlemen'[21] came forward in 1734 not only to propose a bridge, but to offer to contribute to the cost of surveying the area to ascertain if, and where, such a bridge could be built. From its very origins, the building of Westminster Bridge is presented as a result of a public engaging with government to help shape the modern city. Once a convenient location for the bridge was agreed upon, advertisements were placed in newspapers for a meeting of the inhabitants of Westminster, at which a bill was drawn up and agreed to and a petition to parliament signed. In 1736 *An Act for Building a Bridge cross the River Thames, from the New Palace Yard in the City of Westminster, to the opposite shore in the County of Surrey* was passed, declaring that 'it will be advantageous, not only to the City of *Westminster*, but to many of his Majesty's Subjects, and to the Publick in general that a Bridge be built cross the River *Thames*' in Westminster.[22] The law named over two hundred peers and commoners as the commissioners for building the bridge and outlined plans for a lottery to be held as a means to finance its construction. At the same time, the act declared that it would not 'prejudice, or take away any Right, Property, or Jurisdiction' of the City of London (which had been granted rather vague administration of the river in 1197).[23] In addition, the act vested powers in the commissioners for the purchasing of houses on either side of the river for widening the approach roads. Various subsequent acts developed the scope of the project and extended the limits of the area that fell under the powers of the commissioners, who were instructed to transform streets surrounding the bridge that were 'so narrow, that they may be incommodious to Coaches, Carts, and Passengers, and prejudicial to Commerce and Trading' and to attend to the 'Beauty, Regularity, or Uniformity of the said Buildings, Streets, Ways, and Passages' surrounding the bridge.[24]

In 1737 the commissioners advertised a competition to decide the design of the bridge, drawing plans from renowned architects including Nicholas Hawksmoor. The advertisement, however, was for a bridge in timber, a detail that sparked much debate. James Thomson, whose poetry celebrated Britain's national identity as one based on maritime trade and continual progress, published a poem, 'On the Report of a Wooden Bridge to be built at Westminster,' in response to the proposal. Published in the *General Evening-Post*, *Fog's Weekly Journal*, the *Gentleman's Magazine*, and the *London Magazine* in 1737, the poem imagines the complaints of the river Thames on learning it is to be adorned with a wooden bridge funded by a lottery. After appealing to the '*British swains*' whose 'fertile Plains' have been dependant on the flow of the river, the Thames then addresses London's merchants:

> Have I, ye *Merchants*, with each swelling Tide,
> Pour'd *Afric*'s Treasure in, and *India*'s Pride?

Lent you the Fruit of every Nation's Toil?
Made every Climate yours and ev'ry Soil?
Yet *pilfer'd* from the *Poor*, by *gaming base*,
Yet must a WOODEN BRIDGE my waves disgrace?

Thomson's poem imagines the Thames as the epicentre of international commerce, pouring exotic goods from across the globe into London, contrasting the 'treasure,' 'fruit,' and 'pride' of other countries with the proposed wooden bridge funded not by a fair form of commerce, but by 'gaming base.' The poem draws together a number of the issues surrounding the bridge and engages with various contemporary debates about the role of luxury and the relationship between the private wealth of individual merchants and the public good. The debate is presented here in the context of London's (and Britain's) public appearance, as Thomson urges that private fortunes should be reflected in London's public architecture. Given the gentlemen-tradesmen readership of the magazine digests in which the poem appeared, Thomson presents the Thames as appealing to the sensibilities of London's merchants, pleading: 'Tell not to foreign streams the shameful tale.'[25]

A pamphlet published in the following year detailing steps taken towards the bridge building summarizes the debate concerning the use of wood as follows:

> No Resolution from a Commission appointed by Parliament, ever met with so great and general a Dislike from the Publick ... Many Persons of Judgement and Taste, were extremely disgusted at the thoughts of a Wooden Bridge, over so fine a River as the *Thames* ... Others took notice (notwithstanding the Use and Necessity of publick Buildings, and Magnificence, with which they ought to be built) what wretched Appearance, most of those in and about *London* made ... what a scandalous Effect a Wooden Bridge would have over the River *Thames* ... and what injurious Reflections would be made on that Occasion.[26]

The author's description of the debate captures two of the key issues that, I have suggested, were at the centre of the building of the bridge: the emergence of a public, one that, as Thomson's poem suggests, included merchants as well as gentlemen, interested in debating these issues, and the self-consciousness of 'injurious Reflections' on London's appearance. After much consideration about the comparative costs of continually repairing a wooden bridge and the building of one of stone, a stone bridge was finally settled on. In later descriptions of the building of the bridge – whether in histories, topographies, or guidebooks – the decision to build a bridge of stone is presented as the commissioners bowing to the pressure of a powerful public, which suggests a metropolitan identity attempting to assert itself as polite, enlightened, and modern.

The decision to build a stone bridge was seen as a triumph over a short-sighted and temporary solution in favour of a long-term response to London's changing needs. The next dilemma facing the commissioners was whether or not a sufficiently qualified British architect and engineer could be found. Among the proposals submitted was one from 'a set of Artificers' who, having heard 'a Report that there were not *English* artificers able to build the intended Bridge,' offered to build a single pier out of their own cost to demonstrate their abilities, an offer that was described as having 'some National Spirit in it.'[27] The fierce competition for the bridge design sparked its own small pamphlet war: after a number of proposals were published, John James, an architect and surveyor who had worked closely with Wren and Hawksmoor, followed with a review of the proposals, criticizing each and offering his own thoughts and plans (due to which, it seems, he fell out of favour with the commissioners).[28] Batty Langley followed with a response.[29] In the end, the various proposals by British architects were rejected in 1738, when the commissioners hired the Swiss architect Charles Labelye to undertake the work.

The work, begun in the same year, had its critics. While the first edition of Defoe's *Tour* had looked forward to the time when such a project would be executed, the third edition of 1742, updated and expanded by Samuel Richardson, registered some of the public's concerns about the bridge works. The edition criticizes the location, which had been altered since the original act, and the disruption caused by the alterations to the street layouts on either side for the new approach roads. At the same time, the author suggests, the 'Great Powers [that] are given ... to Commissioners to agree with Persons who own Houses and Lands on either Shore,' have led to overspending. In addition, the decision to raise money via public lottery, described as 'always of pernicious consequence to Trade,' comes under attack.[30] Concerns about lotteries, including their effects on trade and especially on the lower classes of tradesmen and their families, were nothing new.[31] Even the *Gentleman's Magazine*, despite its support of the bridge works, printed an article in which the author asserts the need 'to examine into the Nature of the present Lottery' to ensure that the 'Unwary' may not be lured 'by false Representations, to purchase Tickets at a Price greatly above their real Value.'[32] In response to various criticisms, Labelye published an update of the work in progress, *The Present State of Westminster Bridge* (1743), in which he aimed to defend himself against those 'People ... [who] did imagine that little or nothing has been done hitherto, and that the Bridge would not be passable there ten Years.'[33]

But while there were critics of Labelye's work, and the various additional acts of parliament, the bridge received significant backing from the *Gentleman's Magazine.* As the first and most successful of the new monthly reviews covering a broad range of subjects, the magazine appealed to a wide audience, including an aspiring middling class of tradesmen, merchants, and mechanics who envisioned themselves as gentlemen.[34] The magazine printed reports and updates on the progress of the bridge throughout its development, from short pieces noting, for example, the driving in of the first pile to longer detailed descriptions of the works to date, as well as engravings of plans and views of the bridge. In both Labelye's report of 1743 and in the pages of the *Gentleman's Magazine,* much is made of 'the many new Machines, Engines, and Mechanical Operations which have been made use of in the Course of these Works.'[35] Indeed, the 'technology' involved in building the bridge fed into what Mark Hallett has described as a 'well established category of graphic culture, one that even more specifically focused on ... the fragments of modern structures rather than the completed whole.'[36] In addition to the published proposals for the building of the bridge, engravings of the machines used in the building were produced, while painters including Samuel Scott and the Venetian Canaletto captured the bridge building at various stages.[37] These images would have appealed to a wide audience, from tradesmen and mechanics to art collectors.

The building, eventually dependent on a number of grants from parliament to supplement the funds raised by lottery, continued without any problems until 1747, when shortly before it was originally due to open, the central pier began to sink, throwing the whole design into jeopardy. While a number of reasons for the sinking were circulated, ranging from the presence of a spring underneath the pier to sabotage on the part of the Thames watermen, who apparently delighted in the possibility that the bridge might fail, Batty Langley took advantage of the opportunity to attack Labelye personally, while others satirized the

*Great Architect!* that projected the Plan;
Who first built a *Bridge* – O forbear now all Laughter! –
And sought a *Foundation* – almost nine Years after.[38]

Langley, perhaps seeing a chance to take over the commission, presented Labelye's work to date, and the current structural problems of the bridge, as a national insult, claiming, in *A Survey of Westminster Bridge, as 'tis now sinking into Ruin* (1748), that Labelye pirated his method of building while at the same time misapplying it (hence the sinking). The pamphlet includes an engraving of Labelye hanging from the bridge, with a caption underneath attacking him as an incompetent villain incapable even of pirating Langley's ideas with success (figure 11.2). The imagined dialogue between the Bridge Master and the Engineer ends with the former proclaiming that 'it would indeed have been happy for ye Publick had You been hanged before Westminster Bridge was thought on. For by Your Ignorance, Impudence, and Villainy, ye Publick is greatly Injured, and every Acting Commissioner, is disgraced to the End of Time.'[39] Throughout the pamphlet, Langley makes much of Labelye's Swiss nationality, claiming that not only is he himself 'very greatly injur'd, but my Country is also very highly disgraced, as it 'twas not able to produce One ENGLISHMAN capable to build the Bridge without the Directions of a *Swiss* PRETENDER.' Langley champions the native artists and workmen whose own designs were rejected in favour of Labelye, making use of the term 'pretender' – an especially loaded and derogatory phrase in the wake of the Jacobite rising of 1745 – to accentuate the engineer's un-Englishness. Throughout the narrative, Langley venomously attacks Labelye as an 'imposter,' a '*Foreign Pretender* [who] is the *Bane of* ENGLISH ARTISTS,' and 'an INSOLVENT, IGNORANT, ARROGATING *Swiss*.'[40]

But while Langley was concerned only with attacking Labelye both personally and professionally, attempting to demonstrate his self-professed superiority as an engineer, others took a more public-spirited view. While Langley made xenophobic pronouncements highlighting Labelye's Swiss nationality, other descriptions either neglected to mention it or stated it only in passing, a strategy, perhaps, in appropriating the bridge as a central image in London's (and Britain's) identity. William Halfpenny, in *A New and Compleat System of Architecture Delineated* (1749) includes, among his other architectural drawings, 'A Perspective View of the Sunk Pier.' Without mentioning Labelye (merely regretting 'an acci-

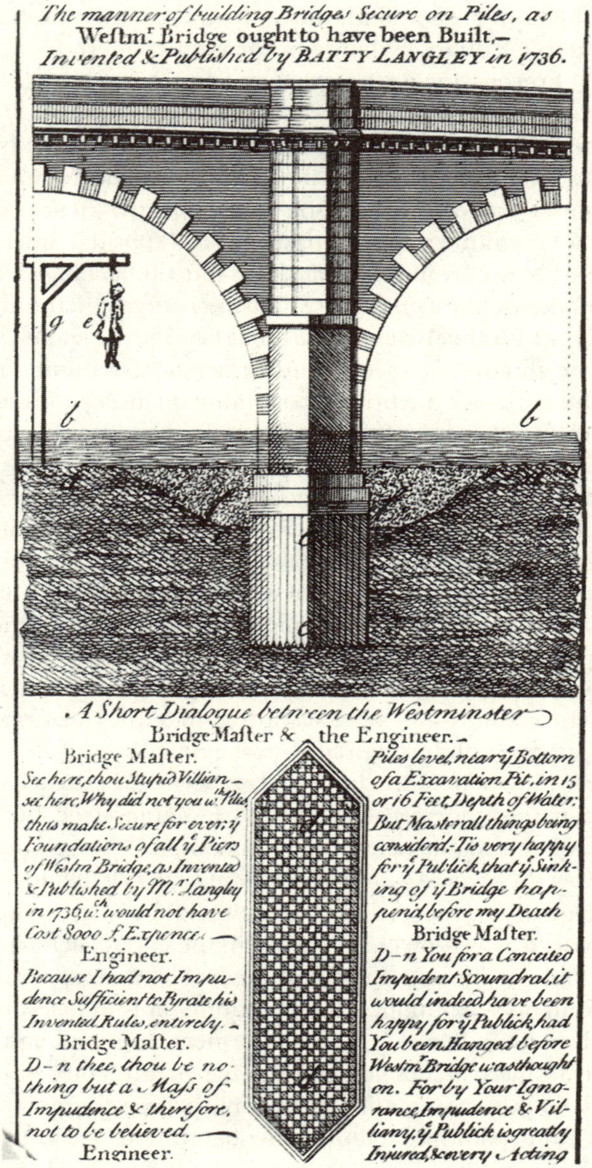


Figure 11.2 Engraving from Batty Langley, *A Survey of Westminster Bridge, as 'tis now sinking into Ruin* (1748), showing Labelye hanging from the bridge. William Andrews Clark Memorial Library.

dent or misfortune of that kind, in so great a structure, and which was erected for publick utility'), Halfpenny proposes a method to secure the pier, and to 'prevent any future mischance to that noble design.'[41] An article reprinted in the *Gentleman's Magazine* in September 1747 argued that the sinking of 'that magnificent fabrick,' as it is 'a national work,' should be looked upon as 'a public misfortune' and, without naming Langley, accuses 'some private persons [of] expect[ing] a benefit of it.'[42] The magazine continued to offer updates on the bridge works, and the methods used to secure the sinking pier; indeed, as Hugh Phillips has pointed out, the public's enthusiasm for the bridge could not be dampened by architectural debates such as that begun by Langley.[43] The correcting of the sinking pier seemed to fit neatly into the interest in the size and engineering of the bridge, becoming, in discussions such as that offered by the *Gentleman's Magazine*, one more feat of engineering to be accomplished.

While artists and print-sellers began to offer topographical views of London and Westminster that included the bridge in the altered cityscape, representations of the bridge itself provided the public with a more enduring image. Canaletto's views, undertaken at various stages of the bridge's development, were the most important of these images executed for patrons, the majority of whom were commissioners for the bridge. His paintings thus actively sought to present the bridge through the eyes of the commissioners as, according to Mark Hallett, the 'monument to civic responsibility' and 'the most explicit site of the modern' in London that they believed it to be. Engravings after Canaletto's paintings provided those less wealthy, but no less interested in the bridge works, with an 'affordable marker of urbane taste.'[44] Canaletto's *Westminster Bridge from the North on Lord Mayor's Day* (1746) is the most important of these images: it was, according to John Eglin, Canaletto's most copied print, and was heavily reproduced through the early 1750s (figure 11.3), during the time of the completion of the bridge and its opening to the public. Canaletto's painting, which presents Westminster Bridge not merely as backdrop but as a significant piece of scenery in the Lord Mayor's Day festivity, reconciles, as Eglin has pointed out, 'tradition and innovation' by showcasing a modern public work alongside traditional festivities.[45] The painting removes any sense of the carnivalesque, a frequent presentation of Lord Mayor's Day in the eighteenth century in, for example, Ned Ward's *The London Spy* (1698-1700) or Pope's *Dunciad* (1743). The image not only draws together a modern feat of engineering, a work that required, in the words of John Summerson, 'neither

Figure 11.3 B. Cole (after Canaletto's *Westminster Bridge from the North on Lord Mayor's Day*), *The South East Prospect of Westminster Bridge* ([1756?]). William Andrews Clark Memorial Library.

mason nor architect, but a professional man which England had not yet learned to produce,' alongside the City's various traditional guilds.[46] It also glosses over, in the very act of bringing together the bridge, the City guilds, and the Thames watermen, who are pictured ferrying passengers over the river, the various tensions that prevented an act for building a bridge in the 1720s, and that resurfaced again when an act was finally passed in the 1730s.

Canaletto's painting, executed shortly before the sinking pier had been discovered, shows the bridge works as complete (including statues of the river gods Thames and Isis that were proposed but never executed), but not yet open to the public. The City guilds and the Thames watermen, the two strongest opponents to the bridge, are represented as thriving alongside it. As shown in Benjamin Cole's engraving, while a number of watermen steer passengers, polite spectators of both the bridge and the procession, along the river, a queue forms at Westminster Bridge stairs for available boats across the Thames (figure 11.4). The larger and more ornate boats ferry the City guilds, which accompany the Lord Mayor in the City Barge to the Exchequer for his swearing-in ceremony. The painting emphasizes the way in which, as Eglin has noted, Canaletto's views of English scenes often focus less on the relationship between individuals than on the relationship between individuals and their environment, through the representations of a thriving river culture, one made up of both traditional ceremonies and the demands of everyday life.[47] In later engravings of this image produced after the pier was stabilized and the bridge opened to the public, such as Cole's *South East Prospect of Westminster Bridge* ([1756?]), both wheeled and pedestrian traffic remain curiously absent. Subsequent engravings of Canaletto's image also provide the viewer with a means of reorientation to this altered cityscape, where the ancient buildings surrounding the modern structure are lettered and identified in a key. The index helps the viewer to place the bridge within recognizable surroundings while at the same time discreetly articulating the importance of the structure by simultaneously suggesting that these ancient structures need to be redefined in relation to it. Despite, for example, its unmistakable towers, Westminster Abbey requires identification in relation to the new bridge, as do St Margaret's Church, Westminster Hall, St John's Church, the House of Commons, and Lambeth House.

Westminster Bridge opened to the public in November 1750 amid much fanfare. The public's excitement for the new bridge was such that a crowd, ignoring public notices, attempted to cross the bridge twelve

Figure 11.4 Detail of B. Cole (after Canaletto), *South East Prospect of Westminster Bridge* ([1756?]), showing watermen ferrying passengers along the river, and a queue at Westminster Stairs for available boats.

days before its opening; the Justice of the Peace had to read the Riot Act before the crowd would disperse – a curious instance in which this icon of rational, Enlightenment modernity became the centre of a popular entertainment, celebrated in almost mob fashion.[48] The bridge's finishing touches, according to the *Gentleman's Magazine*, were completed a few days later, including lamps and the assignment of watchmen, selected from the company of watermen as part of their compensation for loss of business. It was officially opened at midnight on 18 November 1750, with 'a Procession of several Gentlemen of that City, the chief Artificers of the Work, and a Crowd of Spectators, preceded by Trumpets, Kettle-drums, &c. and Guns firing during the ceremony.'[49] The crowd was so large that the new bridge could not cater to all comers: the *Gentleman's Magazine* reported on 20 November that '*Westminster* was all day like a fair with people going to view the bridge and pass over it, and at evening the crowd was so great that multitudes were obliged to cross the water in boats.'[50]

The public's excitement about the bridge in part stemmed from the fact that Westminster Bridge and, in the 1760s, Blackfriars Bridge, provided entirely new views of London. Whereas vantage points from London Bridge were limited to spaces between buildings, Westminster and Blackfriars Bridges opened up the view of London, allowing the viewer to take in broad, sweeping views of the metropolis. As the bridge became hailed as 'an exceeding great Ornament to this Metropolis,'[51] one that, according to the *Gentleman's Magazine*, 'will be looked on with pleasure or envy by all foreigners,'[52] the view *from* the bridge became equally important as the view *of* the bridge. The manner of description of the views *from* the bridge drew heavily on aesthetic ideologies, especially Joseph Addison's formulation of the 'Pleasures of the Imagination,' one of the key features of which is the '*Greatness*' or 'Largeness of a whole View':

> The Mind of Man naturally hates every thing that looks like a Restraint upon it, and is apt to fancy it self under a sort of Confinement, when the Sight is pent up in a narrow Compass, and shortened on every side by the Neighbourhood of Walls or Mountains. On the contrary, a spacious Horison is an image of Liberty, where the Eye has Room to range abroad, to expatiate at large on the Immensity of its Views, or to lose itself amidst the Variety of Objects that offer themselves to its Observation.

Westminster Bridge provided a unique opportunity in London to view a 'spacious Horizon' and survey an extensive view. Other aspects of Addison's formulation, including Beauty, Variety, that which is '*new* or *uncommon*,' and scenes that are 'perpetually shifting, and entertaining the Sight every Moment with Something that is new,' are frequently invoked in descriptions of the bridge.[53] For example, in his *History of the Principal Rivers of Great Britain* (1794) William Combe, after providing details of the history and building of the bridge, states:

> But it would be doing an injustice to this bridge, as well as to the subject of this work, if we were not to give some faint idea, which is all that words can give, of the magnificent prospect that presents itself to those who pass over it; combining such an happy intermixture of water and buildings, of permanent grandeur and varying scenery, as is not to be found in the view of any other river in the world.[54]

The novelty and excitement of the view from the bridge was commented on by many visitors to London. The young James Boswell, eager to improve the view, described in his journal how he picked up a prostitute and 'conducted her to Westminster Bridge, and then in armour complete did I engage her upon this noble edifice. The whim of doing it there with the Thames rolling below amused me much.'[55]

Guidebooks and topographies, in their descriptions of and from the bridge, transform a bridge built primarily as a convenience into a new site of polite leisure, offering a diverting and constantly shifting scene to dazzle the viewer. *London in Miniature* (1755) explains:

> [W]hether viewed from the Water, or by the Passenger who walks over it, it fills the Mind with an agreeable Surprize. The semi-octangular Towers, which form the Recesses of the Foot-way, the Manner of Placing the Lamps, and the Height of the Balustrade, are at once the most beautiful, and, in every other Respect, the best contrived.
>
> The surprizing Eccho in the Arches brings much Company to entertain themselves under it in Summer; and none of the public Walks, or Gardens, can stand in Competition with it for an agreeable airing.[56]

No longer merely a convenience for the inhabitants of Westminster, the bridge is now described as a pleasure ground, complete with promenade and pleasure boats to entertain Londoners on a summer's day. Histories, topographies, and guidebooks repeated superlative descriptions of the bridge, celebrating it as a feat of engineering and a jewel of modern architecture that provided those that pass over it with sweeping views of the celebrated cities of London and Westminster.

The success of Westminster Bridge and the related improvements to its approach roads instigated a number of major public improvements in the City of London and Westminster.[57] As early as 1753, discussions were being held in the City regarding the building of a third bridge and the transformation of London Bridge. While London Bridge had been held up as perfectly satisfactory for the needs of the metropolis when the bridge at Westminster was originally proposed, by 1756 petitioners in favour of repairing London Bridge argued 'that it was absolutely necessary' to widen the arches for the safer passage of sailing vessels and to remove the buildings to create a wider thoroughfare.[58] These improvements, and the building of Blackfriars Bridge, were carried out in the following decade. In 1762 the Westminster Paving Acts set up, for the first time, commissioners responsible for seeing to the paving and cleaning of the streets. Following on from the success of these acts – indeed, in the words of the eighteenth-century historian John Entick, 'imitating' them – a systematic plan was implemented to transform the face of the City. A number of paving, lighting, and cleansing acts were passed, and more strictly enforced, with an eye to making the streets healthier, safer, and more 'commodious.' In addition, Entick explains, steps were taken to make the city 'more airy, the buildings more commodious, and the commerce amongst the inhabitants more easy and agreeable.' The first step in this direction was to obtain an act of parliament 'to pull down what obstructed the free air, and to make and widen such streets &c. as should be found necessary.'[59] This included the removal from posts stretching across the streets of shop signs, which were to be placed flat against the building (and eventually replaced with house numbers); the removal of a number of the City gates; and the dismantling of sections of the City's medieval wall.

Construction of Blackfriars Bridge, stretching from Fleet-Ditch, which had been covered over since 1737, to Southwark, began in 1761.[60] But despite the fact that Robert Mylne's design was, according to John Summerson, more architecturally accomplished, spanning the river in nine rather than twelve arches, it did not capture the public's imagination in the same way as Westminster Bridge did.[61] Fewer artists sought to represent Blackfriars Bridge. Engravings such as those executed of the bridge under construction by Piranesi sought to exaggerate and emphasize the scale of the project, but those that seek to reorient the viewer in the altered landscape by offering a table explaining the various buildings in relation to the bridge were rare if not non-existent. One of the reasons for a different schema of representation for Blackfriars Bridge was that the most significant structure in its vicinity, and in front of which it is almost always depicted, was St Paul's Cathedral, constructed between 1675 and 1710 and thus itself a relatively recent building, clearly recognizable in the City's landscape. The bridge's central location provided a more truly 'panoramic' view, one that, for Thomas Malton, was so spectacular as to resist description.[62] The 'newly invented Panorama,' developed by Robert Barker, created a new, three hundred and sixty degree view of the metropolis from an elevated vantage point.[63] Nevertheless, despite Blackfriars Bridge's superior prospect view and its architectural advantages (not to mention the fact that it was completed on time and within budget), Westminster remained singular for the adulation it received.

With new, and clearly modern, bridges at Westminster and Blackfriars, London Bridge became the subject of extensive criticisms. The spacious arches of the two newest bridges contrasted sharply with the nineteen irregularly shaped arches of London Bridge. It became regarded more as an architectural curiosity, a medieval relic out of place in an Enlightenment city. As late as 1751, only months after the opening of Westminster Bridge, the author of *Gephyralogia: An Historical Account of Bridges, Antient and Modern* described London Bridge as a curiosity 'admired by foreigners, and, in most respects, as remarkable as any in Europe.' Although admitting the 'inconveniency' of the passage, the author felt it was unlikely to be altered, as the rents from the buildings on the bridge paid for its repairs.[64] Even after its houses were torn down and its central pier widened during the improvements of the 1760s, London Bridge remained out of place alongside Westminster and Blackfriars Bridges. By the end of the century it was described as little more than a 'gothic pile' unworthy of the river Thames and its associations with wealth, commerce, liberty and international trade. Writing in 1790, Thomas Pennant described the bridge as it had appeared at mid-century:

> The number of arches was nineteen, of unequal dimensions, and greatly deformed by the sterlings, and the houses on each side, which overhung and leaned in a most terrific manner ... I well remember the street on *London-bridge*, narrow and darksome, and dangerous to passengers from the multitude of carriages: frequent arches of strong timber crossed the street, from the tops of the houses, to keep them together, and from falling into the river.

The inhabitants of the houses, Pennant notes, 'soon grew deaf to the noise of the falling waters, the clamors of watermen, or the frequent shrieks of drowning wretches.'[65]

At the end of the eighteenth century, Westminster Bridge was still being called upon as the point from which to view London's politeness, commerce, and modernity. By the time of Wordsworth's poem 'Composed upon Westminster Bridge, 3 September 1802,' the view from the bridge had become canonized as *the* means of viewing the splendour of London in topographies and histories of the metropolis. In the poem as elsewhere, the view from the bridge draws heavily on the privileged perspective of the figure of the gentleman who, standing at a distant, elevated point, can see the order and harmony in that which lies before him.[66] Such an idea seems, on the surface, at odds with Wordsworth's presentation of London's 'Babel din' in book 7 of *The Prelude*, but even in *The Prelude* Wordsworth admits to a sense of harmony embedded in what seems, on the surface, chaotic, stating that one 'who looks / In steadiness' can see 'the parts / As part, but with a feeling of the whole.'[67] The view from the bridge in the 'calm,' 'silent, bare,' and 'smokeless' morning offers the poet one such opportunity. 'Earth has not anything to show more fair,' he proclaims in the opening line as he surveys the metropolis, a panorama of 'Ships, towers, domes, theatres, and temples' that vie with 'valley, rock, or hill' in their beauty as they 'Open unto the fields, and to the sky.' 'Dull would he be of soul,' Wordsworth proclaims, 'who could pass by' this view without feeling its beauty. The poet captures a unique moment in London, one that inspires in him a feeling of 'calm so deep' as he senses the stillness of the city's 'mighty heart.'[68] The parts do indeed appear for the poet as parts, but with a feeling of the mighty heart of the whole, afforded by the distanced, elevated, and privileged perspective of the viewer.

The building of Westminster Bridge reveals an important moment in London's development, signalling an increasing concern with London's appearance to the growing number of foreigners who visit the metropolis. In its transformation from a planned timber bridge to a stone monument to a metropolis that wanted to be perceived as polite, enlightened, and commercially and scientifically modern, the bridge responded to these concerns. The commissioners, and indeed the public, saw its building as an opportunity not just to articulate London's identity, but to carve its new self-image, the repercussions of which were felt through various transformations to the urban environment across the city. Celebrated in word and image over the course of the second half of the eighteenth century, Westminster Bridge's status was usurped in the early nineteenth century when old London Bridge was replaced with an iron bridge that articulated a new ideology of the urban – that of an industrial modernity. When the current was released from the narrow arches of old London Bridge, the gravel beds and structural foundations of both Westminster and Blackfriars Bridges were damaged.[69] Structural work was carried out to strengthen the foundations of Westminster Bridge, but it was eventually replaced with a new bridge in the 1850s, while Blackfriars Bridge was replaced a decade later. The new London Bridge replaced Westminster Bridge as the emblem of the metropolis's, and nation's, new identity as the centre of industrial advancement.

## NOTES

1 *A Short Narrative of the Proceedings of the Gentlemen, Concerned in Obtaining the Act for Building a Bridge at Westminster* (London, 1738), 45.

2 Richard Sennett, *Flesh and Stone: The Body and the City in Western Civilization* (New York and London: W.W. Norton & Co. 1994), 256.

3 See, for example, John Spranger, *A Proposal or Plan for An Act of Parliament for the Better Paving, Lighting and Cleansing the Streets* (London, 1754); *An Essay on the Many Advantages Accruing to the Community, from the Superior Neatness, Conveniencies, Decorations, and Embellishments of Great and Capital Cities* (London, 1754); John Gwynn, *London and Westminster Improved, Illustrated by Plans* (London, 1766); *Schemes Submitted to the Consideration of the Publick* (London, 1770); and James Stuart, *Critical Observations on the Buildings and Improvements of London* (London, 1771).

4 Miles Ogborn, *Spaces of Modernity: London's Geographies 1680-1780* (New York and London: Guilford Press, 1998), 1–28.

5 Jurgen Habermas, *The Structural Transformation of the Public Sphere*, trans. Thomas Burger (Cambridge: Polity Press, 1989).

6 Ogborn, *Spaces of Maternity*, 75.

7 John Summerson, *Georgian London* (1945; New York: Charles Scribner's Sons, 1946), 95.

8 Thomas Malton, *A Picturesque Tour through the Cities of London and Westminster*, 2 vols. (London, 1792), 1:1–4.

9 *A New Critical Review of the Publick Buildings, Statues, and Ornaments, in and about London and Westminster ... The Second Edition* (London, 1736), 2–3.

10 John Gwynn, *London and Westminster Improved, Illustrated by Plans* (London, 1766), 1.

11 Ibid., 10.

12 Ibid., 1–4.

13 Elizabeth McKellar, *The Birth of Modern London: The Development and Design of the City, 1660–1720* (Manchester: Manchester University Press, 1999), 12.

14 Daniel Defoe, *A Tour thro' the Whole Island of Great Britain* (1724–5), 2 vols. (London: J.M. Dent and Sons, 1962), 1:314.

15 Nicholas Hawksmoor, *A Short Historical Account of Westminster Bridge; with a Proposal for a New Stone-Bridge at Westminster* (London, 1736), 15.

16 According to Mary Dorothy George, 'Improvements in Westminster began with the building of Westminster Bridge (1737–50); many wretched houses were cleared away and thoroughfares were opened.' George, *London Life in the Eighteenth Century* (1925; Chicago: Academy Chicago Publishers, 2000), 107.

17 'Westminster Bridge. Petition of the Inhabitants of Westminster, also of the counties of Kent, Sussex, Surrey, and Southampton. 1721/01/11/.' In *Journals of the House of Commons* (1803 reprint) 19:708.

18 *Reasons against Building a Bridge from Lambeth to Westminster* (London, 1722), 28 and passim.

19 The author asserts that the establishment of competing markets in Westminster would lead to the closure of markets in the City, including those that provide affordable provisions for the poor. The loss of livelihood for the Thames watermen would bring ruin on their families from poverty, but also, he argues, on the nation as the river would no longer be a training ground for naval seamen (ibid., 7, 17–19).

20 Defoe, *Tour*, 2:330.

21 *A Short Narrative*, 3. As this term suggests, those supporting the bridge, and the laws pertaining to its construction and the relating improvements, also used the language of public use, benefit, and good to further their arguments.

22 Public General Acts, 1736, 9 Geo. II c.29. *An Act for Building a Bridge cross the River Thames, from the New Palace Yard in the City of Westminster, to the opposite shore in the County of Surrey* (London, 1736), 431.

23 Ibid., 458.

24 Ibid., 437, and Public General Acts, 1744–1745, 18 Geo. II c.29, p. 437, and 18 Geo. II c.29, *An Act for granting further Powers to the Commissioners for building a Bridge cross the River Thames from the City of Westminster to the opposite Shore, in the County of Surrey; and for the better enabling them to finish the said Bridge, and to perform the other Trusts reposed in them* (London: Thomas Baskett, 1745), 649. In 1749 *An Act for making a Free Market for the Sale of Fish in the City of Westminster* (Public General Acts, 1748–1749. 22&23 Geo. II c.49) was passed (London, 1749), designating a parcel of land that had been purchased by the commissioners for this purpose.

25 James Thomson, 'On the Report of a Wooden Bridge to be built at Westminster' in *Liberty, The Castle of Indolence, and Other Poems*, ed. James Sambrook (Oxford: Clarendon Press, 1986), 300–1.

26 *A Short Narrative*, 22–4.

27 Ibid., 13–14.

28 John James, *A Short Review of the Several Pamphlets and Schemes, that have been offered to the publick, in relation to the building of a bridge at Westminster* (London, 1736). The author of *A Short Narrative of the Proceedings of the Gentlemen, Concerned in Obtaining the Act for Building a Bridge at Westminster* noted that John James's 'Design, together with a Pamphlet published by him, criticizing all others that had appeared before, relating the Bridge at *Westminster*, has very much lessened the good Opinion most People entertained of his Learning and Skill in Numbers, Geometry, and Architecture' (34).

29 Batty Langley, *A Reply to Mr. John James's Review of the several pamphlets and schemes, that have been offered to the publick, for the building of a bridge at Westminster* (London, 1737).

30 Daniel Defoe, *A Tour thro' the Whole Island of Great Britain* 4 vols. (3rd ed., London 1742), 2:109–10.

31 See, for example, *Reasons Humbly Offer'd against Raising Money by a Lottery, for the Service of the Publick* (London, [1718?]), in which the author argues that '*Raising Money* by *Lotteries*, has been always very *dear* and *expensive* to the *Nation*, and a very great Prejudice to *Trade*' (1). Mary Dorothy George states that those tradesmen who were most dependant on the lower class of labourers were most affected during lotteries, as men and women would pay out a large portion of their wages in the purchase of a lottery ticket. George, *London Life*, 306.

32 'Loss and Disadvantage of the Bridge Lottery, and Chance of a Ticket,' *Gentleman's Magazine* 7 (June 1737), 367–8.

33 Charles Labelye, *The Present State of Westminster Bridge. Containing a Description of the said Bridge, as it has been ordered into execution by the Right Honourable, &c. the Commissioners appointed by Parliament, and is now carrying on. With a true account of the time already employed in the Building, and of the works which are now done. In a letter to a friend* (London, 1743), 3.

34 The magazine's original alternative title, 'Trader's Monthly Intelligencer,' was soon abandoned as 'too mechanical,' increasing its appeal to tradesmen who, in the words of Paul Langford, 'did not want to think of themselves as tradesmen but as gentlemen.' Langford, *A Polite and Commercial People: England 1727–1783* (Oxford and New York: Oxford University Press, 1989), 65.

35 Labelye, *The Present State of Westminster Bridge*, 10.

36 Mark Hallett, 'Framing the Modern City: Canaletto's Images of London' in Michael Liversidge and Jane Farrington, eds, *Canaletto and England* (London: Merrell Holberton, 1993), 46–54.

37 For Canaletto's various paintings of the Thames, and Westminster Bridge in

particular, see Hallett, 'Framing the Modern City' and J.G. Links, *Canaletto* (Oxford: Phaidon, 1982), esp. chap. 9, 'London, That Spacious City,' 145–80.

38 *The Downfall of Westminster-Bridge; or My Lord in the Suds. A New Ballad* (London, n.d.), 3.

39 Batty Langley, *A Survey of Westminster Bridge, as 'tis now sinking into Ruin* (London, 1748), n.p.

40 Ibid., iv–vii, 1.

41 William Halfpenny, *A New and Compleat System of Architecture Delineated* (London, 1749), n.p.

42 'On the Sinking of Westminster Bridge,' *Gentleman's Magazine* 17 (September 1747), 433.

43 Hugh Phillips, *The Thames about 1750* (London: Collins, 1951), 136.

44 Hallett, 'Framing the Modern City,' 47, 50.

45 John Eglin, *Venice Transfigured: The Myth of Venice in British Culture, 1660–1797* (New York: Palgrave, 2001), 132.

46 Summerson, *Georgian London*, 98.

47 Ibid., 132.

48 Phillips, *The Thames about 1750*, 136.

49 *London in Minature* (London, 1755), 116.

50 'Historical Chronicle,' *Gentleman's Magazine* 20 (November 1750), 523.

51 *London in Miniature*, 111.

52 'Of Westminster Bridge,' *Gentleman's Magazine* 20 (December 1750), 586.

53 Joseph Addison and Richard Steele, *The Spectator*, ed. Donald F. Bond, 5 vols. (Oxford: Clarendon Press, 1965), no. 412, 3:540–2.

54 William Combe, *A History of the Principal Rivers of Great Britain* (London, 1794), 158.

55 James Boswell, *Boswell's London Journal, 1762–1763*, ed. Frederick A. Pottle (New Haven: Yale University Press, 1950), 255–6.

56 *London in Miniature*, 111.

57 For overviews of the improvements to London and Westminster, see George, *London Life*, 107–15, and Ogborn, *Spaces of Modernity*, 75–115.

58 Quoted in John Entick, *A New and Accurate History of London, Westminster, Southwark, and Parts Adjacent*, 4 vols. (London, 1766) 3:120–1.

59 Ibid., 3:289–90.

60 Many of the same issues were raised about the building of Blackfriars Bridge as had been first voiced with the building of Westminster Bridge. See *The Expediency, Utility, and Necessity of a New Bridge, at or near Blackfryars* (London, 1756).

61 Summerson, *Georgian London*, 106.

62 Malton, *A Picturesque Tour*, 1:60.

63 Richard Altick, *The Shows of London* (Cambridge, MA: The Belknap Press, 1978), 129. According to Altick, Canaletto's paintings of the Thames, with their broad sweeping views, were a large influence on the development of the panorama.

64 *Gephyralogia: An Historical Account of Bridges, Antient and Modern* (London, 1751), 63–4.

65 Thomas Pennant, *Of London* (London, 1790), 296–7.

66 John Barrell, *English Literature in History: An Equal, Wide Survey* (London: Hutchinson, 1983).

67 William Wordsworth, *The Prelude* (1805), ed. Jonathan Wordsworth, M.H. Abrams, and Stephen Gill (New York and London: W.W. Norton & Co., 1979), 7:11–14.

68 William Wordsworth, 'Composed upon Westminster Bridge, 3 September 1802,' in *The Poetical Works of William Wordsworth*, Ed. E. De Selincourt, 5 vols. (Oxford: Clarendon Press, 1946), 3:38.

69 Phillips, *The Thames about 1750*, 137.

chapter twelve

# Here Comes the Son: A Shandean Project

MANUEL SCHONHORN

If, as Walter Benjamin has said, 'The hero is the true subject of modernism,' the search for that hero can be considered a dominant theme of the early modern period, on the Continent but especially in England.[1] Hanoverian England was a dismembered country whose citizens sought a deliverer of their nation – be he prince, general, admiral, even poet – who would lead them out of the swamp of misery, failure, and despair. How could it be otherwise, when, as Paul Langford explained in his *New Oxford History of England*, 'the traditional structures of corporate and communal life were either absent or inappropriate for the full range of contemporary conditions and aspirations.'[2]

We tend to overlook the historical conditions that gave birth to Laurence Sterne (1713) and *Tristram Shandy* (1759).[3] The thaumaturgic monarch of Samuel Johnson's youth, Queen Anne, had degenerated into 'King Turd' and 'the turnip hoer,' George I and George II of England.[4] When Defoe concluded in 1708 that 'Soldiers fight, and Schollars read, and Parsons preach, 'tis all for Money,' he had assessed the fall from grace of Britain's 'animating centers of society.'[5] Her major national institutions – the army, the church, and the university – had begun their descent that demoralizing cultural change would accelerate.[6] Sterne's century was one of disintegration, fragmentation, and revolution – revolution moral, sexual, social, financial, military, economic, and institutional.[7] John Richetti adds 'epistemological and ontological.'[8] British society had been undermined by specialization, effeminacy, and corruption.[9] Authority had been compromised and complicated. Gibbon's Oxford, that Jacobite haven, was 'engaged in debate over the sources of authority.'[10] Internally and externally the 1750s were deplorable years, 'a decade of one prolonged crisis.'[11] Sterne was begot and his fiction engendered in a nation threatened by domestic violence from within and by foreign oppression from without. He was fully aware of the 'torrent of corruption' that had overwhelmed Great Britain.[12] By 1758, in the face of military losses, humiliating retreats, rudderless ministries, breakdowns in public order, food riots, and invasion scares, a twenty-six-year-old shopkeeper in Sussex and the Earl of Chesterfield both bemoaned a nation that had lost its institutional vigour. Thomas Turner wrote in his diary: 'Oh my country, my country! Oh Albion! I doubt you are tottering on the brink of ruin and desolation.' Chesterfield wrote in a private letter: 'Whoever is in or whoever is out, I am sure we are undone both at home and abroad. We are no longer a nation. I never yet saw so dreadful a prospect.'[13] It is true that 1759 was to be a memorable year, the *annus mirabilis* of the Seven Years' War. It began with Admiral Hawke's sea victories at Lagos and Quiberon Bay and ended with General Wolfe's epical victory on the Plains of Abraham in Quebec.[14] In December Sterne published the first two volumes of *Tristram Shandy* at York. But while church bells were ringing in York Minster all through that celebratory year, Tristram records no victorious moments, nor do any church bells ring in his parallel narrative.

Who, then, would be able, in Linda Colley's phrase, to reforge the nation?[15] Who had the art, as Stephen Sondheim wrote, of 'making connections, of putting it together,' of bringing order to the whole?[16] For many, Britain's 'deplorable state of feebleness' could only be repaired by a dominant leader possessing supreme power.[17] Defoe, in an earlier and subversive program, had hoped to establish a magnipotent minister who would dominate the business of government.[18] One of the conclusions that can be drawn from J.H. Plumb's magisterial biography of Walpole is that one-party government, like non-party government, can devolve into one-man rule.[19] Bolingbroke's *Idea of a Patriot King*, written in 1735 though not published until 1749, preached for a benevolent autocrat, 'a standing miracle.'[20] In 1757 Rev. John Brown, in his popular and oft-reprinted *Estimate of the Manners of the Times*, fervently hoped that 'some GREAT Minister,' 'some leading *Mind*,' 'some superior *Intelligence*,' might arise to reform his fractured nation.[21] And as late as 1787, Thomas Day interrupted his seminal work intended for the use of children, *Sandford and Merton*, to ask for 'some legislator' who could draw up a 'code,' some 'general rules of conduct,' by which all could live in harmonious equality.[22]

Unlike his Continental antecedents, who proclaimed the visual artist and the inspired poet as luminaries of distinction and grace, Sterne, true to his idiosyncratic and parochial heritage, projected the writer-artist as the exemplary figure of his English culture.[23] Tristram Shandy, I will suggest, is the new authority, the reigning voice that was to repair the fragmentation of English society. His 'talismanic power of metaphor' (239)[24] coheres the discordant and potentially destructive elements of family life. How often we find him putting things together – upstairs/downstairs, chords/cords, parlour/kitchen, wit/judgment; even an unexpected sermon on heart, mind, and conscience arrives linked to a treatise on science and engineering; and the recurring allusions to Don Quixote and his horse fuse Tristram with his father, uncle, and Yorick, and each with all. Despite the recurring tensions in his household, Tristram is an agent of reconciliation who establishes a myth of community among jarring and mutually exclusive political, social, and religious interests that contemporaries and historians proclaimed among the discordant strains of the period.

Tristram's book begins with the truth that his three fathers – a scholar, a soldier, and a priest – are dead. (Three fathers, curiously, appear to have provided the English hero in myth and fiction with the biological, psychological, and moral foundations sustaining his role as redeemer and saviour.)[25] In Tristram's world, birthed in 1718 and published in 1759, these status personalities are dead, but Tristram, artist that he is, has absorbed all their vocabularies, has assimilated each of their voices. He has become, as Arthur Cash presciently observed, 'the fit moral heir of Shandy Hall.'[26]

The story of Sterne, Tristram, art, and the creative imagination has been told too often and too well to need more than a summary treatment here.[27] Sterne, we know from his letters and from the letters of those who met him, encouraged identification with Yorkshire Parson Yorick. But Sterne also saw and shaped himself in other ways. Queried by a French admirer, Sterne replied that he was of 'three elements, one inborn, two acquired,' the acquired two being the Scriptures and John Locke.[28] Lewis Curtis, editing his letters, came to this conclusion: 'He unconsciously watched his own reflection and supposed it to be variously Walter or Tristram or Parson Yorick or even Uncle Toby.'[29] In 1761 Sterne was already wearing a sword in Paris – unlike Pope, whose pen alone was wielded like a sword. In 1763 he wrote of himself as a 'miles emeritus.'[30] And in 1765 he sat for a portrait in Florence as the acclaimed author of Tristram Shandy. He is dressed in clerical black, and 'curiously' is wearing a sword when the clergy were prohibited by their profession to 'wear a Sword, or any Arms, (their habit being alone their

Defence).'[31] Kenneth Monkman has helped us to see *Tristram Shandy*, the book and the person, 'as the successful fusion of a number of Laurence Sternes.'[32]

When John Traugott writes that 'Tristram speaks as a preacher,' he overlooks the fact that Tristram has defined the vocabularies of all his fathers, incorporated their voices, assimilated their multiple temperaments.[33] Eric Rothstein's observation, that 'the bases for Tristram's life and opinions are found in them,' has been insistently repeated in the scholarship.[34] Through them Tristram has moulded his identity and projected his image. Examining his Lockean language, Bernard Harrison concluded that 'the figures who people Tristram's narrative imagination make him what he is.'[35]

But above all, Tristram is an artist, as Boswell immediately recognized.[36] His artistry is evident on almost every page of his life. He paints – paints his journey and paints his book; and in his opinion 'to write a book is for all the world like humming a song' (315). A whole chapter is but a fiddle solo in words (371–2). He pencils, he chisels, he wields the fiddle stick, prints a black and marbled page, and is both inventor and engraver of his volumes' trajectories (100, 473). Reading him, Voltaire and others were reminded of the paintings of Rembrandt, the sketches of Callot, and the talents of Vandyke and Teniers.[37] He surprises us with his references to Guido Reni.[38] Also, 'music is everywhere in *Tristram Shandy*,' William Freedman has written.[39] Sermons are dramatic performances. He sculpts his language. Like all narrative artists, Tristram recreates time, but also, like Parson Yorick and seemingly a discovery of modern fiction – he is able to stop time in its tracks; 'centaur-like,' 'labour stood still as he pass'd, – the bucket hung suspended in the middle of the well, – the spinning-wheel forgot its round, – even chuck-farthing and shuffle-cap themselves stood gaping till he had got out of sight' (19).[40] Diderot observed, coincidentally after reading *Tristram Shandy*, that the true poet fuses gesture and music, painting his words.[41]

From the beginning of his narrative, Tristram has been the artist of coherence in his house. Significantly in his fiction, his conscious amalgamation of all the arts can perhaps be seen as his successful attempt to reconcile them. In his time, Britain was in the midst of the same debate that earlier had preoccupied Renaissance artists and humanists concerning the hierarchy of the arts. Steven Copley, exploring the controversy, revealed how the positive analogy between literature and the other forms of aesthetic representation was coming under attack.[42] Copley does not allude to it, but just as opera earlier had been seen as a sign of cultural decadence, so now the instigator was painting. Sculpture and music too were seen to have a debilitating effect on the new impressionable public. Goldsmith was one among many who wrote that the division between literature and the other arts remained unbridgeable. For others, to enjoy those politer arts was to have one's Protestant faith called into question. In 1760, the Anglican cleric Benjamin Newton warned of the 'suspected loyalty' of sculpture and painting, seeing them as the "handmaids to Idolatry."'[43] Tristram refuses to delimit narrowly the artist to the 'literary.' Though he is writing his memoirs, he steadfastly refuses to be a 'man of letters,' and is as much a 'man of music' or a 'man of art' as he is a 'man of writing.'

And Tristram is an artist because he fits the now-recognized profile of the artist that we can track from the earliest records of achievement till the present day. I want to consider the seemingly eccentric life of Sterne together with his creation, comforted by what the hero of Bernard Malamud's novel *Dubin's Lives* observed : 'There is no life that can be recaptured wholly, as it was. Which is to say that all biography [and autobiography] is ultimately fiction.'

In 1934 Ernst Kris and Otto Kurz published their study *Die Legende vom Künstler: Ein geschichtlicher Versuch,* translated in 1979 as *Legend, Myth, and Magic in the Image of the Artist.* Its wealth of ideas, range of documentation, and boldness of speculation have made it one of the most influential texts in art history. From it, and the studies that it has generated, we learn 'that from the moment when the artist made his appearance in historical records, certain stereotyped notions were linked with his work and his person – preconceptions that have never entirely lost their significance and still influence our view of what an artist is.'[44] These preconceptions, or anecdotes, as Kris and Kurz called them, encompass Greek poets, painters, and dramatists, and apply to Renaissance painters, sculptors, and architects. Biographies and autobiographies of artists are characterized by topics and formulistic phrases distinguishing them from all other kinds of biographies and autobiographies. These impart a legendary status to the artist. What are these themes?

Quite naturally, they deal with the artist's conception, birth, and youth. First, he is given uncertain origins, or parents, at times suggesting that he is an illegitimate child. In childhood, his special gifts are immediately apparent. And he is singled out, recognized, and confirmed by miraculous rescues that set him apart from ordinary individuals. If a poet, his poetry images the imagined self at times as a monarch, a soldier, or a priest. In every possible manner, it is clear, his attributes and accounts parallel those of the mythological hero.[45] These features in the life of the artist occur with recognizable symmetry in the records of Sterne and Tristram. Sterne's *Memoirs* and Tristram's *Life and Opinions* incorporate anecdotes that, throughout history, have revealed artists constructing their own contemporary myths, or having them constructed for them. In the former, Sterne tells of 'that wonderful Escape, in falling thro a Mill Race whilst the Mill was going – and of being taken up unhurt. – The Story is Incredible – But known for Truth in all that part of Ireland, where hundreds of the Common people flocked to see me.'[46] This event, and a later tale of a fall from a church steeple, discounted by an early biographer, exhibit a young hero of an incredible adventure. And while Professor Cash surmised that 'the boy imagined the incident, appropriating to himself a similar myth about Archbishop Sterne,' the man of the *Memoirs* may just as likely be recalling the miraculous escape of Horace, who himself was appropriating a classical formula.[47]

Sterne later recalled for his daughter 'this anecdote of myself, and school-master' while he was a student at Halifax: 'He had had the ceiling of the school-room new white-washed – the ladder remained there – I one unlucky day mounted it, and wrote with a brush in large capital letters, LAU. STERNE, for which the usher severely whipped me. My master was very much hurt at this, and said, before me, that never should that name be effaced, for I was a boy of genius, and he was sure I should come to preferment.'[48] Now this may not have anything to do with a somewhat similar anecdote in Vasari's life of Giotto – any study of Vasari in the eighteenth century would have to consider Sterne – but we do know this: Sterne set aside his *Memoirs* in 1759.[49] As Kenneth Monkman deduced with the discovery of the manuscript of the first part of the *Memoirs*, this anecdote dates from 1767. He wrote: 'Sterne takes up the narrative with the anecdote about him as a 'boy of genius,' which obviously could not be decently told before his genius had been established; but which is the first thing he "cannot omit mentioning" when it has been.'[50]

Like Pope before him, whose portrait at age seven reveals his little hands already holding the laurels of the soon-to-be-acclaimed poet-artist, and like Picasso in our day, we have genius, in their projections of childhood, dramatizing themselves according to legend. From Spence's *Anecdotes* we learned that the laurels were added later by Pope's friend the painter Jervas, 'and it is a fair guess that the instigation for the addition came from Pope.'[51] And from John Richardson's first volume of his acclaimed biography of Picasso we learned that every childhood anecdote of this seemingly precocious and untutored genius – he never walked like a child, he never drew like a child, he presented himself as 'a prodigy of both intuitive genius and also an artistic virtuoso who sprang fully developed from his father's head' – derives from Picasso himself, who spent his whole life creating his own myth, 'fantasizing, embroidering, and reordering the truth.' Richardson thoroughly dispatched Picasso's 'Andalusian embellishments' and finally dispelled the twentieth-century's grandest myth of origin.[52]

Like Sterne and Pope and Picasso, Tristram too was born to travel out of the common road. Recall *his* anecdote of the top: his father,

> observing a most unaccountable obliquity, (as he call'd it) in my manner of setting up my top and justifying the principles upon which I had done it, – the old gentleman shook his head, and in a tone more expressive by half of sorrow than reproach, – he said his heart all along foreboded, and he saw it verified in this, and from a thousand other observations he had made upon me, That I should neither think nor act like any other man's child (6).

Even brother Bobbie's death contributes in a mythic way to Tristram's ascension, for his inheritance, and thus his rise in social standing, suggests another typical motif in the life of the artist.[53] E.H. Gombrich, in his admiring introduction to the Kris and Kurz study, writes that 'the stories told about artists [and that artists tell about themselves] in all ages and climes reflect a universal human response to the mysterious magic of image-making.'[54] That image-making magic is Tristram's 'talismanic power of metaphor.'

Tim Parnell, analysing Sterne's wanton sentimentalism, parenthetically remarked that 'his theological underpinnings are almost entirely lost to us.' His unexplored insight is worth detailed interrogation.[55]

Matthew Hole, a venerable and respected Church of England priest, published in 1716 a series of *Practical Discourses* elaborating on all the Collects, Epistles, Gospels, and Saints Days in the Church's calendar. His commentaries appear to have retained their popularity. The text from Matthew 4:13, 'Deliver us from Evil,' he expounds on thus: 'By the Mystery of thy Holy Incarnation, by thy holy nativity, and circumcision ... Good Lord deliver us.'[56] Hole's conjoining of Christ's conception, birth, and circumcision came naturally to other Anglicans, from Joseph Hall in 1661 to Samuel Johnson's Lenten sermon in 1740.[57] The 'theological underpinnings,' these three mysteries in the life of Sterne's 'small HERO' (10), deserve our consideration.

Tristram Shandy 'was begot in the night, betwixt the first *Sunday* and the first *Monday* in the month of *March*, in the year of our Lord one thousand seven hundred and eighteen, *I am positive I was*' (8). It is that intimate evening of that set-aside day when Mr Shandy takes care of a family concernment, as much a matter of business as amusement – but surely one not driven by lust – for with this activity, a prime family duty is gotten out of the way so that he would be no more plagued and pestered with it for the rest of the month. About his time of conception, Tristram is very precise, supported as it is by 'anecdote' and memorandum book (8). It is the first Sunday in Lent. As Septimus Hodson reminded his parishioners in the first of his five Lenten sermons, 'We have commenced that solemn season, which the church hath in its wisdom set apart for a more especial enquiry into the state of our spiritual concerns, and which it hath accompanied with suitable holy exercises and ordinances.'[58] The biblical text selected by the clerics dictated the dominant themes. Some preachers, for example, remarked about the calendric significance – forty days and forty nights – of the penitential season. Some sermons were simply moral disquisitions; some said nothing about the unique significance of the season. But the predominant and prevailing theme of the sermons I have read, from Bishop Sherlock's in 1692 to Hodson's in 1792, is the season's occasion, and its requirement of fasting, prayer, and self-examination. Elias Sydall discoursed for many when he wrote that the 'Use and End of Religious *Fasting* is to Subdue and Mortifie our Carnal Lusts, and Affections, and bring the *Flesh* into subjection to the *Spirit*.'[59] Yet some preachers caution excessive zeal in mortification and denial. 'This Church hath limited a certain Number of Days to be employed in such Acts of Humiliation and Abstinence as are likely to help forward our Repentance, leaving it to every Person to practise it, or dispense with it, as his Bodily Health, his worldly Business, or the Occasions of his Soul require,' Edward Littleton wrote in 1749.[60] Or as Richard Lucas preached, discoursing '*Of the due Moderation of our Pleasure*,' 'That God has not interdicted all Pleasure is manifest: for, in the *Old Testament* we are taught, that *the ways of wisdom are ways of pleasantness, and that all her paths are peace*.'[61] Thomas Berriman continued the theme of moderation and restraint, cautioning his readers that 'the *Wednesdays* and *Fridays* through the whole Season [are] the more solemn days of Prayer [to] be observed with such Retirement and Abstinence as may assist our Devotion ... excepting the *Sundays*, which are always festival.'[62] Some, like Berriman and Coney, began and concluded their presentations to their parishioners by reciting the Collect of the day, continuing with the reading from

2 Cor. 6:1: 'I have heard thee in a time accepted, and in the day of salvation have I succoured thee; behold, now is the accepted time; behold, now is the day of salvation.'[63] Finally, Bishop Joseph Hall, in his final contemplations of the annunciation, birth – 'The Purification' – and circumcision of Jesus, with the rare wit and imagination of an earlier day, reminded his readers that 'Christ was conceived in the Spring, born in the Solstice. He in whom the World received a new life, receives life in the same season wherein the World received his first life from him.'[64] Fasting and prayer, yes, but the beginning of a season that promoted piety and virtue and spiritual preparation.

There is, of course, no way to prove that Sterne examined a calendar to affirm that date for Tristram's momentous conception. But we do know that Lent was of special, sacred significance for the Reverend Sterne. The holy season is unusually prominent in his calendric church history and in his correspondence. He preached annually in York Cathedral every sixth Sunday in Lent. And in response to queries from his archbishop in 1743, he writes, 'I Catechise every Sunday in my Church during Lent, But explain our Religion to the Children and Servants of my Parishioners in my own House every Sunday Night during Lent, from six o'clock till nine.' Canon Ollard, who compiled the collected *Queries* from respondents in the diocese, commented: 'In all the hundreds of returns which I have examined, this is unique and stands alone.'[65]

Begotten in Lent, Tristram Shandy also has an auspicious birthday: 'On the fifth day of *November* 1718, – was I, *Tristram Shandy*, Gentleman, delivered into this scurvy and disasterous world of ours' (9-10).[66] It is, of course, Guy Fawkes Day, as the editors of the Florida edition of the Works of Laurence Sterne note, 'and the inauguration of the Revolution of 1688 by the landing of William III. in Torbay.'[67] It too had a special Form of Prayer and Service, 'for the happy Deliverance of King JAMES I ... And also for the happy Arrival of His late Majesty on this Day, for the Deliverance of our Church and Nation.' It is a rare sermon that does not celebrate this 'double deliverance,' and nearly always with special recognition of William, 'Our great Deliverer,' 'a glorious Deliverer,' '*Him*, whom God was pleas'd to make his *chief Instrument* in one of our *Great Deliverances*.'[68] Others linked William with 'God, our Saviour and mighty Deliverer,' making it difficult for the reader to distinguish the two.[69] Following his death, many dutifully concluded with thankful references to Queen Anne or George I. The day was a sacred festival day for Henry Grove, who, in 1747, proceeded to proclaim it similar to the deliverance of the children of Israel, 'a type of redemption of mankind by Jesus Christ.'[70]

Tristram's birth day was eight months from his Lenten conception, a figure that has provoked arguments and debates over Sterne's attentiveness and Tristram's illegitimacy. Yet Sterne's correspondence and the judgment of Kenneth Monkman should caution an argument for error. Examining the early editions of *Tristram Shandy*, Monkman concluded: 'There is ample evidence that [Sterne] took a meticulous interest in his words and how they were printed.'[71]

No one as far as I know has connected the interval with William himself, who was known to be a premature child, having been born 'at the *Hague, Nov.* 4, 1650, ten days after his Father's Death, and two Months before his Time, which made him weak and infirm.'[72] In fact, *The Clockmakers Outcry*, in May 1760, five months after the London publication of Sterne's celebrated fiction, reaffirmed the rumours of William's impotency.[73] Impotency, of course, has been one persistent theme in Tristram's story and with Tristram's critics. Thus, as the Florida editors note, 'It is almost assuredly no accident that Sterne chose November 5,' but not only for its national significance, and not assuredly for the anti-Catholic diatribes that abounded in some sermons, common though they were. For, paradoxically, Tristram's birthday is William's birthday, for the latter's was 'joyn'd' to the King's landing and 'the Gunpowder Treason,' as Thomas Hearne explained, for purely political reasons to solemnize two – really three – disparate events.[74] Annually, in his *Review* essays on or close to 4 November, Defoe did not distinguish between the two days, the birthday of his nation's Moses or the day of his nation's deliverance, when 'all these great Things [have] been directed by Heaven to meet on the Conjunctures of this Day ... On this Day he was born, on this Day he was marry'd to the Daughter of *England* ... and on this Day he marry'd the Nation.'[75] In 1749 Walter Harris observed '4 Festival Days to [our great Deliverer's] memory,' linking 4 and 5 November to William's victories in Ireland.[76] The two days were also linked in the nation's political calendar by processions, bonfires, and, of course, sermons.[77] And the two festival days were linked also in the theatrical calendar, for both 'on the fourth and fifth of November, the anniversary of the landing of William III (1688) was celebrated by performing Rowe's *Tamerlane*.'[78] By 1760 we can imagine Sterne's readers envisioning Tristram as a domestic reincarnation of England's 'Christian Hero,' as Thomas Comber wrote in York in 1757; the redeemer of Israel-England, 'a name that will ever be precious to all who in Religion are Protestants, and in the State Lovers of Liberty.'[79] Though the king's reputation as a war leader had been dimmed by mid-century, he remained 'Glorious King William,' 'our

Protestant Deliverer,' as one admirer proclaimed in 1759.[80] His birth, and the birth of his literary progeny, heralded 'the return of fundamental life-giving values in danger of extinction.'[81] As another prayer book reminds us, 'In the child God gives humanity a chance to make good its mistakes and to fulfill its fondest dreams.'[82]

Another 'great Deliverer' was honoured by the nation, as the late Bishop of Exeter preached, in a sermon published in 1747, and readable again in a second edition of his works published in 1755.[83] He was the circumcised Christ, who, 'in this state of weakness, pain, and sorrow,' revealed the wisdom and power of God 'to effect the deliverance of his people,' as George Horne preached before the University of Oxford in 1763.[84] This festival day was celebrated on the first of January in commemoration of the circumcision of the Saviour, which brought 'a mighty Blessing to the World ... even the glorious Liberty ... that *there* should be for the future *neither Jew nor Greek, neither circumcised nor uncircumcised,* but *one Fold and one Shepherd.*'[85] It is the last of those three mysteries enunciated by Matthew Hole, when Christ, suffering for humanity, obtained favor for all. Recognized as a day of initial sorrow, it was proclaimed a day of joy in the conclusion of many of the sermons.

Let me now comment, if only sketchily, on what the English wrote, preached, and read, about circumcision in the century before Tristram Shandy's. John Downame's *Annotations upon all the Books of the Old and New Testament* appears to have provided the foundation for all the commentaries that followed:

> The uses of circumcision were, partly, to distinguish Gods people from other nations, and to separate them from them; so that, without they would yeeld to that, they were to have little communion with them ... Partly to signifie the cutting off of concupiscence, and the extirpation of sin ... And partly, to betoken the purity of the promised seed, who was to have mans nature, but without any corruption, which in ordinary generation is derived from it; and withal, not onely to be a signe of Gods Covenant with Abraham, and his seed, especially the godly, but to be a Seal of the righteousness of faith, and remission of sins by the blood shedding of the promised seed.[86]

The main points of the doctrine were as follows:

1 Circumcision was of divine appointment, institutionalized by God, 'a Sacrament of Initiation,' but initially a ritual of exclusion, separation, enjoined upon God's 'Peculiar People.'[87]
2 Circumcision recognized man's imperfect and carnal nature, and

served as 'a token and Memorial of Restraint upon fleshly Lusts,' as William Tilly preached before the University of Oxford on New Year's Day, 1711; commanded by God, as Roger Altham preached, 'to subdue, to kill, to crucify the flesh, with its affections and lusts.'[88]

3 Circumcision guaranteed typologically the final deliverance from sin, signalling the beginning of redemption not only for the seed of Abraham but for all mankind. 'You must see in it,' Bishop Weston told his audience, 'a mighty blessing to the World, which it commemorates on this great Day; even the glorious Liberty, and Title, of the Children of God, made universal.'[89]

4 Circumcision Sunday, New Year's Day, was the 'Sweet Earnest of an Happy Year.'[90] As Thomas Horne exuberantly told his audience at the University of Oxford in 1763, with an exuberance shared by many, though the day began with sorrows, it was 'the beginning of joy ... because the beginning of redemption.'[91]

Perhaps in a pattern we cannot fully fathom, Tristram is again linked with an agent of salvation and deliverance, one who took upon himself the 'beginning of redemption,' on this naming day on which Christ embraced his human, and thus double, nature. In many of the sermons of Sterne's time, the event was proclaimed a blessing, 'not a useless or indifferent Ceremony'; and 'the day that opens, as it were, a fresh Scene of Life, by letting us into a New Year.'[92] What began as a physical sign singling out Abraham and the chosen people, when later undergone by Christ it prefigured the spiritual circumcision of the heart. Salvation was guaranteed not for one primitive tribe but for all mankind.

Conceived during the first Sunday evening in Lent, birthed on the fifth of November, and circumcised when five years old – 'thousands suffer by choice what I did by accident' (379) – such are the elements of the prodigious beginnings of Tristram Shandy.[93] To assert conscious artistry for Sterne is no longer an uphill battle for his admirers; but to suggest that his dizzy fiction sacralizes the earthly legend of the artist is to risk scholarly excommunication. Arguments about 'ambiguity' and 'uncertainty' still abound concerning Sterne's fiction.[94] Tristram Shandy is difficult to accept as a culture-hero, the totemic figure of the artist. Perhaps one of the many possible reasons is that, despite his seemingly modern stances, he is not defined by his alienation from his society. While he may recognize, and even give generous voice to, its incoherence and instability, he does not remove himself from it. But we should be aware that his creator is not a lone voice dismantling old ideologies. The years immediately before the publication of *Tristram Shandy* reveal a new assertiveness by writers challenging the age's perception of them as hacks or scapegoats for the failures of their time. James Ralph, in 1758, published an impassioned defence of authors, 'the Dispensers of Knowledge, [which] is the Light of the World.'[95] One year later Goldsmith sought to rescue literary genius 'from the shackles of pedantry and criticism.'[96] That same year, before he felt the enthusiastic fit, Imlac proclaimed the poet as 'the legislator of mankind.'[97] And Gibbon also responded to the temper of the time. His earliest work, *Essai sur l'étude de la littéraire*, written in 1761, is more than 'a philosophical defense of erudition against a philosophy he thought to denigrate it,' as John Pocock has determined.[98] It is also a young man's calm, reasoned, but dedicated recognition of *belles lettres*, of 'Great men of letters,' and of 'Poets and Oraters' in the ancient world and the present, who were responsible for the progress of civilization.[99]

Sterne's fiction epitomized, and Tristram's voice resolved, the dislocation of the social order of their time. At least, to Sterne's credit, we should admit that he is the author of the first fictional autobiography of the artist in his time, and that Tristram Shandy is the first fictional autobiographer of whom the only work is the autobiography itself.[100]

## APPENDIX

All the following are London publications unless noted.

### Lent

Altham, Roger. On Lent, in *Sermons Preached on Several Occasions*. 2 vols. 1732. In vol. 1.

Ashton, Thomas. *Sermon XIX. Sermons on Several Occasions*. 1770.

Bayley, Benjamin. *Sermon I* (9 March 1708 ). *Fourteen Sermons on Various Subjects*. 2 vols. 1721. In vol. 1.

Berriman, William. *Sermon IX. Christian Doctrine and Duties Explained and Recommended; in Forty Sermons*. 2 vols. 1751. In vol. 2.

Clarke, Samuel. *Sermon XCIV. The Works of Samuel Clarke. In Four Volumes. Volume the First. Containing Sermons on Several Subjects*. 1738.

Coney, Thomas. *Sermon XXIII. Twenty Five Sermons Preached upon Several Subjects and Occasions*. 2 vols. 1750. In vol. 2.

Dawes, William. *A Sermon Preach'd before the Queen, in Lent, 1704*. 2nd edition. 1707.

Fleetwood, William. *A Sermon Preached before the King the First Sunday in Lent, 1717.* In *A Compleat Collection of the Sermons, Tracts, and Pieces of all Kinds, That were written by The Right Reverend William Fleetwood, late Lord Bishop of Ely.* 1737.

Fletcher, William. Lent Sermon. *Twenty Sermons by the Late Reverend William Fletcher.* Dublin, 1772.

Hayter, Thomas. *A Sermon Preached before the King. On Sunday, March 22, 1752.* 1752.

Hodson, Septimus. *Sermon I. Sermons on the Present State of Religion In This Country, and On Other Subjects.* 1792.

Jeffery, John. Lent Sermon. *A Complete Collection of the Sermons and Tracts.* 2 vols. 1751. In vol. 1.

Johnson, Samuel. Lent Sermon. *Thirty-six Select Discourses Doctrinal and Practical.* 2 vols. 1740. In vol. 1.

Littleton, Edward. Lent Sermon. *Sermons upon Several Practical Subjects.* 3rd edition. 1749.

Lucas, Richard. *Sermon X. A Lent Sermon. Twelve Sermons Preached on Several Occasions.* 2 vols. 2nd edition. 1729. In vol.2.

Newlin, Thomas. *Sermon III. A Sermon Preach'd before the University of Oxford, at St. Peter's, in Lent.* In *English Sermons on Several Occasions.* Oxford, 1720.

Orr, John. *Sermon X. Preached in Lent.* In *Sermons of the late Reverend John Orr.* 3 vols. 1772. In vol. 2. The Preface is dated 28 October 1765.

Pearse [Pearce], Robert. *The Reasonable, the Ease, the Pleasure of the Christian Life. On the Third Sunday in Lent, 1717.* Oxford, 1717.

Reading, William. *Sermon XXI. On the First Sunday in Lent. Morning.* In *One Hundred and Sixteen Sermons Preached Out of the First Lessons At Morning and Evening Prayer, for all Sundays in the Year.* 4 vols. 2nd edition. 1736. In vol. 1.

– *Sermon XXII. On the First Sunday in Lent. Evening.* In vol. 1.

Smalridge, George. *Sermon X. An Enquiry Into the Rise and Antiquity of the Lent-Fast.* In *Sixty Sermons Preach'd on Several Occasions.* Oxford, 1724.

Sydall, Elias. *Of the True Uses and Ends of Religious Fasting: With a Brief Account of the Original of Lent.* 1713.

Tillotson, John. *A Sermon Preached before the King and Queen. Being the First Sunday in Lent.* 1694.

Warren, Robert. *Discourse for the First Sunday in Lent.* In *Practical Discourses on Various Subjects. Proper for all Families.* 2 vols. 1723. In vol. 1.

Wheatly, Charles. *Sermon XI. Christ tempted.* In *Fifty Sermons on Several Subjects and Occasions.* 3 vols. 1753. In vol. 1.

**November 5**

Atterbury, Francis. *A Sermon Preach'd before Her Majesty. On Sunday November the 5th, 1704.* 1704.

Beveridge, William. *A Sermon Preach'd before the House of Peers, Sunday, November the 5th. 1704.* 1704.

Bilstone, John. *Sermon IX. A Divine Providence Considered and Vindicated.* In *Thirteen Sermons Preached before the University of Oxford.* Oxford, 1749.

Bourne, Samuel. *Popery a Craft and Popish Priests The Chief Crafts-Men. A Sermon Deliver'd On the Fifth of November.* 1735.

Bradford, Samuel. *The Reasonableness of Standing Fast in English and in Christian Liberty. A Sermon Preach'd before the Lord Mayor On Thursday, November 5th, 1713.* 3rd edition. 1713.

– *A Sermon Preach'd before the Lord Mayor and Aldermen, and the Citizens of London, On Sunday, November 5th, 1704.* 1704.

– *A Sermon Preach'd before the Lord Mayor and Aldermen, and the Citizens of London, On Tuesday, November 5th, 1700.* 1701.

Bundy, Richard, *Sermon XXIV. Preach'd on the Fifth of November.* In *Sermons on Several Occasions.* 2 vols. 1740. In vol. 1.

Burscough, William. *The Revolution Recommended to Our Memories. A Sermon Preach'd on November the Fifth, 1715.* 2nd edition. 1722.

Butler, Lilly. *A Sermon Preach'd before the Lord-Mayor, Aldermen, and Citizens of London, on the Fifth of November, 1710.* 1710.

Chandler, Edward. *A Sermon Preach'd at the Cathedral Church of Worcester; On the 5th of November, 1714.* 1714.

Clarke, Samuel. *Sermon CI. Providential Deliverance from Slavery. Preached on the 5th of November.* In *Sermons on Several Subjects.* 4 vols. 1738. In vol.1.

Cornell, William. *A Sermon Preached to the Congregation of Protestant Dissenters in Suffolk, November 5, 1757.* 1758.

Crowe, William. *A Sermon Preach'd before the Lord Mayor, on November the Fifth, 1734.* 1734.

Dalton, John. *A Sermon Preached before the University of Oxford, On the Fifth of November, 1747.* 1747.

Dawes, William. *The Continual Plots and Attempts of the Romanists. A Sermon Preach'd in Cambridge, On the Fifth of November, 1705.* 1705.

Devis, James. *The Principles of the Church of Rome Exploded. In a Sermon Preached at St. Peter's Sandwich, Nov. 5, 1755.* 1756.

Fleetwood, William. *A Sermon Preached before the House of Commons Thursday the 5th of November 1691.* In *A Compleat Collection of the Sermons, Tracts, and Pieces of all Kinds.* 1737.

Free, John. *A Sermon Preached at St. Mary's in Oxford Nov. 5th 1745.* In *A Volume of Sermons Preached before the University of Oxford.* 1750.

Goldwin, William. *The Protestant Religion. A Sermon Preach'd in Bristol, on November the 5th, 1726.* 1726.

Grove, Henry. *The Grounds of Anniversary-Days, both Festivals and Fasts, November 5, 1717.* In *The Works of Mr. Henry Grove of Taunton.* 4 vols. 1747. In vol. 1.

Kimberly, Jonathan. *A Sermon Preach'd before the Lower House of Convocation, November the Fifth, 1702.* 1702.

Knaggs, Thomas. *Divine Providence. A Sermon before the Lord Mayor, and Aldermen and Sheriffs of the City of London; Preach'd on November 5th. 1702.* 1702.

Lambe, Henry. *Christian Zeal Display'd. A Sermon Preach'd before the Lord-Mayor of the City of London, on Tuesday, November the 5th, 1723.* 1723.

Macro, Thomas. *Charity of Temper. A Sermon Preach'd in Great Yarmouth, November the 5th 1731.* 1732.

Newlin, Thomas. *Sermon XII. A Sermon Preach'd on the Fifth of November before the University of Oxford.* In *English Sermons on Several Occasions.* Oxford, 1720.

Postlethwait, Matthew. *The Moral Impossibility of Protestant Subjects Preserving Their Religious or Civil Liberties Under Popish Princes. Briefly Shewn In a Sermon Preach'd at the Cathedral Church of Norwich. On the 5th of November 1718.* Norwich, 1719.

Silvester, Tipping. *Extraordinary and Particular Vows Consider'd. A Sermon Preach'd before the University of Oxford, On Sunday in the Afternoon, November the Fifth, 1732.* Oxford, 1732.

Stanhope, George. *A Sermon Preach'd before the Queen, November the 5th 1706.* 1707.

Stubs, Philip. *The Church of England, Under God, An Impregnable Bulwark Against Popery. A Sermon Preach'd at St. Paul's Cathedral November the 5th. 1703.* 1703.

Sutton, William. *A Sermon Preach'd at the Cathedral-Church of Norwich, On Thursday Nov. 5. 1724.* Norwich, 1725.

Sydall, Elias. *The Insupportable Yoke of Popery. In a Sermon Preach'd at the Cathedral-Church of Canterbury, On Saturday, Nov. 5, 1715.* 1715.

Sykes, Arthur Ashley. *The Consequences of the Present Conspiracy to the Church and State, Considered. In a Sermon Preach'd at the Chapel in King-Street; Upon November 5, 1722.* 1722.

Weston, William. *The Safety and Perpetuity of the British State, Under the Influence of Political and Religious Zeal. Being the Substance of Several Discourses*

*Preach'd before the University of Cambridge During the Late Rebellion and Present War.* Cambridge, 1759.

Whincop, Thomas. *A Sermon Preach'd before the Honourable House of Commons, on Thursday the 5th of November, 1702.* 1702.

Willis, Richard. *A Sermon Preach'd before the Honourable House of Commons, on Monday, the 5th of November, 1705.* 1705.

Wrench, Jonathan. *The Spirit of Christianity, and the Spirit of Popery, Compared Together. In a Sermon Preached in the Parish-Church in Norfolk, Nov. 5th. 1721.* 1721

Wright, John. *A Sermon Preach'd before the Honourable House of Commons, Saturday, November the 5th, 1715.* 1715

Wright, Samuel. *Salvation from Popish Counsels, and a Popish Prince the Joy of Britain: A Sermon Preach'd on the Fifth of November, In the Year 1717.* 1717.

– *A Sermon Preach'd on the Fifth of November In the Year 1719. Wherein the Greatness of our Salvation by King William III (of Glorious Memory) is represented as an Occasion of Continual Joy to these Nations.* 1719

**Circumcision**

Altham, Roger. See *Lent*, above.

Bradford, Samuel. *The Unprofitableness of External, Without Internal Religion. A Sermon Preach'd before the King, January 1st, 1715/6.* 1716.

Bragge, Francis. *Sermon III. The Circumcision of the Heart. Of Undissembled and Persevering Religion: In Several Sermons Upon the Following Subjects.* 1713.

Bundy, Richard. *Sermon IX. Preached on New-Years's Day.* In *Sermons on Several Occasions.* 2 vols. 1740. In vol. 2.

Burkitt, William. *Expository Notes, with Practical Observations, on the New Testament. The Eleventh Edition Carefully Corrected.* 1739 [1749?].

Bury, Arthur. *The Danger of Delaying Repentance: Set Forth in a Sermon Preached to the University in Oxford on New-Years Day, 1691/2.* 1692.

Clarke, Samuel. *Sermon XCIII.* See *Lent* above.

Downame, John. *Annotations Upon All the Books of the Old and New Testament; This Second Edition Enlarged.* 1651.

Hall, Joseph. *The Contemplations Upon the History of the New Testament, Now Complete. The Second Tome.* 1651.

Hele, Arthur. *The Four Gospels Harmoniz'd, and Reduc'd Into One.* Reading, 1750.

Hole, Matthew. *Practical Discourses On All the Parts and Offices of the Liturgy of the Church of England.* 4 vols. 1714–16. See also below, note 56.

Horne, George. *Discourse XI. The Circumcision. Preached before the University of Oxford, Jan 1. 1763, Being the Festival of the Circumcision.* In *Discourses on Several Subjects and Occasions.* 4 vols. 4th edition. Oxford, 1793. In vol. 1.

Hough, John. *A Sermon Preach'd at the Church of St. Mary-le-Bow, before the Societies for Reformation of Manners, On Monday, Jan. 1, 1704.* 1705.

Reading, William. *Sermon I. On the Circumcision. Morning.* In *Fifty Two Sermons for Every Sunday of the Year MDCCXXVII.* 2 vols. 1728. In vol. 1.

Sharp, John. *A Sermon on Romans II. 28, 29.* In *The Theological Works. A New Edition in Five Volumes.* Oxford: University Press, 1829. In vol. 4.

Stackhouse, Thomas. *A Compleat Body of Speculative and Practical Divinity.* 3rd edition. 1743.

Stanhope, George. *The Circumcision of Christ. A Paraphrase and Comment Upon the Epistles and Gospels, Appointed to be Used in the Church of England.* 2 vols. 1705. In vol. 1.

Tilly, William. *A Sermon Preach'd before the University of Oxford, January 1, 1710/1.* In *Sixteen Sermons. All before the University of Oxford, At St. Mary's, Upon Several Occasions.* 1712.

Warren, Robert. *Discourse XIV. Discourse for the Circumcision of Christ on New-Years Day.* In *Practical Discourses on Various Subjects. Proper for All Families.* 2 vols. 1723. In vol. 2.

Weston, Stephen. *Sermon II. Of the Circumcision of Christ, Preached on New-Year's Day.* In *Sermons on Various Subjects, Moral and Theological.* 2 vols. 2nd edition. 1755.

## NOTES

1 The Editors, *Aperture* 110 (Spring, 1988), 3. For a brilliant history of the writer-hero in France that parallels some of my arguments below see Paul Benichou, *The Consecration of the Writer, 1750–1830 (Le Sacre de l'ecrivain)*, trans. Mark Jensen (Lincoln: University of Nebraska Press, 1999), introduction, chapter 1, and some illuminating notes on pp. 347, 348.

2 *A Polite and Commercial People: England, 1727–1783* (Oxford: Clarendon Press, 1989), 100.

3 Morris Golden, 'Periodical Context in the Imagined World of *Tristram Shandy*,' in *The Age of Johnson: A Scholarly Annual*, ed. Paul Korshin (New York: AMS Press, 1987), 1:237–260, considers 'suggestions,' 'hints,' and 'likely

impressions' that Sterne could have received from some periodicals during the 1750s. The absence of any historical context of the age that begot him, other than the narrowly literary, continues with all the essays in *Approaches to Teaching Sterne's 'Tristram Shandy,'* ed. Melvyn New (New York: MLA, 1989); see especially the editor's four questions suited for 'Classroom Presentation,' 22–7. Sterne's latest biographer, Ian Campbell Ross, *Laurence Sterne: A Life* (Oxford: Oxford University Press, 2001), continues to deny Sterne his full historical context. I have not yet read Tom Keymer's favourably reviewed *Sterne, The Moderns, and the Novel* (Oxford: Oxford University Press, 2002). But see his 'Horticultural Wars: *Tristram Shandy* and *Upon Appleton House,*' *The Shandean* 11 (1999–2000), 38–46, which 'reads' Sterne, Uncle Toby, and the Seven Years' War through Marvell's great poem.

4 Marc Bloch, *The Royal Touch,* trans. J.E. Anderson (London: Routledge & Kegan Paul, 1973), and Nicholas Rogers, *Crowd, Culture, and Politics in Georgian Britain* (Oxford: Clarendon Press, 1998), 55 ('which did much to erode respect for the monarchy').

5 *The Review of the Affairs of France,* ed. Arthur W. Secord, 22 vols. (New York: Columbia University Press, 1938), 5:406 (20 Nov. 1708); Clifford Geertz, 'Centers, Kings, and Charisma: Reflections on the Symbolics of Power,' in *Rites of Power,* ed. Sean Wilentz (Philadelphia: University of Pennsylvania Press, 1985), 15.

6 The scholarship on these subjects is overwhelming; I cite authors and the dates of their pertinent books only. For the army: Alan Guy (1985), Tony Hayter (1978), J.A. Houlding (1981), R.E. Scouller (1966), J.R. Western (1965), and Sir John W. Fortescue, *A History of the British Army,* 13 vols. (London: Macmillan, 1899–1930), 2:32, and his essay 'The Army,' in *Johnson's England,* ed. A.S. Turberville, 2 vols. (Oxford: Clarendon Press, 1933), 1:66–87. For the universities: John Gascoigne (1989), L.G. Mitchell (1986), W.R. Ward (1958), D.A. Winstanley (1935), and *History of the University of Oxford: The Eighteenth Century,* ed. L.S. Sutherland and L.G. Mitchell (Oxford: Clarendon Press, 1986). For the church: C.F.A. Best (1964), V.H. Green (1964), Peter Virgin (1989), and some dispiriting contemporary accounts in Spencer Cowper, *Letters,* ed. Edward Hughes (1956), Thomas Turner, *Diary,* ed. David Vaisey (1984), and John Hildrop, *The Contempt of the Clergy Considered,* 2nd ed. (London, 1756).

7 See the studies by Michael McKeon (1987), Dudley Bahlman (1957), John Barrell (1983), Ian Christie (1984), P.G.M. Dickson (1967), Brian McCrea (1998), Nicholas Rogers (1998), and Kathleen Wilson (1995). Many of these studies are indebted to the work of J.G.A. Pocock, *The Machiavellian Moment* (1975) and *Virtue, Commerce, and History* (1985).

8 *The English Novel in History, 1700–1780* ( London: Routledge, 1999), 3.

9 See John Sekora, *Luxury: The Concept in Western Thought, Eden to Smollett* (Baltimore: Johns Hopkins University Press, 1977).

10 J.G.A. Pocock, *Barbarism and Religion, Volume One: The Enlightenments of Edward Gibbon, 1737–1764* (Baltimore: Johns Hopkins University Press, 1999), 27.

11 Gerald Newman, *The Rise of English Nationalism, A Cultural History, 1740–1830* (New York: St Martin's, 1997), 26, citing E.P. Thompson. See also W.E.H. Lecky, *A History of England in the Eighteenth Century*, 8 vols. (New York: 1888–90), 2:492–3 ('the years 1756 and 1757 were among the most humiliating in her history'); also 1:14–15.

12 *Letters of Laurence Sterne*, ed. Lewis Perry Curtis (Oxford: Clarendon Press, 1935), 126.

13 Thomas Turner, *The Diary of Thomas Turner, 1754–1765*, ed. David Vaisey (New York: Oxford University Press, 1984), 75; Chesterfield's despair can be read in Lecky, *History of England*, 2:531–2.

14 The invasion scares that had continually agitated the English public until 1759 are noted in Horace Walpole, *Letters*, ed. Paget Toynbee, 6 vols. (Oxford: Clarendon Press, 1903), 4:266, 268; and David Hume. *Letters*, ed. J.Y.T. Greig, 2 vols. (Oxford: Oxford University Press, 1969), 1:307. See also Alan McNairn, *Behold the Hero: General Wolfe and the Arts in the Eighteenth Century* (Montreal: McGill-Queens University Press, 1997), 7.

15 *Britons: Forging the Nation, 1707–1837* (New Haven: Yale University Press, 1992).

16 *Sunday in the Park with George*, act 2, scene 12.

17 *Memoirs of a Royal Chaplain, 1729–1763*, ed. Albert Hartshorne (London: John Lane, 1905), 271. The historical moment is explored by Leonard Krieger, *Essay on the Theory of Enlightened Despotism* (Chicago: University of Chicago Press, 1975) and in his contribution to the *Dictionary of the History of Ideas*, 4 vols. (New York: Scribner's, 1973), 1:141–62.

18 Manuel Schonhorn, *Defoe's Politics* (Cambridge: Cambridge University Press, 1991), 99–109.

19 J.H. Plumb, *Sir Robert Walpole: The King's Minister* (Boston: Houghton Mifflin, 1961).

20 *The Works of Lord Bolingbroke*, 4 vols. (Philadelphia 1841), 2:396.

21 2 vols. (London 1757–8), 1:25.

22 Newman, *Rise of English Nationalism*, 108. Brian McCrea, *Impotent Fathers: Patriarchy and Demographic Crisis in the Eighteenth-Century Novel* (Newark: University of Delaware Press, 1998), leaves unanswered the essential question, 'Who is to replace the decentered, marginalized patriarch?' (12).

23 See Ernst Kantorowicz, 'The Sovereignty of the Artist,' in *De artibus Opuscula XL: Essays in Honor of Erwin Panofsky*, ed. Millard Meiss (New York: New York

University Press, 1961), 261–79; also Catherine Soussloff, *The Absolute Artist* (Minneapolis: University of Minnesota Press, 1997). Old but useful commentary can be read in Rudolph Wittkower, 'The Artist,' in *Man versus Society in Eighteenth-Century Britain*, ed. James L. Clifford (Cambridge: University Press, 1968), 70–84. See the essays collected in *The English Hero*, ed. Robert Folkenflik (Newark: University of Delaware Press, 1982), and Robert Folkenflik, 'The Artist as Hero in the Eighteenth Century,' *Yearbook of English Studies* 12 (1982), 91–108. Sterne gets short shrift in both.

24 I will be quoting from *Tristram Shandy*, ed. James A. Work (New York: Odyssey Press, 1940). All further quotes will be noted in the text.

25 I am concluding a study of this seemingly unique English trope, examining the pattern from Spenser to Fanny Burney. Commentary on *Tristram Shandy* is replete with readings that Tristram has two surrogate fathers – Uncle Toby and Parson Yorick – in addition to his biological one.

26 *Laurence Sterne: The Later Years* (London: Methuen, 1986), 199.

27 See the early comments in Dorothy Van Ghent, 'On *Tristram Shandy*,' in *The English Novel: Form and Function* (New York: Rinehart, 1953), 83–98; and the more recent John Preston, *The Created Self* (New York: Barnes and Noble, 1970). A useful summary of the subject is in Arthur Cash, *Laurence Sterne: The Early & Middle Years* (London: Methuen, 1975), 208–13. See also Peter de Voogd, 'Lawrence Sterne, the Marbled Page, and "the use of accidents,"' *Word & Image* 1 (1985), 279–87, and his '*Tristram Shandy* as Aesthetic Object,' *Word & Image* 4 (1988), 383–92.

28 M. Sward, *Memoirs*, quoted in Wilbur L. Cross, *The Life and Times of Laurence Sterne*, 2 vols., 3rd ed. (New Haven: Yale University Press, 1929), 1:275–7.

29 *Letters*, xxx.

30 Ibid., 196.

31 Cash, *Sterne: The Later Years*, 196; John Chamberlayne, *Magnae Britanniae Notitia; Or, The Present State of Great-Britain* (London, 1727), 145.

32 'Sterne, Hamlet, and Yorick: Some New Material,' in *The Winged Skull*, ed. Arthur Cash and John M. Stedmond (Kent, OH: Kent State University Press, 1971), 122.

33 *Tristram Shandy's World* (Berkeley: University of California Press, 1954), xiii.

34 *Systems of Order and Inquiry in Later Eighteenth-Century Fiction* (Berkeley: University of California Press, 1975), 63.

35 'The Defence of Wit: Sterne, Locke, and the Particular,' *Comparative Criticism: An Annual Journal* 10 (1988), 117.

36 'A Poetical Epistle to Doctor Sterne, Parson Yorick, and Tristram Shandy' (1760), in *Boswell's Book of Bad Verse*, ed. Jack Werner (London: White Lion, 1974), 133.

37 Walter Sichel, *Sterne, A Study* (London: Williams and Norgate, 1910), 79; Edward Mangin, *A View of the Pleasures Arising from a Love of Books* (London: Longman, 1814), 92.

38 Karina Williamson, '*Tristram Shandy*: An Allusion to Guido Reni,' *N&Q* 35 (1988), 187–8.

39 '*Tristram Shandy* and the Art of Literary Counterpoint,' *MLQ* 32 (1971), 269; see his *Laurence Sterne and the Origins of the Musical Novel* (Athens: University of Georgia Press, 1975).

40 See the ground-breaking essay by Joseph Frank, 'Spatial Form in Modern Literature,' in *The Widening Gyre* (New Brunswick: Rutgers University Press, 1963). It first appeared in the *Sewanee Review* 53 (1945). His first mention of Sterne appeared in his 'Some Further Reflections,' in *The Idea of Spatial Form* (New Brunswick: Rutgers University Press, 1991), 115–20 ('sporadic anticipation' and 'abnormal' in its day but a harbinger of the present, 115, 120).

41 Alice Green, 'Diderot's Fictional World,' *Diderot Studies* I, ed. Otis Fellows and Norman Torrey (Syracuse: Syracuse University Press, 1949), 1–26; see also P.N. Furbank, *Diderot* (New York: Knopf, 1992), 267, 280.

42 'The Fine Arts in Eighteenth-Century Polite Culture,' in *Painting and the Politics of Culture: New Essays in British Art*, ed. John Barrell (New York: Oxford University Press, 1992), 13–37, esp. 29–31.

43 James Gregory, 'Anglicanism and the Arts: Religion, Culture and Politics in the Eighteenth Century,' in *Culture, Politics and Society in Britain, 1660–1800*, ed. Jeremy Black and James Gregory (Manchester: Manchester University Press, 1999), 86.

44 New Haven: Yale University Press, 1979, 4. Though Kris and Kurz primarily considered the graphic artist – the painter, sculptor, and architect – they discovered these preconceptions, or 'anecdotes,' in the writings of poets (37–8). For myths of development in the autobiographies of classical poets, see the essays collected in Mary Lefkowitz, *Lives of the Greek Poets* (Baltimore: Johns Hopkins University Press, 1980). For modern artists, see the comments by James Turrell in *The New Yorker*, 30 July 1990, 75; and the review of Ada Louise Huxtable, *Frank Lloyd Wright*, in *The New York Times Book Review*, 27 November 2004, 16.

45 Compare Soussloff's 'Schematic Structure of the Artist's Biography' (2) with Lord Raglan's pattern of the hero in *The Hero*; The Thinker's Library (London: Watts, 1949), 178–9. I would like to thank Professor Soussloff for her correspondence, her articles published and unpublished, and her support.

46 Kenneth Monkman, *Sterne's Memoirs: A Hitherto Unrecorded Holograph Now Brought to Light, Introduction and Commentary* (Coxwold: Laurence Sterne Trust, 1985), 9.

47 Cash, *Sterne: The Early & Middle Years,* 19; *Tristram Shandy,* ed. Charles Whibley, 2 vols. (Chicago, 1895), 1:xxiv. See also Horace, *Odes,* 2.13.

48 *Memoirs,* 17–18.

49 See Giorgio Vasari, *The Lives of the Artists,* trans. and ed. Julia Conaway Bondanella and Peter Bondanella. The World's Classics (Oxford: Oxford University Press, 1991), 16–17. The legendary artist, Kris later wrote, as a child 'is discovered in a 'childish' activity'; *Psychoanalytic Explorations in Art* (New York: International Humanities Press, 1952), 72. Note Tristram's play with his top, below. See Paul Barolsky, *Giotto's Father and the Family of Vasari's 'Lives'* (University Park: Pennsylvania State University Press, 1992), 103; also his *Michelangelo's Nose* (University Park: Pennsylvania State University Press, 1990). I wish to thank Professor Barolsky for his correspondence and support.

50 *Memoirs,* xxxi.

51 David Piper, 'The Development of the British Literary Portrait Up to Samuel Johnson,' *Proceedings of the British Academy* 54 (1968), 68. Published for the British Academy (London: Oxford University Press, 1970).

52 *Picasso: A Life* (New York: Random House, 1991), 33, 3, 28. Richardson's destruction of Picasso's fantasies makes for fascinating reading.

53 Could Mr Shandy's demand for his son's Caesarian birth suggest a relationship to the hero's father's attempt to kill his unborn or infant son? Otto Rank in his writings on the hero and the family romance notes that in the competition for primacy a father or a brother has to be removed or killed off. See *The Myth of the Hero* (New York: Nervous and Mental Diseases Publishing, 1914), 88.

54 *Legend of the Artist,* xii.

55 'A Story Preached to the Heart? *Tristram Shandy* and Sentimentalism Revisited,' *The Shandean* 9 (1997), 126.

56 *Practical Discourses on all the Parts and Offices of the Liturgy of the Church of England,* 4 vols. (London, 1714–16), 1:57. Volume 4 has discourses on the Circumcision of Christ (142–57) and for the First Sunday in Lent (345–62). My appendix will list the sermons and texts cited. The following notes provide only the author's name and page number of the relevant works. All material was read at the Huntington Library, where the once unique copy of John Gordon Spaulding, *Pulpit Publications, 1660–1782,* in 6 volumes, was available, listing sermons by authors' names, year of publication, and biblical text, and with a superb index volume. It has been printed and edited with introductory matter: New York: Norman Ross, 1996.

57 Hall, 7; Reading, Sermon 22, 283; Johnson, 214; Clarke, 587. Bishop Joseph Hall was one of Sterne's favourite authors; see Ross, *Sterne,* 108, 176, 217,

and 242, where he notes his indebtedness to Melvyn New for Sterne's borrowing for his sermons.

58 Hodson, 1.

59 Sydall, 59.

60 Littleton, 110.

61 Lucas, 267.

62 Berriman, 180.

63 Coney, 389; Berriman, 182.

64 Hall, 6. Coincidentally, Tristram directs the reader's attention only three paragraphs after his conception to March 25, Lady Day, the celebration of the Feast of the Annunciation (9).

65 *Letters*, 22–3, and n. 4; see also *Archbishop Herring's Visitation Returns, 1743*, ed. S. L. Ollard and P.C. Walker, Yorkshire Archaeological Society, Record Series 75 (n.p.: Yorkshire Archaeological Society, 1929), 3:92–3.

66 Shandy of course wrote 'brought forth' (10).

67 *Tristram Shandy*, vol. 3: The Notes, ed. Melvyn New, Richard A. Davies, and W.G. Day (Gainesville: University Presses of Florida, 1984), 52–3.

68 Burscough, 21; Walter Harris, 500; Dawes, 22, among dozens. Johnson, in his *Dictionary*, under 'Deliverer,' cites *Paradise Lost*, 12:149: 'By that seed / Is meant thy great *deliverer*, who shall bruise / The serpent's head.'

69 Cornell, 6; Dalton, 7; Bradford (1713), 13; S. Wright (1719), 5. Cf. Cornell, preaching on 1 Samuel 12:24, 'the great things He hath done for you' (6, 19).

70 Grove, 33.

71 'The Bibliography of the Early Editions of *Tristram Shandy*,' *The Library*, 5th ser. 21 (March 1970), 31. Monkman remarks on Sterne's 'genius' commenting on the ceiling episode (12).

72 Halliday, 28.

73 Anne Bandry, 'The First Reactions to *Tristram Shandy* in the Oates Collection,' *The Shandean*, 1 (1989), 30.

74 *The Remains of Thomas Hearne, Reliquiae Hearnianae*, rev. John Buchanan-Brown (London: Centaur Press, 1966), 263–4, 411–12.

75 *Review* 6:114 (4 November 1707). See also Defoe's *Commendatory Sermon Preach'd November the 4th, 1709. Being the Birth-day of King William, of Glorious Memory.*

76 *The History of the Life and Reign of William-Henry, Prince of Nassau and Orange* (Dublin, 1749), 501.

77 Kathleen Wilson, *The Sense of the People: Politics, Culture and Imperialism in England, 1715–1785* (Cambridge: Cambridge University Press, 1995), 91–2.

Wilson footnotes events in York (92n); Nicholas Rogers, *Crowds, Culture, and Politics in Georgian Britain* (Oxford: Clarendon Press, 1998), 23–4.

78 In Garrick's Drury Lane calendar; see George W. Stone, Jr and George M. Kahrl, *David Garrick: A Critical Biography* (Carbondale: Southern Illinois University Press, 1979), 92.

79 Thomas Comber, *A Vindication of the Great Revolution in England. In A.D. MDCLXXXVIII. And of the Characters of King William and Queen Mary* (London, 1758), 27. It should be noted that Comber concludes his eulogy of his '*Christian Hero*' at York, 10 April 1758 (149); Grove, 33, 61.

80 William Weston, 13.

81 Dorothy Norman, *The Hero: Myth/Image/Symbol* (New York: World, 1969), 113.

82 Note to the Torah reading on the first day of Rosh Hashanah, Genesis 21, which tells of the birth of a son to Abraham and Sarah in their old age.

83 Weston, 133.

84 Horne, 320.

85 Weston, 132.

86 While of multiple authorship, the text is usually indexed under Downame's name. I am quoting, from the second enlarged edition (London, 1651), the annotations under Genesis 17:10, 11.

87 Stackhouse, 642; Stanhope, 358.

88 Tilley, 219; Altham, 197.

89 Weston 1:132.

90 Joseph Beaumont, *Original Poems in English and Latin* (London 1749), 157. Beaumont's dates are 1616–99. His manuscript poems were printed in this selective edition. The quote is from his song for 'Newyear Day.' His poems are of some 'metaphysical' interest. He also preached a sermon on 5th November before the University of Cambridge.

91 Horne, 320.

92 Hough, 5; Stanhope, 365. Commenting on the Gospel verse for the day – Luke 2:15 – Stanhope writes: 'For he delivers us from the heaviest of all Bondages, from the most formidable of all Enemies' (375).

93 I am fully aware of the scholarly debate about Tristram's circumcision; see Robert Darby, '"An Oblique and Slovenly Initiation": The Circumcision Episode in *Tristram Shandy*,' *Eighteenth-Century Life* 27 (2003), 72–84. I would only remark that Sterne's eighteenth-century reader would not have been fully cognizant of the anatomical subtleties. Note also that only eight years earlier the Jewish Naturalization Bill (1753) had provoked pamphlets, prints, popular songs and ballads, and newspaper commentary in London and the provinces with circumcision – not medically considered – the focus of the vitriol;

see Frank Felsenstein, *Anti-Semitic Stereotypes: A Paradigm of Otherness in English Popular Culture* (Baltimore: Johns Hopkins University Press, 1995), 187–215, and his bibliography, 319–37. No critics appear to have considered, or looked at, the sermons.

94 Duncan Largen, 'Nietzsche and Sterne,' *The Shandean* 7 (1995), 12.

95 *The Case of Authors by Profession or Trade Stated*, ed. Philip Stevick (Gainesville: Scholars' Facsimiles and Reprints, 1966), 73.

96 *An Enquiry Into the Present State of Polite Learning in Europe*, in *The Works of Oliver Goldsmith*, ed. Peter Cunningham, 4 vols. (London 1878), 2:6 n.1.

97 *The History of Rasselas*, in *Samuel Johnson*, ed. Donald Greene, The Oxford Authors (Oxford: Oxford University Press, 1989), 353.

98 J.G.A. Pocock, *Barbarism and Religion*. Vol. 2: *Narratives of Civil Government* (Cambridge: Cambridge University Press, 1999), 5.

99 *An Essay on the Study of Literature* (London, 1764), 73, 17.

100 To give due credit, I owe this insight to Professor Barolsky.

chapter thirteen

# Science, Projects, Computers and the State: Swift's Lagadian and Leibniz's Prussian Academy

MARTIN GIERL

> This I am told by a very skillful Computer, who hath given a full Demonstration of it from Rules of Arithmetick.
>
> Jonathan Swift, *A Tale of a Tub* (1704)[1]

> I invented by the agency of artis combinatoriae some items in mathematics and mechanics of not low importance, especially in arithmetic a device which I call Living Calculator's bench [Lebendige Rechenbanck].
>
> Gottfried Wilhelm Leibniz,
> Letter to Duke Johann Friedrich, 1671[2]

> 1001 4922 [= God wants]
>
> Caspar Rödecke (1726)[3]

> One; One = $\text{Two}^0$; $\text{Two}^n + \text{Two}^n = \text{Two}^{n+1}$

Neither individuality, nor wit, nor knowledge and education, nor free speech made up enlightenment and modernity, but rather a potentially universal, fast-growing linking of information and organization. That is the thesis I want to pursue with this paper. The idea is doubtlessly very general and matches, in a wry sense, the goals of the members in Swift's famous Academy of Lagado, who were obsessed with turning the world upside down with their projects and research.[4] Accordingly, I was less proud than worried about the claim, until I realized that the idea is in fact not only already old, but the core of what was called science or, better, modern knowledge by Thomas Sprat in his famous *History of the Royal Society* (1667).[5]

What does 'linking of information and organization' mean? The linking of information and organization is a sound definition of knowledge, but – interestingly enough – it is also a quite precise definition of an institution. An A becomes a letter as part of the alphabet. It becomes a mark as part of the system of marks. Following the meanwhile broadly accepted description given by the social sciences, one has to keep in mind that not only the government, the church, universities, and hospitals are 'institutions,' but that abstract social units – the law, the family, and (not least) language itself, from grammatical structures down to single words – may also be regarded as institutions.[6] Thus, to achieve a potentially universal linkage of information and organization, we must establish institutions, whether they are an element of knowledge, finance, or administration, in any possible relation. The underlying logic is in a way the conversion and complement of Murphy's Law, from 'everything that can go wrong will go wrong' to the analogous proposition 'anything that allows the linking of information and organization can be invented.'

At this point, projects come into play, and with them my second thesis. The continuous attempt to connect units of organized information is furthered by ingenuity and creativity. At its fundamental level, however, it is sufficient to have a defined pattern of trial and error to keep the process going. Thus, a project may be defined as a distinct process of connecting possibilities. Projects lie between utopia and realization. They start with the condensing of plans and are followed by planning and the search for acceptance. Projects adjust utopia towards reality. They are basic tools for history, accompanying its process from the beginning. But at a certain point, something new comes into play: the organization of planning itself, the organization of organization.[7] This point was reached by 1700 – the age of projects in the eyes of the contemporary observers. And the incarnation of the attempt to organize organization was the academy.

My argument will run as follows. I will start with a short outline of Sprat's ideas about the Royal Society and modern learning. The point here is to give at least some evidence that the linking of information and organization actually constitutes the centre of Sprat's conception. But a personal reading of Sprat is certainly not enough to mark the organization of organization as the starting point of modernity and the founding of academies as its sign. Therefore, I will submit my thesis to a double-check. I will confront Sprat and his supporters, who regarded modern learning as a means to organize organization, on the one hand with their sharpest critic and on the other hand with an attempt to realize the concept, which – in the eyes of contemporary historians – completely failed.[8] I will confront them with Jonathan Swift and his famous tale of the Academy of Lagado in *Gulliver's Travels*, which consists of crazy bunglers who torture their country with mad projects, and with the Prussian Academy in Berlin, founded 1700, whose membership first existed only on paper and which issued not more than one volume of its learned journal (in Latin) during the first twenty years of its existence.[9] At first glance, there is nothing surprising in this for us, given that we believe in individuality, structures, or systems as the agencies and actors behind the process of history. Misguided individuality results in foolishness; insufficiently conceptualized undertakings result in failure. Instead of stumbling into the void, however, Swift's criticism ended in accurately depicting the future, whereas the Berlin Academy failed in its attempt at building basic instruments for the future. In fact, the bungler's undertakings in Lagado and Berlin should be regarded as ultimately successful. They should be considered as basic elements of the present. Lagado and Berlin point to an ancient but then more and more prevailing agent besides individuality and structure on the stage of history: they refer to institutionalization. Let me then make the concept of an all-encompassing linking of information and organization more concrete by telling the story of how Lagado and Berlin invented and furthered the design of the two most powerful historic realizations of that concept. Let me tell how Lagado and Berlin invented the concept of the computer and furthered the concept of the modern national state, that universal administrative organizer of social affairs.

## I. Thomas Sprat, the Royal Society, and the Universal Organization of Knowledge

Thomas Sprat's *History of the Royal Society*, published only five years after the Society's foundation, is less a report than a manifesto. It was – as manifestos sometimes are – a highly influential book that defended not so much the new method of experimenting as the unprecedented phenomenon of an autonomous organization outside the governmental and, especially, the ecclesiastical framework by attacking the traditional forms of learning as useless quibbling of words.[10] The Royal Society became the symbol of a new form of learning. In fact, the passages of Sprat's book directly concerning the Royal Society are relatively small.[11] The main part treats the history of learning, its old and its new conceptions, and foreshadows the battle between the moderns and the ancients, as Sprat calls them, long before Charles Perrault did in his famous speech in front of the Académie française 1687.[12] Defending the moderns, Sprat followed Bacon's idea of useful practical knowledge, but it was not Bacon's scientific factory based on the division of labour he had in mind. Sprat propagated the linking of information and organization as a general cooperation between representatives of the existing social bodies. The Royal Society is where the organization of organization allegedly had begun: 'Thus they have form'd that Society, which intends a Philosophy, for the use of Cities, and not for the retirements of Schools, to resemble the Cities themselves: which are compounded of all sorts of men, of the Gown, of the Sword, of the Shop, of the Field, of the Court, of the Sea; all mutually assisting each other.'[13]

For Sprat, the organization of organization should be based on the universal repository of information: 'For these, they have begun to settle a correspondence through all Countrys; and have taken such order that in short time, there will scarce a Ship come up the Thames, that does not make some return of Experiments, as well as of Merchandize.'[14] 'Nothing shall be so remote,' Sprat wrote, 'as to escape their reach.' Obviously, *universality* of organized information lies at the core of Sprat's proposition. Besides the concept of experimenting, universality is in fact the term Sprat uses most to characterize modern learning. He talks about 'universal intelligence,' 'universal philosophy,' 'universal inspection,' 'universal inquiry,' and 'universal light' to qualify the new project of learning.[15] Not any longer would universality be an exclusively divine attribute of knowledge beyond the reach of humans. In the same way, though a wider frame, in which Addison's *Spectator* was to claim to have brought knowledge and education down from Olympus into the coffee houses of contemporary London, Sprat boasted that he had brought

> Philosophy down again to men's sight, and practice, from whence it was flown away so high: the Royal Society has put it ... into one of the Arts of Life, ... that cannot ... be extinguish'd, at the loss of a Library, at the overthrowing of a Language, or at the death of some few Philosophers: but men must lose their eyes, and hands, and must leave off desiring to make their Lives convenient, or pleasant; before they can be willing to destroy it.[16]

The practical nature of modern philosophy basically involved the empirical investigation of nature, but it had its utilitarian side as well. As Sprat asserted:

> A New City is to be built, on the most advantageous Seat of all Europe, for Trade, and command. This therefore is the fittest Season for men to apply their thoughts, to the improving of the materials of building, and to the inventing of better models, for Houses, Roofs, Chimnies, Conduits, Wharfs, and Streets: all which have been already under the consideration of the Royal Society.[17]

## II. Gulliver's Travels to the Academy of Lagado: The Academy as Allegory of the Organization of Organization and the Computer as Metaphor of Nonsense

Forty years after Sprat had published his vision of what the Royal Society ought to represent as an institution, in the last days of February 1708, Gulliver reached Lagado, the capital of Balnibarbi. He arrived in Balnibarbi after a visit to Laputa, the flying island and seat of the court and monarch, an island that, owing to the power of a huge loadstone, was able to hover over its dominions below and exert power over them. The stay in Laputa was interesting for Gulliver, but rather dissatisfying. He felt neglected because only Mathematics and Music were valued there. Laputa is the Olympus of pure science and Balnibarbi the reign of applied learning. Both, of course, serve as partial representations of England in Swift's 1726 classic.[18] Gulliver finds Lagado-London and the land about it in a devastated condition. The houses are decayed, the fields lie fallow. And all this is the fault of the only thing that flourishes – the huge Lagadian Academy.[19] The Academy consists of several departments with more than five hundred rooms. There is a political school, a school concerned with practical discoveries, a school of languages, and the school of speculative learning. To provide the perfect state, the political school proposes to raffle for political positions, to tax self-assessed virtues as well as the vices of the others, and, in the case of clashing political parties, to open their members' heads, to merge and then redistribute their brains in equal portions. In the 'Applied Sciences' section, there is a man who wants to let hogs grub the fields in order to save ploughs and labour. Another tries to extract sunlight out of cucumbers. Someone hopes to convert excrement back into food. Others try to turn hailstones into gunpowder and spiders' webs into silk. There are plans to construct houses beginning with the roof and working down and to pump water uphill into a reservoir to supply the mill, instead of using the nearby river.

Swift combined a variety of literary representations with contemporary, actually existing, scientific ideas, for his satire.[20] But on the basis of these representations, the Lagadian academy is first and foremost a direct answer to Sprat.[21] Gulliver is taught that the 'Sum' of the 'Discourse' was

> That about Forty Years ago, certain Persons went up to Laputa ... and after five Month Continuance, came back with a very little Smattering in Mathematicks, but full of Volatile Spirits acquired in that Airy Region. That these Persons upon their Return, began to dislike the Management of every Thing below; and fell into Schemes of putting all Arts, Sciences, Languages, and Mechanicks upon a new Foot. To this End they procured a Royal Patent for erecting an Academy of PROJECTORS in *Lagado* ... [Here] the Professors contrive new Rules and Methods of Agriculture and Building, and new Instruments and Tools for all Trades and Manufactures, whereby, as they undertake, one Man shall do the Work of Ten; a Palace may be built in a Week ... All the Fruits of the Earth shall come to Maturity at whatever Season we think fit to chuse, and increase an Hundred Fold more than they do at present; with unnumerable other happy Proposals. The only Inconvenience is, that none of these Projects are yet brought to Perfection; and in the mean time, the whole Country lies miserably waste.[22]

'Projectors' – the only word completely written in capital letters in the whole text – is the decisive term. Projects may bridge utopia and the intended reality, even if they fail at last, as long as they carry on in a steady process, testing their practicality and viability with every step. As imagined by Swift, however, projectors are doomed to failure even if their ideas are feasible in the end, because they are fixated on their ideas. Projectors attempt to fit nature to ideas, while true scientific experimenters fit ideas to nature.

These are the poles for which Sprat and Swift stand. The ultimate proof for Swift's position is the ridiculous plan of the Lagadians to rationalize language: the project to base discourse exclusively upon nouns, the idea of a universal language, and the contrivance to write texts automatically (see Figure 13.1). Swift depicted this gadget in his book.[23] It is, in fact, the only engraving besides the maps of Gulliver's voyages, and it resembles a microchip not completely by chance. The machine consists in wooden bits quadratically ordered within a frame. The whole vocabulary of the English language is dynamically fixed on wires connectable to the bits with the help of forty-five handles. Every turn of the handles produces a new text by changing the order of the words.

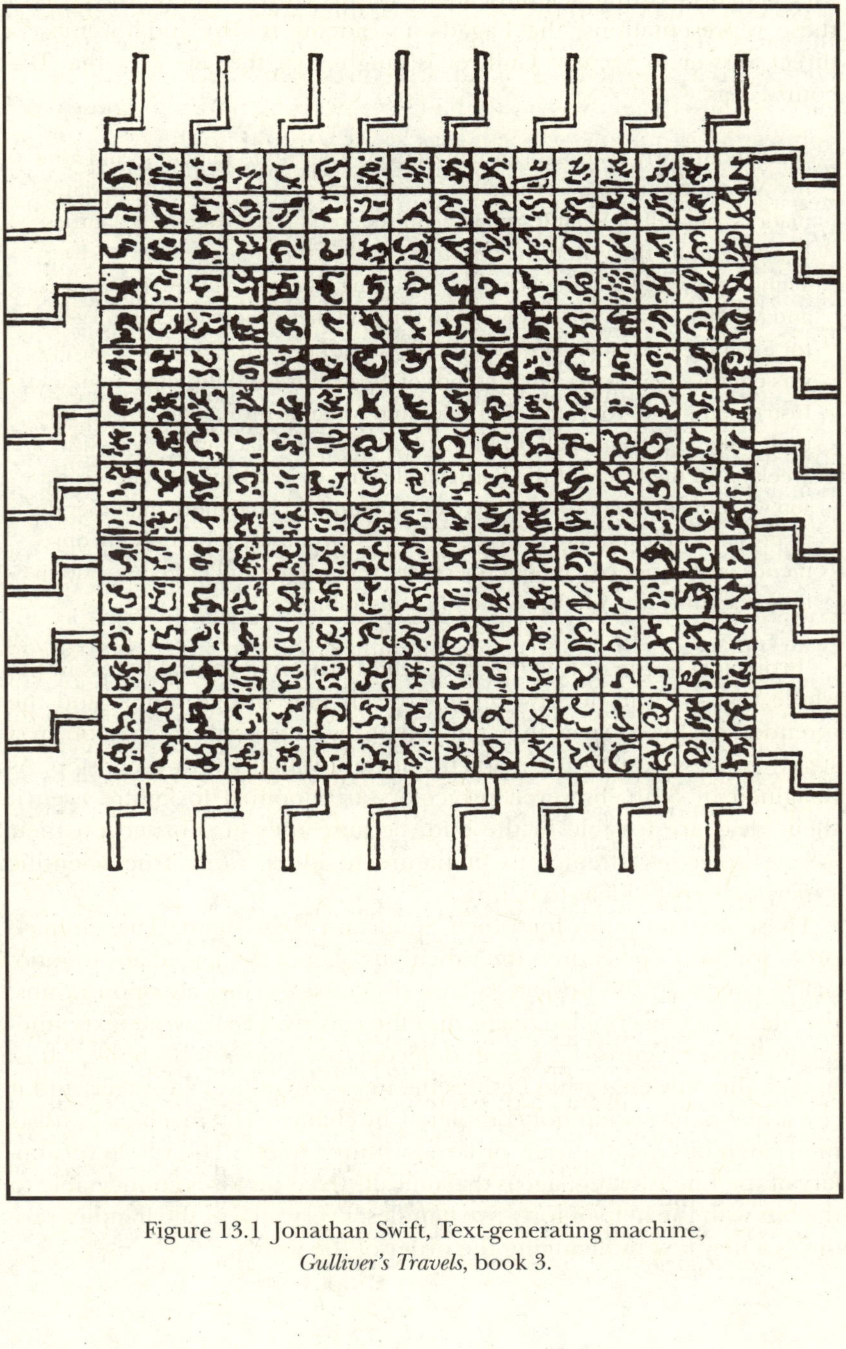

Figure 13.1 Jonathan Swift, Text-generating machine, *Gulliver's Travels*, book 3.

No doubt, if there were no other operating powers in history than, on the one hand, individuality with intellect, language, violence, and discourses, and, on the other hand, structures with forces and traditions, such as the efficiently working mill on the river, Swift's computer as a connection of mindlessness and uselessness would be the most foolish thing one could conceive. Two questions arise: Why then could it happen that Swift was so entirely wrong? And how could it happen that somebody who wanted to describe complete nonsense actually described the future? The answers to these questions are interrelated. For Swift, in opposition to Sprat, modern learning was the business of individuals, of non-cooperative representatives of certain ideas, sitting alone in their offices. Sprat, by contrast, stressed the communication of ideas. In his eyes, getting information organized was the basis of an experiment. In short, while Swift believed in individuality and structures, Sprat believed in institutionalization. The more Swift neglected the impact of institutionalization, the better he could ridicule the misbehaviour of his allegedly free-acting contemporaries. Behind the tale, however, Swift's book itself answered perfectly to the requirements of institutionalization, for *Gulliver's Travels* was far from being solely his individual invention. Discussing contemporarily debated positions, it organized information by bringing literary traditions, the ongoing debate on the moderns and ancients, and scientific proposals and projects together. Swift's book was an actualized, carefully ordered variety of ideas and practices based on previous authors and literary traditions, Rabelais, Cervantes, and the satiric pattern of the world turned upside down. It was also a response to the contemporary scientific thought of Bacon, Sprat, Newton, Boyle, Wren, Hooke, and perhaps Leibniz, too.[24]

## III. Leibniz, His Academy, the Concept of a Computer, and the Emergence of the Modern State

The Prussian Academy, founded 1700, was not a five hundred-room institute. On the contrary, its first lodging had been a royal stable. Nevertheless, the new academy brought diverse institutions together: economic, social, political, cultural, and scientific ones.[25] It was a new element of the state administration, which controlled its budget, and part of court life led by a recently crowned monarch who wanted an academy because the French and English had one already. The official task of the Academy was to reform the calendar, but it was also supposed to house a *Theatrum anatomicum*, an observatory, a botanical garden, a library, and a

Figure 13.2 Wilhelm Gottfried Leibniz's calculator of 1693.

collection of biological specimens. Even if some of these institutes were not much more than projects on paper in the beginning, the so constituted Academy illustrates the organization of information as a process of tying institutions together. Its founder, Gottfried Wilhelm Leibniz, the Hanoverian minister, philosopher, engineer, librarian, historian, lawyer, mathematician, architect, and pioneer of the early media age, organized journals and communicated with 1100 correspondents.[26] Leibniz embraced an institutionalized network in its still personalized form. And he tried to realize core ideas of Swift's later tale of Lagado to a remarkable extent. 'Theoria cum praxis' was the motto of his life: 'The art of practice consists in forcing chance under the yoke of science.'[27] As a mining engineer, he actually designed a windmill that would pump water uphill. He drafted a universal language, constructed calculators, and invented the binary number system – the basis of a computer.

On his 'living calculator's bench' (see figure 13.2) one finds the handles of Swift's automatic text machine again. Both devices consist – as a microchip does – in representations of a completely embodied system of information with the ability to reorganize and memorize these representations.

The difference between the two projects is that Swift's computer acted on the basis of mindless chance, that is, nonsense, whereas Leibniz planed to translate logic into the movements of a machine. Leibniz's binary system is an outstanding example of the hope of 'forcing chance under the yoke of science.'[28] The system bears on the possibility of constructing a complete representation of all numbers using less or more digits than the ten of the decimal system. The 'trick' is to express the digital positions by powers of the respective basic number of the system and to add the resulting figures. The number 84 in the decimal system, for instance, equals $8{\cdot}10^1{+}4{\cdot}10^0$. The number 100 equals $6{\cdot}10^2{+}0{\cdot}10^1{+}0{\cdot}10^0$. The powers keep the respective digital positions. Consisting in additions and the ongoing multiplication of the basic digit with itself, number systems resemble crystals or the petals of an endlessly unfolding calyx. Based on these characteristics, the core idea of Leibniz's number system was as ingenious as it was simple. The dyadic system confines the digits to one and zero and operates on the basis of powers to two.

*Omnibus ex nihilo ducendis SUFFICIT UNUM* – one is enough to draw everything out of nothing – was Leibniz's motto for his dual system. This system invokes two decisive advantages: Because two plus two equals two times two, the equation $2^n + 2^n = 2^{n+1}$ is valid. And whereas the single digital positions in the decade system enhance the numbers from 0 to 9, the digital positions in the binary system are confined to 0 and 1.

Number systems

| | Number | Notation | Written in powers to the basis | Written as sum |
|---|---|---|---|---|
| Decade system | 84 | *84* | $8 \cdot 10^1 + 4 \cdot 10^0$ | 80+4 |
| (*0,1,2,3,4,5, 6,7,8,9*) | 600 | *600* | $6 \cdot 10^2 + 0 \cdot 10^1 + 0 \cdot 10^0$ | 600+0+0 |
| Basis 10 | 109 | *109* | $1 \cdot 10^2 + 0 \cdot 10^1 + 9 \cdot 10^0$ | 100+0+9 |
| Three-way system | 9 | *100* | $1 \cdot 3^2 + 0 \cdot 3^1 + 0 \cdot 3^0$ | 9+0+0 |
| (*0,1,2*) Basis 3 | 11 | *102* | $1 \cdot 3^2 + 0 \cdot 3^1 + 2 \cdot 3^0$ | 9+0+2 |
| Binary system | 4 | *100* | $1 \cdot 2^2 + 0 \cdot 2^1 + 0 \cdot 2^0$ | 4+0+0 |
| (*0,1*) Basis 2 | 9 | *1001* | $1 \cdot 2^3 + 0 \cdot 2^2 + 0 \cdot 2^1 + 1 \cdot 2^0$ | 8+0+0+1 |

Leibniz's binary system

| Decade system | Binary system | Written in powers to two | Written as sum |
|---|---|---|---|
| 1 | 1 | $1 \cdot 2^0$ | 1 |
| 2 | 10 | $1 \cdot 2^1 + 0 \cdot 2^0$ | 2+0 |
| 3 | 11 | $1 \cdot 2^1 + 1 \cdot 2^0$ | 2+1 |
| 4 | 100 | $1 \cdot 2^2 + 0 \cdot 2^1 + 0 \cdot 2^0$ | 4+0+0 |
| 5 | 101 | $1 \cdot 2^2 + 0 \cdot 2^1 + 1 \cdot 2^0$ | 4+0+1 |
| 6 | 110 | $1 \cdot 2^2 + 1 \cdot 2^1 + 0 \cdot 2^0$ | 4+2+0 |
| 7 | 111 | $1 \cdot 2^2 + 1 \cdot 2^1 + 1 \cdot 2^0$ | 4+2+1 |
| 8 | 1000 | $1 \cdot 2^3 + 0 \cdot 2^2 + 0 \cdot 2^1 + 0 \cdot 2^0$ | 8+0+0+0 |
| 9 | 1001 | $1 \cdot 2^3 + 0 \cdot 2^2 + 0 \cdot 2^1 + 1 \cdot 2^0$ | 8+0+0+1 |
| 10 | 1010 | $1 \cdot 2^3 + 0 \cdot 2^2 + 1 \cdot 2^1 + 0 \cdot 2^0$ | 8+0+2+0 |
| 11 | 1011 | $1 \cdot 2^3 + 0 \cdot 2^2 + 1 \cdot 2^1 + 1 \cdot 2^0$ | 8+0+2+1 |
| 12 | 1100 | $1 \cdot 2^3 + 1 \cdot 2^2 + 0 \cdot 2^1 + 0 \cdot 2^0$ | 8+4+0+0 |

That means that the whole system can be easily transformed into a system of additions and subtractions and in consequence, into a system of geometrical operations: $2^n + 2^n = 2^{n+1} \rightarrow I(2^n) + I(2^n) = I(2^{n+1})$. Added, two 'I's at one digital position ($2^n$) result in one 'I' on the next higher digital position ($2^{n+1}$). Thus, Leibniz's binary system can easily be trans-

Addition: Examples

| Decade system | Binary system | Written in powers to two | Written as sum |
|---|---|---|---|
| 1 | I | $I\cdot 2^0$ | 1 |
| | | * | |
| +1 | I | $I\cdot 2^0$ | 1 |
| 2 | I0 | $I\cdot 2^1+0\cdot 2^0$ | 2+0 |
| | | | |
| 2 | I0 | $I\cdot 2^1+0\cdot 2^0$ | 2+0 |
| +2 | I0 | $I\cdot 2^1+0\cdot 2^0$ | 2+0 |
| 4 | I00 | $I\cdot 2^2+0\cdot 2^1+0\cdot 2^0$ | 4+0+0 |
| | | | |
| 5 | I0I | $I\cdot 2^2+0\cdot 2^1+I\cdot 2^0$ | 4+0+1 |
| +5 | I0I | $I\cdot 2^2+0\cdot 2^1+I\cdot 2^0$ | 4+0+1 |
| 10 | I0I0 | $I\cdot 2^3+0\cdot 2^2+I\cdot 2^1+0\cdot 2^0$ | 8+0+2+0 |

Addition in the binary system and its mechanical transformation: the example of Leibniz's medal

| Dec. system | Dya. system | M. trans. Step 1 | Step 2 | Addition Dig. pos. 1 | Dig. pos. 2 | Dig. pos. 3 | Dig. pos. 4 |
|---|---|---|---|---|---|---|---|
| 1 | I | ////* | ////* | ///*/ | //*// | ///// | ///// |
| 3 | II | ///** | ////* | ///*/ | //*// | ///// | ///// |
| 5 | I0I | //*/* | //*** | //**/ | //*// | /*/// | ///// |
| 7 | III | //*** | //*** | //**/ | //*// | /*/// | ///// |
| 16 | I0000 | | | / | // | /// | *//// |

formed into a mechanical system, for instance, by using a ball for an 'I' and a space for the zero at digital positions – at least in principle (see first table above).

Imagine pipes for each digital position, balls (*) for the 'I's and spaces (/) for the '0's in the second table above. Then, the transformation of an addition into a mechanical operation would run as illustrated in the table. Leibniz actually planned to realize the corresponding machine with pipes, cans, and little balls. But, as we know, the possibilities to process information and organization are limited in mechanical carrying-out, when signs have to be transferred into clumsy mechanical parts.[29]

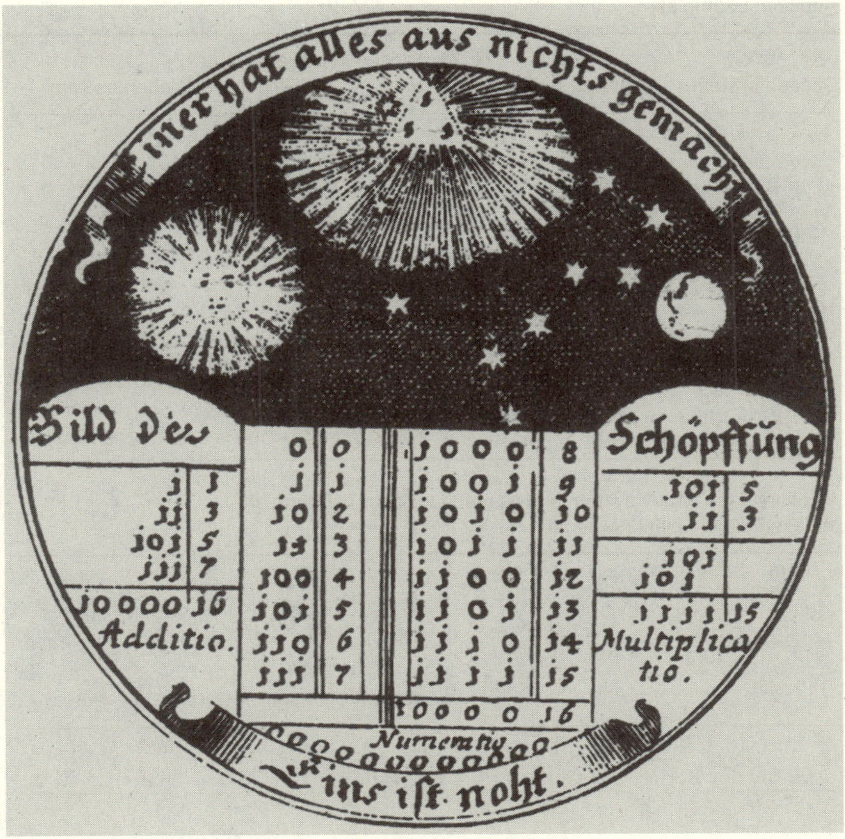


Figure 13.3 Draft of a medal praising Leibniz's dyadic system. Popp, *Leibniz*, 36.

Nevertheless, Leibniz was so proud of his system that he drafted two medals in its praise. Even more, Leibniz believed he had discovered the key of god's creation, beauty, and nature. He engraved his medals (see figure 13.3) with the aphorism: 'One is enough to draw everything out of nothing.'[30] His system showed that God created everything out of nothing and also how beautifully and well everything had been done. He argued, 'For now the wonderfully beautiful order and unity of the number system which is impossible to improve is proven against the common view maintaining that numbers possess neither regularity nor certain sequences.'[31] Even if we would not subscribe to Leibniz's aesthetic and

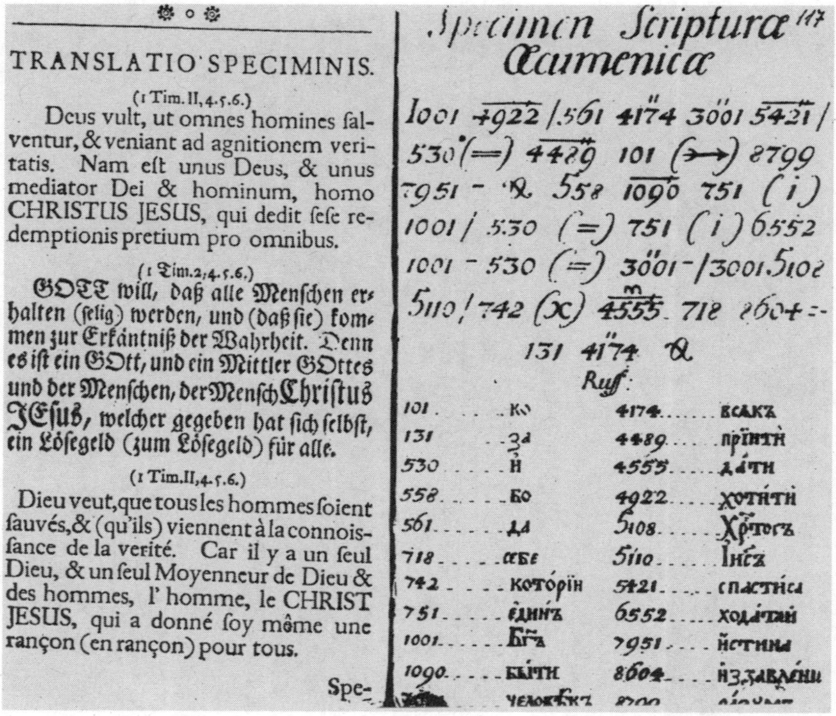
TRANSLATIO SPECIMINIS.

(1 Tim. II, 4.5.6.)

Deus vult, ut omnes homines salventur, & veniant ad agnitionem veritatis. Nam est unus Deus, & unus mediator Dei & hominum, homo CHRISTUS JESUS, qui dedit sese redemptionis pretium pro omnibus.

(1 Tim. 2, 4.5.6.)

GOTT will, daß alle Menschen erhalten (selig) werden, und (daß sie) kommen zur Erkäntniß der Wahrheit. Denn es ist ein GOtt, und ein Mittler GOttes und der Menschen, der Mensch Christus JEsus, welcher gegeben hat sich selbst, ein Lösegeld (zum Lösegeld) für alle.

(1 Tim. II, 4.5.6.)

Dieu veut, que tous les hommes soient sauvés, & (qu'ils) viennent à la connoissance de la verité. Car il y a un seul Dieu, & un seul Moyenneur de Dieu & des hommes, l' homme, le CHRIST JESUS, qui a donné soy même une rançon (en rançon) pour tous.

Spe-

Specimen Scripturæ Œcumenicæ 117

1001 4922 / 561 4174 3001 5421 /
530 (=) 4489 101 (→) 8799
7951 - 558 1090 751 (i)
1001 / 530 (=) 751 (i) 6552
1001 - 530 (=) 3001 - / 3001 5108
5110 / 742 (x) 4555 718 8604 =-
131 4174

Ruß:

| | | | |
|---|---|---|---|
| 101 | ко | 4174 | всакъ |
| 131 | за | 4489 | прїити |
| 530 | и | 4555 | дати |
| 558 | бо | 4922 | хотити |
| 561 | да | 5108 | Хр͠тосъ |
| 718 | себе | 5110 | Іисъ |
| 742 | которїи | 5421 | спастися |
| 751 | єдинъ | 6552 | ходатай |
| 1001 | Бг͠ъ | 7951 | истина |
| 1090 | быти | 8604 | изъзбавлении |

Figure 13.4 Rödeke's Universal Language, examples from Rödecke, *Probe*, II, Archiv der Berlin Brandenburgischen Akademie der Wissenschaften, I–V, 2.

religious enthusiasm, his concept gives analytical evidence and formulates, in terms of creation, the possibility and practical potential of the linking of information and organization. Leibniz had found the principle of modernity.

Both Leibniz's ingenious achievement and his mechanical failure bring us back to his academy in Berlin and to Swift's language machine and his reference to a 'skillful computer.'[32] Actually, not only Leibniz's projects, but also his academy resembled the Lagadian academy. There were plans to implant chestnut trees and sunflowers in Berlin. There was the idea and actual attempt to fund the academy by producing silk. There was a design of a new threshing machine constructed to save labour. There were proposals for silver bullets as well as the idea to set lamps on horses' heads to light the street.[33] And there was the draft of a universal language, submitted by Caspar Rödecke (see figure 13.4). His idea to translate language into numbers was, in a way, the next step from Leibniz's dyadic number system to a modern computer.[34]

What, then, distinguished the Berlin Academy and the Lagadian Academy? Essentially two things, I believe. First, the 'Lagadian projects' were not pursued by the Prussian Academy itself. They were submitted to the Academy to be reviewed and fostered. So, the Academy played a controlling role. In short, it organized organization. Second, the Academy possessed a concrete official task and privilege. It was responsible for the production and distribution of calendars in Prussia. Calendars bear on astronomy, and the then still old-style Prussian calendar had to be attuned to the solar year. That was the scientific background of the task. Thus, the Academy became the warden of time, but more, as monopolist of calendars, it became an organizing instrument and main propagator of life and world as a coherent administrable environment. The Academy's calendar represented not only the run of the stars, but also represented – as calendars always do – much more the run of social life. It fixed holidays, market days, and paydays in the now officially distributed Prussian timetable. Yet, the then highly popular literary genre 'calendar' was not confined to data of time. It was popular because it was an important source of information about basic lines of social life, containing weather information, tables of prizes, measures and weights, and ready reckoners. The Berlin-based Prussian Academy had published more than just calendars of that kind since 1704; it had issued the so- called *Berliner Adreßkalender*: an official inventory of Berlin institutions, from the court to the schools, with their members' names and addresses. The inventory was probably based on similar productions in London and Paris, and it was the forerunner of the later state manuals and municipal mailing list. More than 100,000 copies a year of the various calendar types were issued by the Academy, and printed and distributed by eight publishers all over Prussia.[35]

Besides the calendars, prizes, measures, weights, institutions, and addresses, the Academy mapped the state's parishes with all its churches and schools. Addressing, mapping, inventorying: introducing the matrix of and for administration and further organization, the Academy was a focal point in the transformation of a territory into a modern state.[36]

For as a social reality beyond government, the state consists essentially in the transformation of social life into an ordered process beyond local boundaries. It consists in regulations expressed as interrelated scales. Bringing life itself into coherent interrelated registers, the Academy itself was the 'skillful computer' Swift had been talking about. Indeed, Swift coined the term not for a machine, but for modern scholars who would take some handful of information out of the more and more common registers of books in order to write new texts instead of reading the books first.[37] Swift's observation hit and missed the point even more than his Lagadian text machine. For registers were in fact crucial parts of an upcoming machine of communication, which would be commonly called the Republic of Letters. The scientific backbone of this network – 'scientific' taken in the sense of 'methodical production of knowledge' and 'backbone' in the sense of 'a device to guarantee coherence' – was a huge enterprise that Bacon had already proposed: the *Historia literaria.*[38] Innovations depend on access to the already given. The idea was to make up a complete and steadily supplemented inventory of all texts and their authors attached to the respective field of study. This registry was not just a list. Creating the complete bibliographical body of knowledge, it created as well the virtual universal library and transformed on its basis the real libraries from closed book collections into interrelated repositories as basic instruments for the coherent increase of knowledge. *Historia literaria* was not a job for five people and a pair of scissors. The project implied completeness, actuality, participation, and authority. It meant the organization and reorganization of the production of knowledge, and was achieved by the interplay of a variety of media: histories of scientific institutions and disciplines, manuals, dictionaries, book lists – several thousand alone for Germany in the eighteenth century – but also, and not least, review journals – some two hundred in Germany – published and authorized foremost by learned societies, especially the academies. The result was not a new process, but a now systematically performed process. Knowledge stored in its completeness as texts in the hardware of the libraries was taken and processed by authors and restored afterwards by the *Historia literaria* for further use.[39] Sprat was right. The computer had started to work.

Obviously, Swift's Lagadian Academy corresponds closely with Leibniz's ideas. Did Swift have Leibniz in mind when he wrote his satire? I have gone through the literature and Swift's correspondence. There is no hint of Leibniz. Even in a letter of 1714 in which Swift asked John Gay, then in Hanover for diplomatic reasons, to look something up in one of the libraries directed by Leibniz, Leibniz is not mentioned personally.[40] Thus, I am quite sure that a direct connection did not exist between Leibniz and Swift. But then, suddenly, I had the idea to ask Swift's 'skillful computer' itself in its contemporary form – I searched not Yahoo, but Google about the connection and I immediately got fifty hits. Most of them stemmed from computer projects, and projects that pursue the idea of automatic text machines and especially translation devices or computer philology. Many name Swift as well as Leibniz as founders of their enterprise, stressing of course not so much Swift's disdain of 'computers' as his foresight in attempting to outline and – to a certain extent – to construct and invent a personalized story and history of the field.

Paradoxically, then, the Lagadian Academy was a debate of Leibniz's ideas. And following Norbert Wiener, who argued this Leibniz-Swift connection in his autobiography, the websites support this notion.[41] Who is Norbert Wiener? Not a member of a literature department, but one of the founders of cybernetics. And is cybernetics? Let me cite the Deutsche Gesellschaft für Kybernetik here: 'According to ... Norbert Wiener, cybernetics shall be understood as not confined to the theory and technique of regulation alone, but also as the pursuit of transferring and processing information based upon analytical, modeling, measuring, and calculating methods for the purpose of prognoses.'[42]

The history I have been tracing is the story of institutionalization. Sprat's and Leibniz's hope to organize the future methodically has grown up into an organized discipline called cybernetics. 'The past is our present' is the slogan of the historians. And the cyberneticists seem to answer, 'The present is our future.'

## NOTES

1 Jonathan Swift, *A Tale of a Tub*, ed. A.C. Guthkelch and D. Nichol Smith (Oxford: Clarendon Press, 1920), 146.

2 Gottfried Wilhelm Leibniz, *Sämtliche Schriften und Briefe*, ed. Deutsche Akademie der Wissenschaften zu Berlin, series 1, vol. 1 (Berlin: Akademie Verlag, 1923), 160.

3 Caspar Rödecke, *Probe der allgemeinen Schrifft nebst den Schlüsseln von 19 Sprachen* (Salzwedel [For the Author], 1725), II.

4 Jonathan Swift, *Gulliver's Travels* (1726), ed. Paul Turner (Oxford: Oxford University Press, 1986), 171–84.

5 Thomas Sprat, *The History of the Royal Society of London, For the Improving of Natural Knowledge* (London: Printed by T. R. for J. Martyn and J. Allestry, 1667), 61–123.

6 On institutions, see Helmut Schelsky, 'Zur soziologischen Theorie der Institutionen,' in *Zur Theorie der Institution*, ed. H. Schelsky (Düsseldorf: Bertelsmann, 1970), 9–26, esp. 16–22. See also Ephrem Else Lau, *Interaktion und Institution: Zur Theorie der Institution und der Institutionalisierung aus der Perspek-*

*tive einer verstehend-interaktionistischen Soziologie* (Berlin: Duncker & Humblot, 1976), 18–24; Lynne G. Zucker, 'Institutional Theories of Organization – Conceptual Development and Research Agenda,' in *Institutional Patterns and Organizations: Culture and Environment*, ed. Lynne G. Zucker (Cambridge, Mass.: Ballinger, 1988), xiii–xix; Mary Douglas, *How Institutions Think* (London: Routledge & Kegan Paul, 1986); Peter L. Berger and Thomas Luckmann, *The Social Construction of Reality: A Treatise in the Sociology of Knowledge* (New York: Anchor, 1967); and John R. Searle, *The Construction of Social Reality* (New York: Free Press, 1995).

7 For planning, plans, utopia, and organization, see Martin Gierl, *Geschichte und Organisation: Insitutionalisierung am Beispiel der europäischen Wissenschaftsakademien um 1900.* Abhandlungen der Akademie der Wissenschaften zu Göttingen, 233 (Göttingen: Vandenhoeck & Ruprecht, 2004).

8 For the failure of the Berlin Academy in its first decades, see Adolf Harnack, *Geschichte der Königlichen Akademie der Wissenschaften zu Berlin*, vol. 1.1 (Hildesheim: Georg Olms Verlag, 1970), 142–8, 240f.

9 Cf. Harnack, *Geschichte* 1.1:159–65, 236.

10 Sprat, *History*, 25–8, 345–78; for the criticism of Henry Stubbe on the Royal Society, see Henry Stubbe, *Campanella Revived Or an Enquiry into the History of the Royal Society* (London: [Printed for the author,] 1670); Henry Stubbe, *A Censure upon Certain Passages Contained in the History of the Royal Society, As being Destructive to the Established Religion and Church of England* (Oxford: Printed for Richard Davis, 1670); on Stubbe, see James R. Jacob, *Henry Stubbe: Radical Protestantism and the Early Enlightenment* (Cambridge: University Press, 1983).

11 Sprat, *History*, 52–157.

12 Ibid., 4–51.

13 Ibid., 76.

14 Ibid., 86.

15 Ibid., 5, 20, 38, 75, 81, 86, 114, 130, 323.

16 Ibid., 119; cf. *The Spectator*, no. 10, in *Addison & Steele and others, The Spectator in Four Volumes*, vol. 1, ed. Gregory Smith (London: Dent, 1963), 31f.

17 Sprat, *History*, 122f.

18 On Laputa and the Lagadian Academy, see Marjorie Nicolson and Nora Mohler, 'Swift's "Flying Island" in the Voyage to Laputa,' *Annals of Science* 2 (1937), 405–30; Marjorie Nicolson and Nora M. Mohler, 'The Scientific Background of Swift's Voyage to Laputa,' in Marjorie Nicolson, *Science and Imagination* (Ithaca: Great Seal Book's, 1962), 110–54; John N. Sutherland, 'A Reconsideration of Gulliver's Third Voyage,' *Studies in Philology* 54 (1957), 45–52; John Munro, 'Book III of Gulliver's Travels Once More,' *English Studies* 49 (1968), 429–36; Paul Korshin, 'The Intellectual Context of Swift's Flying

Island,' *Philological Quarterly* 50 (1971), 630–46; Pat Rogers, 'Gulliver and the Engineers,' *Modern Language Review* 70 (1975), 260–70; Dennis Todd, 'Laputa, the Whore of Babylon, and the Idols of Science,' *Studies in Philology* 75 (1978), 93–120; Eric Rothstein, 'Gulliver 3; or The Progress of Clio,' *Proceedings of the First Münster Symposium on Jonathan Swift* (1985), 217–31. Robert P. Fitzgerald, 'Science and Politics in Swift's "Voyage to Laputa,"' *Journal of English and Germanic Philology* 87 (April 1988), 213–29; John Christie, 'Laputa Revisited,' in *Nature Transfigured: Science and Literature, 1700–1900,* ed. John Christie and Sally Shuttleworth (Manchester: Manchester University Press, 1989), 45–60; Joseph E. Argent, 'The Etymology of a Dystopia: Laputa Reconsidered,' *English Language Notes* 34.1 (Sept. 1996), 36–40; and Robert Phiddian, 'A Hopeless Project: Gulliver Inside the Language of Science in Book III,' *Eighteenth Century Life* 22.1 (Feb. 1998), 50–62.

19 Swift, *Gulliver's Travels,* 165–84.

20 Cf. Nicolson and Mohler, 'The Scientific Background'; Paul Turner, *Explanatory Notes,* in Swift, *Gulliver's Travels,* 334–9.

21 And indirectly to *New Atlantis* and the concept of science of Francis Bacon, who intended with his Advancement of Learning 'a general and faithful perambulation of learning, with an inquiry what parts thereof lie fresh and waste, and not improved and converted by the industry of men.' Francis Bacon, *Of the Proficience and Advancement of Learning Divine and Human,* part. 2,' in *The Works of Francis Bacon,* ed. James Spedding, Robert Leslie Ellis, and Douglas Denon Heath, vol. 3 (London: Longman and Co., 1859), 328.

22 Swift, *Gulliver's Travels,* 168f.

23 Ibid., p. 174. Cf. Eric A. Weiss, 'Jonathan Swift's Computing Invention,' *Annals of the History of Computing* 7 (1985), 164–5.

24 For example, Shadwell's *Virtuoso* and Rabelais's *Garagantua* and *Pantagruel.* Cf. n. 20, 40.

25 On the early history of the academy in detail, see Hans-Stephan Brather, ed., *Leibniz und seine Akademie: Ausgewählte Quellen zur Geschichte der Berliner Sozietät der Wissenschaften 1697–1716* (Berlin: Akademie Verlag, 1993); Harnack, *Geschichte,* 1.1.

26 Cf. Reinhard Finster and Gerd van den Heuvel, *Gottfried Wilhelm Leibniz* (Hamburg: Rowohlt, 1990); Karl Popp and Erwin Stein, eds, *Gottfried Wilhelm Leibniz: Philosoph, Mathematiker, Physiker, Techniker* (Hannover: Schlütersche, 2000).

27 Leibniz to G. Wagner, 1696, in Carl Immanuel Gerhard, ed., *Die philosophischen Schriften von Gottfried Wilhelm Leibniz,* vol. 7 (Berlin: Weidmann, 1890), 525.

28 See Finster and van den Heuvel, *Leibniz,* 103–8; Erich Hochstetter, *Herrn von*

*Leibniz' Rechnung mit Null und Eins* (Berlin: Siemens Aktiengesellschaft, 1979); and Hans J. Zacher, *Die Hauptschriften zur Dyadik von G. W. Leibniz: Ein Beitrag zur Geschichte des binären Zahlensystems* (Frankfurt am Main: Klostermann, 1973).

29 See Ludolf von Mackensen, 'Die ersten dekadischen und dualen Rechenmaschinen,' in *Leibniz*, ed. Popp and Stein, 85–100, 94; Ludolf von Mackensen, 'Leibniz als Ahnherr der Kybernetik: Ein bisher unbekannter Leibnizscher Vorschlag einer "Machina arithmeticae dyadicae,"' in *Akten des II. Internationalen Leibniz-Kongresses, Hannover, 17.–22. Juli 1972*, vol. 2 (Wiesbaden: Steiner, 1974), 255–68.

30 Quote: Popp and Stein, *Leibniz*, 36. For the medals, see Carl Günther Ludovici, *Ausführlicher Entwurff einer vollständigen Historie der Leibnitzischen Philosophie zum Gebrauch seiner Zuhörer heraus gegeben* (Leipzig: J.G. Löwe, 1737), 132–5.

31 Leibniz to Duke Rudolf August von Wolfenbüttel, 1697, in *Gottfried Wilhelm Leibniz: Sämtliche Schriften und Briefe*, ed. Deutsche Akademie der Wissenschaften zu Berlin, series 1, vol. 13 (Berlin: Akademie Verlag, 1987), 118.

32 The term derives from the Latin *computare* and denominated in the seventeenth century a person doing mathematical calculations. Early references of the noun and of the verb 'to compute' stem from the first half of the century.

33 See Archiv der Berlin Brandenburgischen Akademie der Wissenschaften, 'Wissenschaftliche Verhandlungen,' I–V, 1–25; cf. Brather, *Leibniz*, 123–78.

34 Cf. Rödecke, *Probe*, If. On Rödecke and the Berlin Academy, see Archiv der Berlin Brandenburgischen Akademie der Wissenschaften, 'Wissenschftliche Verhandlungen,' I–V, 2 and Brather, *Leibniz*, xxxvi f., 411.

35 On the *Kalender*, see Archiv der Berlin Brandenburgischen Akademie der Wissenschaften, I–VIII; especially Brather, *Leibniz*, 233–58; Harnack, *Geschichte*, 1.1, 123–7.

36 For further examples, see James E. McClellan III, *Science Reorganized: Scientific Societies in the Eighteenth Century* (New York: Columbia University Press, 1985), 77, 94, 126, 177–8.

37 Swift, *Tale*, 144–9. On Swift's criticism of the moderns, see Miriam Kosh Starkman, *Swift's Satire on Learning in A Tale of a Tub* (New York: Octagon Books, 1968), 3–22.

38 Cf. Bacon, *Proficience*, 329f.

39 On *Historia literaria*, see Martin Gierl, 'Bestandsaufnahme im gelehrten Bereich: Zur Entwicklung der 'Historia literaria' im 18. Jahrhundert,' in *Denkhorizonte und Handlungsspielräume: Historische Studien für Rudolf Vierhaus zum 70. Geburtstag* (Göttingen: Wallstein, 1992), 53–80; and Gierl, 'Compilation and the Production of Knowledge in the Early German Enlightenment,' in *Wissen-*

*schaft als kulturelle Praxis, 1750–1900*, ed. Hans Erich Bödeker, Peter Hanns Reill, and Jürgen Schlumbohm (Göttingen: Vandenhoeck & Ruprecht, 1999), 69–104.

40 *The Correspondence of Jonathan Swift*, ed. Harold Williams, 5 vols. (London: Oxford University Press, 1963), 2:33, 38; on Leibniz and Swift's *Gulliver's Travels* see also Charles McDowell, 'Catastrophism and Puritan Thought: The Newton Era,' *A Symposium on Creation* 6 (1977), 57–90. Grand Rapids, Mich. Pacific Meridian Publishing Co.

41 Norbert Wiener, *I Am a Mathematician* (Cambridge, MA: MIT Press, 1964), 123.

42 See http://www.gesellschaft-fuer-kybernetik.org.

chapter fourteen

# Geographical Projects in the Later Eighteenth Century: Imperial Myths and Realities

CAROLE FABRICANT

While the earlier part of the eighteenth century was littered with the corpses of failed or abortive projects – the Darien Expedition, the South Sea Bubble, Bishop Berkeley's Bermuda Project – the latter part of the century, according to the received wisdom, was filled with great and successful ones. These projects were spurred on by the period's scientific advances, animated by Enlightenment ideas, and enriched by the accumulated experiences of the preceding six decades of global exploration. The air was filled with anticipations of grand discoveries and with visions of vast knowledge in the process of being collected, organized, and systematized – knowledge based on travels throughout the world. Thus, Edmund Burke could proclaim to a friend in 1777, 'now the great Map of Mankind is unrolld at once; and there is no state or Gradation of barbarism, and no mode of refinement which we have not at the same instant under our View.' He proceeded then to enumerate the different degrees of civility and savagery characterizing the various parts of the globe.[1]

The prospect of adding even more detail to 'the great Map of Mankind' inspired a host of geographical projects in the final decades of the century, which were launched in a spirit of confidence and optimism, and bolstered by a belief in the inevitable spread of civilization and the triumphant march of reason and science. My own project, in this essay, is characterized by a very different spirit, and informed by a considerably more sceptical view of the putative march of 'progress' and of its penetration into the furthest nooks and crannies of the earth. For this undertaking, I want to enlist the aid of a seemingly unlikely ally – Daniel Defoe himself, who is generally taken to be an enthusiastic advocate and promoter of projects, especially via the numerous schemes propounded in his *Essay on Projects* (1697), but who, in fact, expressed a rather more ambivalent attitude towards them, as his example of one of mankind's earliest projects, the Tower of Babel, suggests. For all its negative associations, the tower, according to Defoe, was 'a Right Project; for indeed the true definition of a Project, according to Modern Acceptation, is ... a vast Undertaking too big to be manag'd, and therefore likely enough to come to nothing.' Although the image here suggests an ominous hubris and inevitable failure, Defoe does offer some ray of hope, noting that 'if the People of the Old World could have Built a House up to Heaven, they should never be Drowned again on Earth, and they only had forgot to Measure the Heighth, *that is*, as in other Projects, it only Miscarri'd, or else 'twould have Succeeded.'[2] Defoe's attempts to define what a project is are mired in contradiction and culminate in tautology – if only they had not failed, he argues, all projects would be successful. Faced with the alternatives of either tearing down the tower and admitting defeat or obstinately pushing ahead in an effort to turn failure into success, Defoe implies that he would opt for the latter, although he surely realizes that a scheme to raise an edifice high enough to reach the heavens is an impossibility, and likely to exact a heavy toll from those involved in its construction – death and damnation being only the two most extreme examples of penalties incurred.

Defoe's ability to see the different sides of the projecting spirit – its energy and optimism counterbalanced by its recurring failures and by what we can only term a form of madness (for who but a madman would draw up precise measurements to build castles in the air?) – helps us to think about some of the complexities and contradictions of expeditionary projects later in the century. Published accounts of these projects were, in general, avidly consumed by a domestic audience and rapidly absorbed into a story much larger than itself, centred on a saga of progress and empire-building that was often undermined by many details of both the expedition and its narrative. As a means of exploring the conflicting faces of these projects and the ideological contradictions underpinning them, I will focus mainly on Captain Cook's voyages to the South Pacific while also glancing at James Bruce's travels to Abyssinia and concluding with some reflections on Mungo Park's journey into the interior of Africa.

The ideology of bourgeois individualism, in conjunction with the cult of celebrity that began developing in the earlier part of the century with swashbuckling adventurers such as William Dampier and George Anson sailing around the world and then publishing popular accounts of their exploits that turned into the best-sellers of their day, produced a situation in which explorers like Cook, upon their return to England, were exalted as heroic explorers and adventurers and feted for their 'accomplishments,' regardless of the relative success or failure of their undertakings.[3] But while their expeditions were on one level being reduced to extreme individualism, to the portrait of the sea captain or lone journeyer as hero, they were, paradoxically, on another level being transformed into symbols of a national project that far transcended the particulars of their individual journeys. In certain ways, this rendered the particulars irrelevant, making them little more than anonymous components of a vast colonial venture that they were enlisted to promote, regardless of their intentions or stated goals. My point is not that these expeditions were necessarily hostile to or overtly critical of the national project, but rather that their narratives tell a number of different, often conflicting stories – of defeat, humiliation, and vulnerability, as well as of conquest and achievement – and that only one of these stories became the officially acknowledged one, passed down to future generations to help fuel the ongoing project of colonial expansion.

What exactly were these 'other' stories and perspectives woven into the fabric of explorational narratives but then often erased from the collective memory of later generations? A few of the recurring narrative strands that I want to pay particular attention to include the following: First, there was an enormous investment, ideological as well as financial, in the advanced scientific technology that the explorers brought with them on their expeditions, which functioned as symbols of the superiority of European knowledge vis-à-vis the rest of the world no less than as instruments enabling the accomplishment of certain specific scientific goals. Yet, the scientific paraphernalia proudly displayed at the beginning of the expeditions often exacerbated the explorers' vulnerability to the natives they encountered on their travels rather than strengthening their position of dominance over them, and frequently failed to perform their designed function. Adding to this vulnerability was the explorers' dependence on native guides, interpreters, and informants – or on the native population at large – which indicates that far from asserting control in the distant lands they traversed, the Europeans were often rendered confused and helpless on foreign soil, desperately relying on friendly natives for their very survival in conditions that were hostile and inscrutable to them. Third, while the projects proclaimed their commitment to the acquisition and circulation of knowledge about other countries and peoples, the travel narratives often, at least intermittently, tell a different story: of the opaqueness of other cultures, of the explorers' lack of linguistic and other skills needed to 'read' these cultures accurately, the result being that the more they travelled, the less they understood, and the more inconclusive their observations became. Fourth, there is the fact that the obligatory moment of 'discovery' in many cases actually points to its absence, its illusory quality, or the farcical nature of its claims to 'heroic' achievement, especially when viewed through the lens of native observers. Finally, there are the many instances in which the civilization-barbarism polarity is undermined by narrative strands that tell of enlightened British explorers setting forth to bring civilization to savage peoples but eventually finding it hard to differentiate the one from the other, or discovering, along with Conrad's Marlowe, that the 'heart of darkness' was located not outside, in the alien regions traversed, but within, in the hearts and minds of the explorers themselves.

Captain Cook's voyages constitute an especially revealing example of both the ideals and the contradictions of Enlightenment projects in the latter half of the eighteenth century. The official rationale for the first voyage was scientific – to observe the transit of Venus across the sun on 3 June 1769 from St George's Island (Tahiti). It is therefore not surprising that the initial expedition was sponsored by the Royal Society and that the crew included several noted naturalists (for example, Joseph Banks, Daniel Carl Solander, and Herman Diedrich Spöring), the astronomer Charles Green, and two artists (Sydney Parkinson and Alexander Buchan) to make a visual record of the fauna and flora of the lands they would explore. But Cook was also given a commission from the British Admiralty, which included instructions to search for what was then called the 'Southern Continent,' and he was directed 'with the Consent of the Natives to take possession of Convenient Situations in the Country in the name of the King of Great Britain'; or, in the event the country was uninhabited, to 'take Possession [of it] for His Majesty.'[4] Cook's voyage thus exemplifies the extent to which the (presumably) disinterested aim of scientific advancement and the eminently self-interested goal of colonial expansion had become intertwined by the latter half of the eighteenth century. The name of Cook's first ship, *The Endeavour*, may be understood to signify a multidimensioned effort to establish Britain's dominance both in the field of science and in the spheres of political and economic influence in the Pacific.

When Cook and his crew sailed off from Deptford in May 1768, they brought a large store of scientific equipment with them. Shortly after arriving in Tahiti, their most prized scientific instrument, their best quadrant, needed to record the transit of Venus, was stolen from under their noses. On the morning of 2 May 1769 Cook records, in his journal, that when he and a crew member 'went to set up the Quadrant,' which 'had never been taken out of the Packing case,' it was nowhere to be found (49). Its disappearance was the occasion of considerable 'astonishment' and perplexity on the part of Cook and his men, as the quadrant was very heavy and 'a Centinal stood the whole night within 5 yards of the door of the Tent' where the instruments had been placed (49). Eventually they receive information that 'one of the natives had taken it away and carried it to the Eastward.' At first they decide to take a number of Indians into custody as hostages until the instrument is returned, but then think better of it, given the risk of turning the Tahitians against them. Then two crewmen are sent into the woods to inquire of the chief 'which way and where the Quadrant was gone.' The three men head eastward in search of the quadrant and eventually retrieve it after Cook himself leads a party of armed men to help in the search. The men subsequently try to figure out how the instrument was stolen from under the eyes of the Britons, despite the elaborate security precautions put into place, which included Cook himself and a few others remaining within the fort at all times and participating in sentry duty. Despite a scrupulous attempt to reconstruct the theft, the mystery of the quadrant's disappearance is never really resolved; as Cook explains, 'we found it difficult to beleive [*sic*] that a naked Indian frighten'd of f[i]rearms as they are, would have made such an attemp'd at the certain risk of his life' (51). Apparently the most sophisticated scientific equipment of the technologically advanced Britons was no match for the inexplicable powers of an unclothed Indian who had never before seen a gun, let alone a quadrant, and its return was effected not by the explorers' superior wit but only through the direct intercession of another Tahitian native.

This intercession anticipates the indispensable mediation of a string of native guides and interpreters – Tupia and Omai foremost among them – without whom *The Endeavour*'s mission would have been severely compromised, if not completely thwarted. The journals repeatedly show that Cook and his men need the natives far more than the natives need them. Cook places a great deal of weight on promoting trade with the natives, generally presenting these transactions as activities beneficial to the latter and desired by them. But his journals tell another story as well, recording instances when the natives spurn English cloth, preferring their own, and when they rebuff the sailors' attempts to interact with them, as in the case of the Australian tribesmen who 'had not so much as touch'd the things we had left in their hutts on purpose for them to take away' (130). During his final voyage, Cook comments negatively on Omai's nation, the people of Huahine island, noting with some disdain that 'Europeans have visited them at times for these ten years past, yet we find neither new arts nor improvements in the old, nor have they copied after us in any one thing' (525). Cook interprets the Huahines' imperviousness to European influences as an indication of their primitiveness, lack of initiative, and their indifference to ways of bettering themselves. Thus, he overlooks the more obvious and probable explanation that the natives were simply unimpressed with European 'arts' and 'improvements' and wanted to be left alone to continue living their lives without outside interference. This was clearly the one thing that could not be acknowledged or countenanced by the British explorers, whose very existence was predicated on making connections with a people assumed to be open to their presence and willing to enter into a relationship with them.

A month after the inexplicable theft of the quadrant, two different sets of crewmen were sent to different parts of the island, furnished with instruments to observe the transit of Venus, while Cook and several of his officers remained at the fort to carry on their own observations. The day after the sighting, on 4 June, Cook tersely records, in his journal, that 'the gentlemen that were sent to observe the Transit of Venus returnd with success' (55), and this is the last that we hear of the event that was supposed to be the centrepiece of *The Endeavour*'s voyage. But in a journal entry on the preceding day, we learn that despite the fact that 'this day prov'd as favourable to our purpose as we could wish, not a Clowd was to be seen the whole day and the Air was perfectly clear, so that we had every advantage we could desire in Observing the whole of the passage of the Planet Venus over the Suns disk' (54–5), there were problems with the sightings on the part of Cook and two of his men. He recorded: 'We very distinctly saw an Atmosphere or dusky shade round the body of the Planet which very much disturbed the times of the Contacts particularly the two internal ones. Dr. Solander observed as well as Mr. Green and my self, and we differ'd from one another in observeing the times of the Contacts much more than could be expected.' After noting the difference in the magnifying power between his telescope and Solander's, Cook comments that 'the Thermometer expose'd to the Sun about the middle of the Day rose to a degree of heat ... we have not before met with' (55). It would appear that these inconsistencies and unforeseen circumstances cast doubts on the accuracy of the observations; nor does the fleeting allusion to the other crewmen's 'success' in *their* observations do much to reassure us about the value of these astronomical activities. And indeed, while some modern commentators accept the results at face value, others deem them imperfect and of limited value. The primary editor of Cook's journals, J.C. Beaglehole, considered them completely useless.[5]

But whatever their degree of success or failure, it is clear that science fell short of fulfilling its role as a mechanism of authority – and authorization – in the course of *The Endeavour*'s voyage. This disappointment was repeated during Cook's second voyage aboard *The Resolution*, when other efforts to exploit the weight and prestige of science had similarly problematic (not to mention somewhat risible) results. Dissension among several of the scientists who were part of the expedition produced published accounts of it that not only contradicted one another but also engaged in some rather indecorous and decidedly undignified mud-slinging, turning the supposedly objectifying lens of science into a highly subjective arena of battling egos and conflicting interpretations. The mission involved yet another calculation of the transit of Venus, but the main bone of contention in this case concerned the testing of several chronometers as part of an ongoing experiment to solve the problem of the longitude. One of the naturalists aboard *The Resolution*, Johann Reinhold Forster, charged in his printed 'Observations' that his fellow crewman, the astronomer and mathematician William Wales, had forgotten to wind the chronometers and had damaged their cases through mishandling. Wales responded with a peevish account that heaps scorn on Forster's observational skills and calls into question much of his 'scientific' claims.[6] Even the model of John Harrison's ultimately successful chronometer, which was brought along on the voyage of *The Resolution*, proved inaccurate in this instance.[7] Throughout Cook's voyages, the very things designed to elevate the mission above the vagaries and uncertainties of casual observation or mere opinion – scientific calculation and measurement – were less than successful.

Technology played a similarly problematic role in James Bruce's travels. Unlike Cook, Bruce did not have the backing of government and scientific groups when in 1768 he embarked on what turned out to be a six-year quest to locate the source of the Nile River. Nevertheless, he also embodied a perspective that we associate with Enlightenment ideals – one that placed science, along with discovery, at the centre of human activity, and that linked both to the beneficent rule of Britain in the world. In his dedication to the king, which appears at the beginning of his published memoir, Bruce emphatically differentiates himself from earlier generations of travellers, whom he equates with buccaneers and pirates because of their 'immoderate thirst of Conquest,' and situates himself within the 'golden age' ushered in by George III, which 'united humanity and science' and allowed 'men of liberal minds and education' to lead expeditions abroad that could benefit 'mankind in general.'[8] The pre-eminence of this new type of traveller resided to no small degree in his advanced scientific and technological skills. Bruce, himself an amateur mathematician and astronomer, embarked on his journey equipped with 'a large apparatus of instruments,' including 'the completest of their kind' for the observation of the transit of Venus (which he hoped to see in Algiers) and a 'large and expensive' 'Camera Obscura' to assist in 'executing views of ruined architecture' and the like (1:ix).

As in Cook's journals, the emphasis placed on this equipment represents a very significant psychological and ideological (as well as financial) investment. Yet Bruce's state-of-the-art instruments are of limited use on his travels, proving more a liability than an asset. Their sheer unwieldy materiality, severed from their functionality and from the intellectual ideals of progress they are meant to promote, is made clear when Bruce and his entourage of servants and carriers are faced with having to make an arduous journey over the mountains of Taranta, on their way to Abyssinia. As they begin their ascent, they must use a road of 'incredible steepness,' with 'large holes and gullies' made by torrential rains, and 'huge monstrous fragments of [loosened] rocks' (3:76). Since it is only 'with great difficulty [they] could creep up' the path, carrying only their knapsacks and arms, the problem of what to do about the scientific paraphernalia momentarily stymies them: 'it seemed beyond the possibility of human strength to carry our baggage and instruments' (3:76). Bruce explains that the 'quadrant had hitherto been carried by eight men, four to relieve each other; but these were ready to give up the undertaking upon trial of the first few hundred yards.' A number of expedients are proposed but quickly rejected because of the damage they would cause the quadrant.

Bruce's whole project seems on the verge of collapse because of the very instruments he had hoped would help ensure its reputation and success, and on which he banked his stature as a 'golden age' explorer on behalf of 'humanity and science.' After much anxiety and fretting, he decides to take matters into his own hands: 'As I was incomparably the strongest of the company, as well as the most interested, I, and a stranger Moor who had followed us, carried the head of [the quadrant] for about

400 yards over the most difficult and steepest part of the mountain, which before had been considered as impracticable by all.' What follows in the next two pages of description is the remarkable transformation of the quadrant from a scientific instrument into a measure by which manhood is tested. After carefully easing the now-feminized quadrant up the mountain, Bruce and Yasine, the Moor, return to the rest of the company with hands and knees cut and bleeding and clothes torn to pieces from their exertions, ready to 'declar[e] ourselves now without fear of contradiction, and, by the acknowledgment of all, upon fair proof, the two best men in the company' (3:77). 'Shamed' and 'humbled' by this proof of the two men's superior prowess, the rest of the group push themselves to the limit to help carry the two telescopes and time-keeper up the mountain. In the end, then, the scientific instruments *do* demonstrate an ability to confer authority on Bruce, but it is the authority of a machismo ill suited to the new type of traveller he extols in his dedication – not to mention an authority he must share with his Moorish servant.

Cook manages to establish a degree of authority for himself through his detailed descriptions of Polynesian topography and his minute observations of the natives' physical appearance and daily life, which constitute a picture with obvious geographical and ethnographic significance. And yet, the further we read in the journals, the more we become aware of the yawning gap between empirical facts and knowledge, between even the most scrupulous and precise enumeration of observed details and the kind of comprehension needed to contextualize and make sense of these details. The oft-cited Point Venus scene, in which a young Tahitian girl of ten or twelve has sex with a male in public view, is perhaps the best-known example of this lack of understanding. Cook's account notes the presence of several women 'of the better sort' who 'were so far from shewing the least disaprobation that they instructed the girl how she should act her part, who young as she was, did not seem to want it' (53). As Neil Rennie suggests, the very syntax of this passage is 'confused,' with the pronoun 'it' capable of referring either to the part or the instruction, and with the word 'want' perhaps signifying 'lack,' perhaps 'desire.' The description, he concludes, is 'perfectly ambiguous' since it 'could mean one thing or its exact opposite.'[9] I take this syntactical confusion and ambiguity to be a function of Cook's bafflement over the meaning of what he had witnessed. Was it a religious rite or a sexual initiation, or both? Was the young girl already well versed in the mysteries of sex and therefore not in need of the women's instruction, or was she in fact too young and innocent to 'want,' in the sense of 'desire,' the intercourse?

Cook did not know, and the thing he most counted on to obtain information – namely, keen observation of empirical details – was hopelessly inadequate to provide answers in this regard.

This passage points to a recurring motif in the journals. Again and again, Cook reveals the limits of his comprehension, the gap between seeing and knowing. Coming upon the burial of a man killed by the crew for grabbing at a musket, Cook describes the actions of the natives in precise detail but concludes, 'if it is a Religious ceremoney we may not be able to understand it, for the Misteries of most Religions are very dark and not easily understud even by those who profess them' (47). Later, upon leaving Tahiti after a three-month stay, Cook notes that upon the whole his crewmen and the natives have been on good terms, but that 'some few differences have now and than happend, owing partly to the want of rightly understanding one another' (64). While in London before embarking on his third voyage, Cook found himself at a dinner seated next to Boswell and 'candidly confessed to [Boswell] that he and his companions who visited the South Sea Islands could not be certain of any information they got, or supposed they got, except as to objects falling under the observation of the senses; their knowledge of the language was so imperfect they required the aid of their senses, and anything which they learnt about religion, government, or traditions might be quite erroneous.'[10] Despite these caveats, however, the sheer volume and detail of the sensory perceptions recorded in Cook's journals gave readers the illusion that its author was conveying a comprehensive knowledge about all aspects of Polynesian culture and contributed to a growing belief in the ability of British travellers to become authorities on distant cultures and societies – a belief supporting the construction of Edward Said's 'Orientalist,' who uses the nexus of knowledge and power to exert control over native others who are (in more ways than one) his subjects.[11]

Along with questioning the possibility of ever knowing another culture, Cook's journals also, in various ways, question the assumed distinction between civilization and savagery. At times, the blurring of the distinction is humorous, as when a group of New Zealanders need to be given assurances that Cook and his crew are not cannibals and will not eat them (74). In other instances the destabilizing of the distinction has more sinister implications, as when Cook describes an incident in which an unidentified crewman snuck into his clerk's cabin while the clerk was drunk and proceeded to 'cut off a part of both his Ears as he lay asleep in his bed' (131). Since this incident comes only shortly after Cook's description of some native women whose faces, arms, and legs had been cut and scarified as a sign of mourning for husbands killed by their enemies, we cannot help but compare the two passages and wonder who are the real savages. This question recurs in a variety of contexts. When he first arrives in Tahiti, Cook promulgates a set of rules for his crew to follow, the first being '*To endeavor by every fair means to cultivate a friendship with the Natives and to treat them with all imaginable humanity*' (40). Notwithstanding this instruction, some crew members are overly quick to brandish their muskets and shoot at the natives, with little or no provocation. After witnessing the seemingly motiveless killing of a native by one of his sentries, Cook condemns its barbarity: 'I who was present and on the Spot saw not the least cause for the commiting of such an outrage and was astonished beyond Measure at the inhumanity of the act' (389).

An earlier incident, which occurs soon after the Englishmen arrive in New Zealand for the first time, complicates Cook's relationship with these acts of inhumanity. Seeing two canoes filled with Maoris, Cook wants to take them on board *The Endeavour* and establish a connection with them. When they rebuff Tupia's request that they approach the Britons and instead start moving away, Cook 'order'd a Musquet to be fire'd over their heads thinking that this would either make them surrender or jump over board.' Instead, believing themselves attacked, the Maoris prepare to retaliate, whereupon Cook and his men were, as he puts it, 'obliged' to fire on them, killing two or three of them and wounding another (71). Aware that his actions may be subject to condemnation, Cook launches into a strangely ambivalent explanation for his actions, part *mea culpa* and part self-justification:

> I am aware that most humane men who have not experienced things of this nature will cencure my conduct in fireing upon the people in this boat nor do I my self think that the reason I had for seizing upon her will att all justify me, and had I thought that they would have made the least resistance I would not have come near them, but as they did I was not to stand still and suffer either my self or those that were with me to be knocked on the head. (72)

An earlier version of this account – Cook frequently revised passages in his journal – is more damning. To his exculpatory statement that he could not very well do nothing and thus allow the natives to physically assault his crew – a clear plea of self-defence – he adds the words, 'or else retire and let them g[o] off in triumph and this last they would of Course have attribited to their own bravery and our timorousness' (72n). According to this version, the natives had to die, not necessarily to preserve the lives and limbs of the explorers (an understandable enough if lamentable circumstance) but simply to save face: to ensure that the Englishmen maintained a fearsome reputation in the eyes of the Maoris and were not viewed by them as cowards.

As this latter example suggests, there was more to Cook than the picture that has come down to us via tradition and lore, of the humane and compassionate sea captain (the antithesis of a figure like Captain Bligh) who encouraged his men to treat others well and who denounced acts of violence or cruelty towards the natives.[12] He apparently was also someone who was not averse to complicity in such acts when the occasion demanded – that is, when he perceived a challenge to his authority or a threat to the power relationship he enjoyed vis-à-vis the natives. What is particularly interesting in his case is his willingness to participate not only in European forms of barbarism but also in native forms, as we see most strikingly in his fascination with cannibalism, whose existence he not only seems intent on proving to his sceptical countrymen back home, but towards which he displays a rather unseemly eagerness to examine up close and even involve himself in, to some extent. During his first trip to New Zealand, he in effect oversees the staging of a cannibalistic scene when a crewman brings a bone from the forearm of a dead native to a group of Maoris and 'to shew us that they had eat the flesh they bit a[nd] naw'd the bone and draw'd it thro' their mouth and this in such a manner as plainly shew'd that the flesh to them was a dainty bit' (102–3). Later, during his second voyage to the region, he and some of his officers go on land 'to amuse themselves among the Natives' and come upon a scene of cannibalism (319). Cook describes the particulars of the scene with what one is tempted to characterize as a certain relish as he presents a detailed picture of 'the mangled head or rather the remains of it' belonging to 'a youth who had lately been killed' and whose heart was 'stuck upon a forked stick and fixed to the head of their largest Canoe' (319).

After the natives come aboard the explorers' vessel, bringing the mangled head with them, Cook expresses the obligatory 'horror' and 'indignation against these Canibals' that we find repeatedly in both fictional and non-fictional depictions of them (one thinks immediately of Robinson Crusoe's revulsion at the sight of the cannibal feast), but he very quickly sets his disgust aside and 'ordered a piece of the flesh to be broiled and brought on the quarter deck where one of these Canibals eat it with a seeming good relish before the whole ships Company' (319). Ironies abound in this description – not only do many of the ship's crew vomit with revulsion at the spectacle, but Cook's native informant, Oediddee, a native from Bora Bora, is profoundly sickened as well, being '[so] struck with horror at the sight that [he] wept and scolded by turns,' and refused to allow the cannibals to come near him. Only Cook emerges in this passage with a clear head and an apparent ability to accept the Maoris' cannibalism in a matter-of-fact way. He adopts a detached, almost 'scientific' tone, explaining his actions in terms of his desire to be 'an eye wittness to a fact which many people had their doubts about,' but one cannot help suspecting that this re-staging of the scene of cannibalism shows some kind of fascination with the activity that Cook refuses to admit – perhaps a flirtation with the idea of 'going native': a phenomenon seen to be threatening and dangerous (though no doubt also titillating) for explorers of the day, several of whom succumbed to its siren song. George Vason, for example, went to the Pacific islands as a missionary in 1797 and stayed on after the departure of his evangelical group to become a beachcomber, adopting the Tongan language and dress, marrying a young Polynesian woman, and (in his own words) 'enter[ing], with the utmost eagerness, into every pleasure and entertainment of the natives.'[13] There was, of course, never any danger of Cook's 'going native' in this sense, but his willingness to be part of a cannibal feast (even if he himself refrained from tasting human flesh) suggests an outlook that under other circumstances might have led him onto less conventional paths.

Or, perhaps, it was not so much a flirtation with 'going native' as a recognition that the very concept of 'native' as one identified with savagery and opposed to civilization needed to be rethought. This seems to be what is suggested in Cook's remarks immediately following the 'cannibal feast aboard ship' episode, where he insists that the New Zealand cannibals 'are certainly in a state of civilization, their behaviour to us has been Manly and Mild, shewing allways a readiness to oblige us; they have some arts a mong them which they execute with great judgement and unweared patience' (320). He even offers a partial extenuation of their consumption of human flesh: 'This custom of eating their enimies slain in battle (for I firmly believe they eat the flesh of no others) has undoubtedly been handed down to them from the earliest times and we know that it is not an easy matter to break a nation of its ancient customs' (320). Although he adds, 'especially if that nation is void of all religious principles as I believe the new zealanders in general are,' the main effect of his comments is to suggest an affinity between the British and the Maoris, neither group finding it an 'easy matter' to 'break' with time-honoured customs and traditions.

Elsewhere, Cook goes even further in calling into question the respective body of values that presumably separates the European explorer from the natives he encounters. He draws a decidedly mixed picture of European culture and its exemplars as disseminators not only of enlightenment and civilization but also of decadence and disease – as a force capable of debauching the innocent: that is, those indigenous tribes who, as Cook observes about the aboriginal natives of New Holland (Australia), are 'far more happier than we Europeans' thanks to their intimate connection with nature, their ignorance of Western materialism, and the fact that '[t]hey live in a Tranquillity which is not disturb'd by the Inequality of Condition' (174). At various points in his travels, he worries that ships from Europe have brought venereal disease to the Pacific Islands, and, in the spring of 1773, he is appalled by the changes he sees in New Zealand since his last visit. In the past, the Maoris of Queen Charlotte Sound had always seemed more sexually modest than the Indians from other areas, but they now had become more venal and indecent in their sexual activity, with the males regularly prostituting their own wives and daughters in exchange for objects deemed of value. Cook characterizes this change as less a reflection of the Indians' own immorality as 'the concequences of a commerce with Europeans and what is still more to our Shame[,] civilized Christians,' and he proceeds to offer a blistering condemnation of what results from these cultural encounters:

> [W]e debauch their Morals already too prone to vice and we interduce among them wants and perhaps diseases which they never before knew and which serves only to disturb that happy tranquillity they and their fore Fathers had injoy'd. If any one denies the truth of this assertion let him tell me what the Natives of the whole extent of America have gained by the commerce they have had with Europeans. (277)

Not surprisingly, none of Cook's contemporaries were able to come up with a satisfactory answer to this question. But that did not matter, because in the end Cook's doubts about the presumed nobility of the expeditionary voyages became erased through subsumption into a popular and oft-repeated story of nationalistic pride and success. Readers of his journals could accept his portrayal of innocent, happy natives – the idea of the 'Noble Savage' had, after all, come into its own by the latter part of the century and, for all the ideological contradictions it posed, could be accommodated within a Christian and imperial framework. But the picture of the European barbarian who leaves havoc and destruction in his wake was another matter, and could not be entertained except in the most abstract and conventional terms, as it translated into the trope of Old World corruption versus New World innocence. Both the rhetorical conventions of the travel narrative genre and the momentum of the larger project of geographical discovery functioned to bury Cook's doubts and aspersions beneath a resoundingly celebratory and patriotic message symbolized by the flags, inscriptions, and memorials that he continually had his crew members erect on pieces of barren land in the middle of the Pacific Ocean, and by the 'scene of discovery' that he ostentatiously staged on an island near the Endeavour Strait off the mainland of Australia – perhaps in a bid to offset his failure to locate and map the places he was commissioned by the British Admiralty to find.

Although Cook sees a number of natives on the island, it never seems to occur to him that the land is already settled and hence not available to be claimed by 'first possessors' for the British crown. Only the presence of other *Europeans* cancels out the vacancy of the land and renders it off limits to conquest. Thus, while conceding that he 'can make no new discovery' on the western side of New Holland since it has already been 'discovered' by Dutch navigators, he declares himself confident that 'the Eastern Coast from the Latitude of 38° South down to this place ... was never seen or visited by any European before us.' And so, notwithstanding the fact that he had already 'in the Name of His Majesty taken posession of several places upon this coast,' he stages another ceremony of claim-staking, the most ambitious yet in terms of the amount of territory encompassed:

> I now once more hoisted English Coulers and in the Name of His Majesty King George the Third took posession of the whole Eastern Coast from the above Latitude down to this place by the name of *New South Wales*, together with all the Bays, Harbours Rivers and Islands situate upon the said coast, after which we fired three Volleys of small Arms which were Answerd by the like number from the Ship. (170–1)

Despite the formality of the ceremony, his claim in and of itself could not have meant much – Cook and his crew were in no position to take (and certainly not to secure) possession of such a sweeping amount of territory, and the fact that he kept repeating the ritual in miniature throughout the expedition, sometimes (as he admits above) claiming the same piece of land again and again, serves as a tacit recognition of the impossibility of any single, definitive, and enduring act of possession-taking. The latter becomes an infinitely reproducible set of gestures that no doubt played well to the patriotic audience back home and that was perhaps designed to placate a government hungry for territorial expansion and disappointed in Cook's failure to provide any information about the much-hyped (but of course non-existent) 'Southern Continent' and the Northwest Passage, which up to six months before his death Cook was promising to tackle again.

In lieu of the discoveries that the British Admiralty wanted him to make, Cook keeps on obsessively erecting inscriptions and memorials to his crew's presence on even the most isolated and barren pieces of land. He places messages in bottles, carves them into trees, and has his carpenter fashion wooden plaques of commemoration. In one instance, he extracts a promise from a Maori elder that he would never pull down an inscription post set up near his home by patiently explaining to him, through Tupia's mediation, that 'we were come to set up a mark upon the Island in order to shew to any ship that might put into this place that we had been here before' (108). One can only imagine what must have being going through the old Maori's mind on being told that Cook was marking out this territory for himself. There is at times something distinctly farcical in Cook's compulsive inscription-planting on whatever ground he comes upon, much like a dog peeing on every tree he passes in order to mark his turf – hence the difficulty of not giving a perverse interpretation to his remark that 'after we had done our business upon the Island we return'd on board' (109). It can be argued that the less land Cook actually discovers, the more memorials he erects to himself as a discoverer. It is not surprising, then, that a century later Mark Twain would bring a satiric perspective to this memorial-building and its relation to British colonialism. In letters written for the Sacramento *Union* from Kealakekua Bay, Hawaii, the place of Cook's death, Twain mocks the different kinds of monuments and memorials associated with Cook, including a decayed and rather pitiful-looking cocoanut tree stump completely sheathed with discoloured sheets of copper, each with a barely legible inscription from a succession of ships that visited the island after Cook's death.[14] It is this ugly, over-inscribed stump, bearing the faint imprint of the cannon of Cook's ship underneath the scribbled names of subsequent navigators, that constitutes the only monument to Cook's death. Clearly we are to infer that the memorials heralding Britain's supremacy on the high seas will have a similarly ignominious end.

Despite Twain's sardonically demystifying judgment about Cook's death – 'Plain unvarnished history takes the romance out of Captain Cook's assassination, and renders a deliberate verdict of justifiable homicide'[15] – his murder on 14 February 1779 at the hands of Hawaiian natives remains to this day immersed in mystery and myth, having in recent years become the subject of several heated controversies among anthropologists in particular, exemplified by the dispute between Marshall Sahlins and Gananath Obeyesekere.[16] Did the Hawaiians take Cook to be the fertility god Lono or did they see him merely as a powerful chief? Did Cook come to view himself as Lono? Was he caught up in a kingship ritual that centred around a dying god or was he killed for having desecrated a sacred shrine by using its pilings for firewood? Like the Point Venus scene, the death of Cook forces us to ask questions for which there are no clearcut, definitive answers. For my purposes, what is particularly telling about his death is that an expedition launched under the banner of science ended up firmly ensconced in the realm of myth: a scenario evoking Adorno and Horkheimer's argument that 'enlightenment with every step becomes more deeply engulfed in mythology.'[17] Of course, it might be argued that science and myth were inextricably bound together from the very beginning of the voyage. After all, searching for the Great Southern Continent was about as 'scientific' a pursuit as looking for the Loch Ness monster. Still, the acclaim Cook received after his return from the first two voyages was based on the perception that the expedition had been a success both as a voyage of discovery and as a project for advancing science – not to mention a major contribution to Britain's expansionist enterprise.

Bruce's travels also contributed to nineteenth-century efforts to open up hitherto little-known parts of the world to British penetration, despite the fact that the acclaim awaiting Bruce upon his return to Britain was short lived during his lifetime, thanks in part to the aspersions cast on the truthfulness of his account by Dr Johnson and to the general prejudice Englishmen felt towards 'North Britons' – magnified when the latter were as supremely self-confident and boastful as Bruce. The latter, unlike Cook, offers his readers a very specific and well-defined 'scene of discovery,' staged as a highly elaborate and emotionally wrought event. Catching sight of what he takes to be the aim of his quest after an arduous journey, he throws off his shoes and '[h]alf dressed ... ran down the hill towards the little island of green sods, which was about two hundred yards distant; ... I after this came to the island of green turf, which was in the form of an altar, apparently the work of art, and I stood in rapture over the principal fountain which rises in the middle of it' (3:597). Once again, however, the discovery scenario proves to be both problematic and satirical, in ways that cast a sceptical light on the entire expeditionary enterprise. The first and most obvious problem is that Bruce has mistaken his destination – what he finds is not in fact the source of the Nile, as he believes, but rather of its tributary, the Blue Nile. Then there is the fact that, although he sees himself as a triumphant explorer whose discovery amounts to 'a trophy in which [he] could have no competitor, for the honour of [his] country' (3:641), it is difficult to credit his self-proclaimed achievement as a 'discovery' given that he was almost literally taken by the hand and brought to the place by his native guide, Woldo. Indeed, Bruce needed the help of many Arabs in the area (for safe passage, advice, directions) in order to finally make his 'discovery' – which was, after all, a 'discovery' only to him, since it was a thoroughly familiar and quite unremarkable sight to all the inhabitants of Geesh and its environs. As he himself acknowledges at one point, 'All Abyssinia knows the head of the Nile' (3:511).[18]

The insignificant, indeed illusory, nature of his 'discovery' from the natives' point of view is underscored when Woldo refuses to direct him to the riverhead unless he is first given the sash that Bruce promised to bestow upon him later as a reward for showing him the spot. When Bruce asks another Arab why Woldo insists on being given the sash immediately, he is told that Woldo is sure Bruce will be so 'dissatisfied' with what he sees that he will refuse to give the boy the sash afterward: for, as the Arab explains to Bruce, 'you seemed to think little of the cataract at Goutto, and of all the fine rivers and churches which he had had shewn you; except the head of the Nile shall be finer than all these, when, in reality, it will be just like another river' (3:595). That, of course, is exactly the point. The body of water that Woldo brings him to *is* like any other river, not to mention the fact that it *is* an 'other' river – that is, other than the Nile. In fact, when Bruce first looks down upon it from the mountain village of Geesh, it seems 'only a brook that had scarcely water to turn a mill' (3:590). The only thing that confers special status on it is the myth of discovery that Bruce brings with him to Africa – the idea of finding, claiming, taking possession for oneself and one's country, and in the process transforming oneself into the hero of one's own epic adventure. This is precisely what Bruce does as he portrays his transport at the thought of his singular achievement:

> It is easier to guess than to describe the situation of my mind at that moment – standing in that spot which had baffled the genius, industry, and inquiry of both ancients and moderns, for the course of near three thousand years. Kings had attempted this discovery at the head of armies, and each expedition was distinguished from the last, only by the difference of the numbers which had perished, and agreed alone in the disappointment which had uniformly, and without exception, followed them all ... Though a mere private Briton, I triumphed here, in my own mind, over kings and their armies ... (3:597)

Yet the heroic pose he strikes only serves to highlight the *mock*-heroic quality of the episode. Bruce's delusions of grandeur trigger in the reader's mind memories of the very different way he is viewed by many of the Arabs of the region. One chieftain, for example, refuses him passage through his territory, explaining, '"You white people are all effeminate; you are like so many women; you are not fit for going into a province where all is war, and inhabited by men, warriors from their cradle"' (3:512).

Interestingly, Mungo Park's journey to Africa in 1799 was linked to Cook's through the participation of Joseph Banks, the naturalist who played so large a role on *The Endeavour*'s first voyage and who by the 1790s was president of the Royal Society and head of the Association for Promoting the Discovery of the Interior Districts of Africa (also known as the African Association). It was the latter group that sponsored Park's trip, in hopes of obtaining more information about the Niger River and important cities near it, especially Timbuktu, which would be particularly useful for developing British commercial and trading interests in the region. Inextricably connected with this aim was a civilizing mission: to impart 'the conveniences of civil life, the benefits of the mechanical and manufacturing arts, the attainments of science, the energies of the cultivated, and the elevation of the human character ... to nations hitherto consigned to hopeless barbarism and uniform contempt.'[19] Park infused into this mission an evangelical impulse out of which emerged a fervent desire that the minds of the Africans 'could be softened and civilized, by the mild and benevolent spirit of Christianity,' and which produced the hope that 'a short and easy introduction to Christianity, such as found in some of the catechisms for children, elegantly printed in Arabic, and distributed on different parts of the Coast, might have a wondrous effect' on the Arab traders (77; 275).

Like other explorers in the latter half of the century, Park viewed not only religion but also science as an authorizing discourse and an important component of human progress. Having studied medicine at Edinburgh University (where he also indulged his passion as an amateur botanist) and as a member of the Linnaean Society, Park brought to his African travels a zeal to advance the science of geography and what we would now call ethnography through accounts of the social customs of various African and Moorish tribes and through maps of the areas he traversed. His published *Travels* included an appendix of 'Geographical Illustrations' put together by Major James Rennell, which featured a sophisticated map of North Africa and numerous details of the towns and villages encountered by Park on his journey. Rennell's geographical supplement filled obvious gaps in Park's own account, since the pocket sextant and magnetic compasses he took with him on his trek (like other scientific instruments previously looked at) proved useless in charting his course or determining his bearings, so that he was continually getting lost or having to rely on directions from local inhabitants that more often than not proved inaccurate. Far from aiding his progress, these instruments place him in life-threatening situations. When he points to his pocket compass on the ground and begs for its return after being stripped and examined for booty by a group of bandits, 'one of the banditti, thinking I was about to take it up, cocked his musket and swore that he would lay me dead upon the spot, if I presumed to put my hand upon it' (226). Earlier in his journey, he is stripped of one of the compasses at the command of the Moorish chieftain Ali, who views the instrument with 'superstitious curiosity.' When asked by Ali why its needle always points to the Great Desert, Park tells him that it is guiding him toward his mother beyond the desert, thus in effect transforming the compass from a scientific instrument into a magical fetish (151). And indeed, it is only as the latter that the compass is shown to have any power or utility whatever over the course of Park's journey. Similarly, to the extent that Park possesses any authority in the eyes of the natives, it is based not on his scientific paraphernalia or skills (his one attempt to minister medically to an ailing native ends in utter failure) but on the belief that he is a supernatural creature, the fears and superstitions of the 'Negroes' in one instance having 'dressed [him] in the flowing robes of a tremendous spirit' (127).

Critics have tended to characterize Park as a 'sentimental' traveller, ignoring the significant satirical dimensions of his journey and of the role he plays in it.[20] If we were to imagine Swift's Gulliver in Africa, he would surely resemble Park to a remarkable degree, from the projector's zeal as a specimen-gatherer to the absurd positions he finds himself in vis-à-vis his various captors. Like Gulliver under the microscope in Brobdingnag, a strange creature requiring minute analysis and definition, Park is subjected to the intensely curious gaze of the Moors, who 'inspected every part of my apparel, searched my pockets, and obliged me to unbutton my waistcoat, and display the whiteness of my skin; they even counted my toes and fingers, as if they doubted whether I was in truth a human being' (147). When he is later sheltered in a hut with a wild hog, the Moors 'assembled in crowds to behold' him and obliged him to demonstrate how he puts on and takes off his clothes: 'All this was to be repeated to every succeeding visitor ... and in this manner I was employed, dressing and undressing, buttoning and unbuttoning, from noon to night' (148–9). In a particularly Swiftian moment, a party of Moorish women come to his hut 'and gave me plainly to understand that the object of their visit was to ascertain, by actual inspection, whether the rite of circumcision extended to the Nazarenes (Christians,) as well as to the followers of Mahomet' (154). Park manages to extricate himself from this situation only by turning it to jest and offering to satisfy the curiosity of the youngest and handsomest of the women alone. When he asks a Moor for water to slake his thirst, the man, 'fearing that his bucket might be polluted' by Christian lips, dashes the water into a small trough from which three cows were drinking, whereupon Park gives us the image of himself 'kneeling down' and 'thrust[ing his] head between two of the cows' in order to drink his share (163). Episodes like these (and there are many) underscore the extent to which Park plays a mock-heroic as well as sentimental role in his travels.

Park does offer us a conventional 'scene of discovery,' clearly intended to convey a sense of gravity and moment – but here again, the scene suggests hints of farce and anticlimax. With his horse limping beside him like a lame Rosinante as he approaches the Niger, he passes through several villages where he is made 'the subject of much merriment to the Bambarrans,' whose sarcastic comments and questions convince him that 'the very slaves were ashamed to be seen in my company' (193). He finally arrives at the town of Sego, where, as he is 'anxiously looking around for the river,' a native calls out to him '*geo affilli,* (see the water)':

> And looking forwards, I saw with infinite pleasure the great object of my mission; the long sought for, majestic Niger, glittering to the morning sun, as broad as the Thames at Westminster, and flowing slowly *to the eastward.* I hastened to the brink, and, having drank of the water, lifted up my fervent thanks in prayer, to the Great Ruler of all things, for having thus far crowned my endeavours with success. (194)

One wonders exactly what criteria Park was using to measure this 'success' by, especially since he admits that he already knew the answer to the question of which way the Niger flowed through his 'frequent inquiries during [his] progress' as well as through the information collected by his predecessor, Major Houghton (194). In this sense, his journey has been superfluous, yielding information already known to him and failing to answer any of the other questions for which he had been sent abroad. He certainly did not 'discover' the river in the sense of being the first to come upon it, since it was only too familiar and mundane a part of the natives' landscape, one for which they had their own name – 'Joliba, or *the great water*' (193). Moreover, the moment of great (non-)discovery undergoes further comic deflation via the reaction of the natives. Park admits he can understand the Sego king's suspicions about the true motive of his journey because his guide had expressed similar ones: 'When he was told, that I had come from a great distance, and through many dangers, to behold the Joliba river, [he] naturally inquired, if there were no rivers in my own country, and whether one river was not like another' (199).

Soon after his 'discovery,' Park decides to terminate his mission, foreseeing 'inevitable destruction in attempting to proceed to the eastward,' and begins his return to Gambia (207). But despite the relatively meagre results of his expedition and his failure to reach Timbuktu, Park returned to a triumphant reception in Britain, with the publication of his African adventures attaining great popular success. On the strength of this success, Banks sent a memorandum to the Committee of Privy Council for Trades and Plantations in June 1799, laying out an ambitious plan for the dispatch of a military expedition to acquire for the British Crown, 'either by Conquest or by Treaty,' the entire coast of Africa from Arguin to Sierra Leone, the Niger basin, and all inland areas offering access to the river (16). The result, as he saw it, would be the establishment of a government-controlled trading company that would not only oversee the commercial development of Africa but also convert the Negroes to the Christian religion 'by inculcating in their rough minds the mild morality which is engrafted on the tenets of our faith' (16). Here, Banks is obviously building on Park's evangelical enthusiasm, as well as on those sections of the *Travels* that suggest the potential richness of a land still largely unclaimed by the European powers. At the same time, there is much that Banks conveniently ignores in Park's narrative, such as its testimony to the existence of a civilization flourishing perfectly well without the help of either European technological advances or Christian faith. In Sego, the capital of Bambarra, for example, Park describes how 'the crowded population, and the cultivated state of the surrounding country, formed altogether a prospect of civilization and magnificence, which [he] little expected to find in the bosom of Africa,' and he observes that the nearby town of Kabba is 'situated in the midst of a beautiful and highly cultivated country, bearing a greater resemblance to the centre of England, than to what I should have supposed had been the middle of Africa' (195; 201).

Overlooking these and other aspects of Park's published travels, the latter's patrons and readers helped transform the Gulliveresque object of ridicule into a heroic adventurer, and simultaneously helped absorb what was in many ways an ill-conceived and ill-fated expedition into the mythology of global travel and the romance of empire, modifying it into an inspiration and blueprint for subsequent travellers bent on opening up the African continent to commercial profit and religious proselytizing. In Banks's memorandum, we can see how Park's travels contributed to the mad scramble among explorers in the following decades to be the first to reach (and return alive from) Timbuktu, becoming a springboard for the frenzied stampede throughout the nineteenth century to colonize and divvy up Africa for personal power and profit.

There is a revealingly macabre and grisly postscript to this expedition in Park's return to Africa at the head of a military expedition seven years later, in January 1805, where he died under circumstances only a little less mysterious and mythically resonant than those in which Cook met his end. Neither the weapons nor the bibles the Britons carried with them could bring about a successful resolution to their quest to complete the exploration of the Niger River. On the contrary, all forty-five members of the expedition eventually died, most from disease brought on by the advent of the rainy season. According to Jeremy Swift, an earlier editor of Park's *Travels,* as his companions began dying off in ever greater numbers 'Park pressed on with a mounting sense of despair. Once on the river he threw away his normal caution and diplomacy, and instead of negotiating his passage with the chiefs he was soon shooting on anyone who came near him.'[21] In support of this view Swift cites the claim of Major Gordon Laing – who himself later travelled to Timbuktu only to be killed before his return – that Park 'left behind an appalling and savage image of Europeans, with his boat floating hopelessly down the Niger to certain doom, refusing all contacts and firing indiscriminately at anyone who came near him' (x). Here, Park has left behind the land of Gulliver and entered the world of Werner Herzog, resembling nothing so much as a mad Klaus Kinski, drifting down the Amazon in ever-narrowing circles of destruction as Aguirre, the Wrath of God. He has entered the world of Conrad, where a lengthy sojourn on the Dark Continent culminates in a house adorned with shrunken heads and a command to 'Exterminate the Brutes.'

The ideology that promoted expeditions of 'discovery' in the eighteenth century could transform such figures of madness and futility into images of triumph and conquest – could use them to construct a picture of progress capable of inspiring future explorers to continue the project. The compelling force of the ideology allowed people to overlook individual failures and to ignore the lessons either explicitly taught, or implicitly conveyed as subtext, in travel narratives of the period: that discovery is impossible, or at least meaningful only insofar as it signifies a *re*-discovery of lands already familiar to and inhabited by others; that the line separating civilization and barbarism becomes ever more tenuous the lengthier the interactions between Europeans and natives; that genuine knowledge of other cultures is often an illusion, paradoxically receding ever more into the distance as empirical details pile up higher and higher; that science can at any moment collapse back into myth; that the heroic is only a small step away from the mock-heroic; and that moments of triumphant conquest can easily turn into farce, especially when viewed through the ironizing gaze of the native.

Let me conclude by returning to Defoe's definition of projects, about which he says that ''tis certainly true of 'em all, even as the Projectors propose; that according to the old tale, If so many Eggs are hatch'd, there will be so many Chickens, and those Chickens may lay so many Eggs more, and those Eggs produce so many Chickens more, and so on' (8:40). The image can be seen as one of promise and hope (immediate failure giving birth to future success), but it can also be read in a darker way, as a self-perpetuating process that is out of control – that, once started, cannot be stopped regardless of how negative its results. The image can further suggest, perhaps more ominously, a time in the future when the chickens will come home to roost. Sadly, it is these latter meanings that have proved most applicable to the great ongoing project of redrawing the 'great Map of the World' through colonial adventures that gained a special impetus and cachet in the eighteenth century and that continue, in ever more grotesque and delusional forms, to this day.

## NOTES

1 Burke, *The Correspondence of Edmund Burke*, ed. George H. Guttridge, 10 vols. (Cambridge: Cambridge University Press, 1958–78), 3:351.

2 Defoe, *Political and Economic Writings of Daniel Defoe*, 8 vols., ed. W.R. Owens and P.N. Furbank (London: Pickering & Chatto, 2000), 8:40; hereafter cited in the text.

3 For Anson's adventures, see Richard Walter and Benjamin Robins, *A Voyage round the World in the Years MDCCXL, I, II, III, IV*, ed. Glyndwr Williams (London: Oxford University Press, 1974). See also William Dampier, *A New Voyage round the World* (London, 1697); the 6th ed. of 1717 is reprinted in *Dampier's Voyages*, 2 vols., ed. John Masefield (London: E. Grant Richards, 1906). For further discussion and background, see Philip Edwards, *The Story of the Voyage: Sea-Narratives in Eighteenth-Century England* (Cambridge: Cambridge University Press, 1994); and Anna Neill, *British Discovery Literature and the Rise of Global Commerce* (Houndmills, Basingstoke: Palgrave, 2002).

4 *The Journals of Captain Cook*, ed. Philip Edwards (London: Penguin, 1999), 11; hereafter cited in the text.

5 Jonathan Lamb, *Preserving the Self in the South Seas, 1680–1840* (Chicago and London: University of Chicago Press, 2001), 88.

6 See *Exploration and Exchange: A South Sea Anthology, 1680–1900*, ed. Jonathan Lamb, Vanessa Smith, and Nicholas Thomas (Chicago and London: University of Chicago Press, 2000), 99.

7 Lamb, *Preserving the Self in the South Seas*, 88.

8 James Bruce, *Travels to Discover the Source of the Nile, in the Years 1768, 1769, 1770, 1771, 1772, and 1773*, 5 vols. (London, 1790 [repr. Gregg International Publishers, 1972]), 1: n.p.; all other parts of the *Travels* are paginated and are hereafter cited in the text.

9 Neil Rennie, 'The Point Venus "Scene," Tahiti, 14 May 1769,' in *The Global Eighteenth Century*, ed. Felicity A. Nussbaum (Baltimore and London: Johns Hopkins University Press, 2003), 250.

10 James Boswell, *Boswell: The Ominous Years, 1774–1776*, ed. Charles Ryskamp and Frederick A. Pottle (New York: McGraw-Hill, 1963), 341.

11 See Edward W. Said, *Orientalism* (New York: Vintage Books, 1979), esp. Introduction and chap. 1.

12 In this connection, see Neill, *British Discovery Literature*, chap. 6.

13 *Exploration and Exchange*, 167.

14 Ibid., 285–6.

15 Ibid., 281.

16 See Marshall Sahlins, *Islands of History* (Chicago: University of Chicago Press, 1985), and *How 'Natives' Think: About Captain Cook, for Example* (Chicago: University of Chicago Press, 1995); and Gananath Obeyesekere, *The Apotheosis of Captain Cook: European Mythmaking in the Pacific* (Princeton: Princeton University Press, 1992).

17 Max Horkheimer and Theodor W. Adorno, *Dialectic of Enlightenment*, trans. John Cumming (1944 [repr. New York: Continuum, 1987]), 11–12.

18 For an insightful discussion of Bruce's travels that touches upon some of the issues I deal with here, see Lora Edmister Geriguis, 'Bows without Arrows: The Role of "Native Agency" in the Travel Narratives of Daniel Defoe and Other English Texts, 1668–1790' (unpublished PhD thesis, University of California, Riverside, 1997), 270–308.

19 Mungo Park, *Travels in the Interior Districts of Africa*, ed. Kate Ferguson Marsters (Durham and London: Duke University Press, 2000), 9; hereafter cited in the text.

20 E.g., according to Mary Louise Pratt, Park presents himself as 'a sentimental hero' and his *Travels* 'richly exemplifies the eruption of the sentimental mode into European narrative of the contact zone at the end of the eighteenth century'; see *Imperial Eyes: Travel Writing and Transculturation* (London and New York: Routledge, 1992), 75, 74–5.

21 Jeremy Swift, 'Preface' to Mungo Park, *Travels into the Interior of Africa* (London: Eland Books, 1983), ix; hereafter cited in the text.

chapter fifteen

# Forging Figures of Invention in Eighteenth-Century Britain

SARAH TINDAL KAREEM

When a scholar mentions Rudolf Erich Raspe (1736–94) in the same sentence as projection, the writer is usually invoking that quintessentially twentieth-century usage of the term, referring to the process of projecting unconscious fears or fantasies onto others.[1] Raspe's anonymous authorship of *Baron Munchausen's Narrative of his Marvellous Travels* (1785), in which the eponymous baron recounts outlandishly tall tales, has been interpreted as a 'projection' in precisely this sense. In fact, Raspe's twentieth-century biographer, John Carswell, discusses Raspe's composition of the narrative in a chapter of his 1950 biography entitled 'Baron Munchausen: The Projection.' The consensus of scholars who have analysed Raspe's composition of *Munchausen,* most of whom are historians or biographers, is that Munchausen, the fabulous, fairy-tale success story, was the projected fantasy of Raspe, the frustrated failure, who wrote *Munchausen* after his career as an earth scientist fizzled out. While Munchausen, as Carswell puts it, 'lives in his actions alone,' Raspe's 'grandiose plans, hatched with naive enthusiasm, go one after another to make up a dismal chronicle of failure.'[2] For the last fifty years or so, that has been more or less the last word on Raspe, Munchausen, and projection.

There is another story to tell about the projections of Rudolf Raspe. Raspe was a man who seems to have channelled Daniel Defoe's 'projecting spirit.'[3] Raspe's 'brain hummed with projects,' in Carswell's words,[4] and other scholars claim 'the story of Raspe's life reveals a mind fertile for hatching schemes and plots for advancement of all kinds.'[5] Raspe imagined enterprising and wide-ranging 'projects' that ranged from the creative to the downright illegal. 'Imagined' is the operative word here; as one scholar notes of the proposed schemes for advancement of the science of mineralogy laid out in Raspe's final geological publication, 'in spite of such an elaborate planning, none of these projects ever materialized.'[6]

After this final geological publication in 1776, Raspe did not give up on projecting; on the contrary, 'he spent more time among his crucibles and peddling schemes of exploitation.'[7] Carswell's image of his subject 'among his crucibles' pertains to another meaning of 'projection,' which in alchemy referred to the casting of powdered philosopher's stone, known as the 'powder of projection,' into a crucible in order to transmute a base metal into gold or silver.[8] Invoking the alchemical meaning of projection might seem as potentially anachronistic as invoking the psychoanalytic meaning of projection in an eighteenth-century context, but it was an anachronism that was already prevalent in Raspe's lifetime, to his cost. Raspe negotiated his identity as a would-be professional geologist in a world in which such a creature did not yet properly exist, and in which the charge of alchemy, precisely because it was discredited as an explanatory system, was readily hurled at him. The process of alchemical projection is evoked literally and metaphorically by many of the transmutations Raspe was so adept at pulling off, in his manipulation both of different discourses and of his own persona.

The use of the twentieth-century discourse of unconscious projection to analyse Raspe has tended to dredge up familiar clichés about the drudgery of 'science' and the fantasy of 'art.' As Richard Hamblyn cautions us in his essay on the production and consumption of geology in the eighteenth century, the ease with which we talk of 'scientists' and 'artists' today can lead us astray in discussions of eighteenth-century discourses. He notes that 'the shaping of those discourses into what we now recognize, however uncomfortably, as two distinct intellectual cultures is a story that has yet to be adequately told.'[9] As someone who both had a hand in shaping, and was shaped by, these shifting discourses, Raspe lurks in the margins of this untold story. The early-modern discourse of alchemical projection and the eighteenth-century discourse of entrepreneurial projection coax him onto the main page, where he proves, simultaneously, a compelling character in, and charismatic narrator of, a story that recounts how the discursive conventions we have come to associate with art and science emerged in the latter half of the eighteenth century.[10]

Given that Raspe is perhaps an unfamiliar figure to many readers, it may be helpful to outline his career. After receiving his master's degree from the University of Göttingen in 1760, he was appointed to a junior clerkship in the manuscript department of the Royal Library at Hanover. Though it was a modest position, Raspe was perfectly placed to exploit his knack for unearthing forgotten works. In 1763 he published his first and only major geological work, *An Introduction to the Natural History of the Terrestrial Sphere*, which advanced ideas influenced by Robert Hooke. Contemporary readers credited Raspe with having 'discovered' – a term often placed in quotation marks in scholarly accounts of this event – a largely forgotten 1705 work of Hooke that he had found in the library.[11] Raspe's main innovation was to introduce 'islands born from the sea' as a new example of the terrestrial change Hooke had argued was caused by volcanic eruptions. Raspe's publication was an immediate sensation and thrust him into the limelight, and into correspondence with the prevailing international elite of natural philosophers, including Benjamin Franklin and Sir William Hamilton in Naples.[12] His subsequent publication of a neglected pile of Leibniz's unpublished papers he had found languishing in the library solidified his reputation. He supplemented his geological publications with curatorial work and translations of literary works, including Macpherson's *Ossian*, as well as original poetic compositions.[13] His newfound prestige won him the curatorship of a collection of antique medals and gems at Hesse-Cassel in 1767 and, in 1769, fellowship in the Royal Society. It was a serendipitous period for Raspe, during which he seemed to have the Midas touch, gaining the nickname from his friend J.G. von Herder of 'glücklicher Finder,' or fortunate finder.[14]

His giddy success, however, did not last. Facing mounting debts due to his extravagant lifestyle, Raspe began embezzling from the collection entrusted to his care.[15] In 1775 Raspe fled when his actions were discovered. A manhunt ensued. The warrant launched by the Landgrave of Hesse-Cassel noted that Raspe 'walks, in general, hastily,'[16] and he was described in a police advertisement as 'a man with red hair, who usually appeared in a scarlet dress embroidered with gold.'[17] Cutting such an eye-catching figure, Raspe was arrested within four days, but allegedly allowed to escape two days later by an official moved by the charismatic Raspe's confession of his plight. This time Raspe made it out of the country, never to return. He headed for London, but his disgrace followed him, and by the end of 1775 he attained the dubious honour of being the only Fellow expelled from the Royal Society 'due to the infamy of his character.'[18]

In the late 1770s and 1780s, Raspe eked out a living in Britain, seeking to remain afloat as a publishing scholar by selling subscriptions to the ambitious scholarly projects he always had on the back-burner. But increasingly he lived on the brink of destitution; his friend Horace Walpole had to pay his tailor.[19] Without the means to fund scholarly expeditions and publications, he turned to more practical enterprises, working as an assayer and possibly as an industrial spy for the famed Cornwall mining industrialist Matthew Boulton, and authoring a well-received descriptive catalogue of the collection of casts of gems formed by James Tassie.[20] It was during this peripatetic period that Raspe wrote and anonymously published *Munchausen*. In the last five years of his life, Raspe worked mainly as a mineralogical surveyor. He finally succumbed to spotted fever while prospecting in Ireland in 1794.[21]

As his biography attests, Raspe was involved in numerous projects, which he strived to keep distinct. As Carswell notes, 'His surviving correspondence shows how carefully he divided his life into compartments, concealing from one set of friends the very existence of another and possibly conflicting coterie with which he was also dealing.'[22] The fleeting, colourful references to Raspe that occasionally crop up in recent histories of eighteenth-century earth science often make him the subject of a lone sentence that ends in a long list of his miscellaneous occupations and preoccupations.[23] The suspicion invited by Raspe's miscellaneous projects in his own time illustrates how distinctions were increasingly being made between aesthetic, industrial, and scientific invention. Raspe walked a fine line between 'the sciences' and 'the arts' – as understood both then and now – and various amateur and professional identities. His negotiation of these discourses and *personae* are instructive in revealing the emergence, in the late eighteenth century, of disciplinary and occupational divisions that persist in our own times.

Raspe was active during a period in which the terms and concepts 'art' and 'science' were in transition, and he himself had a hand – and a stake – in reshaping these discourses. In the eighteenth century, the term science in its widest sense refers to knowledge in general.[24] Science was also used more specifically to refer to knowledge committed to demonstrative certainty, as opposed to practical arts that proceed by trial and error.[25] Raspe was one of several eighteenth-century natural philosophers ('natural philosopher' was the usual term for 'scientist' until the 1830s) who sought to redefine science as knowledge based on practical findings rather than speculative thinking.

In the latter half of the eighteenth century, earth science was established more firmly as a discipline of empirical enquiry, and demarcated from both other sciences and other forms of discourse, such as theology, the study of man, and the arts.[26] Roy Porter argues that in the period 1775–1815, 'scientific study of the Earth became *geology*.'[27] Hamblyn pinpoints this development even more precisely, arguing that the 'basalt controversy' of the 1760s and 1770s – a debate in which only about five men were major players, Raspe being one of them – 'created the discourse of modern geology from out of a kind of growing connoisseurship of the earth.'[28] Such an assessment relocates the apparently marginal figure of Raspe at the epicentre of seismic shifts in the intellectual landscape of the Enlightenment.

Newly emerging distinctions in the persona of the natural philosopher accompanied the subdivision of natural philosophy. Early-eighteenth-century natural philosophers were not specialists, but were often equally devoted to other interests, such as antiquities or Celtic studies.[29] The tendency towards specialization intensified in the 1790s with the appearance of distinctively scientific periodicals and societies, such as the Geological Society of London. While geology scholars rarely put their knowledge to practical use, those whose occupations demanded an extensive knowledge of the earth, including coal prospectors, mineral assayers, and quarrymen, sought to parlay their professional know-how as 'scientific knowledge.'[30] However, the word 'scientist' was not coined until 1835, and Porter argues that it is not until that same year that 'an Englishman of any rank makes a regular and secure living out of his skills at geological fieldwork.'[31]

Sub-fields such as volcanology erupted within the burgeoning new field of geology and acquired distinctive cultural identities. Perhaps because research meant travelling far afield – at least before the volcanic origin of rocks within the British isles was discovered – British volcanists tended to be wealthy aristocrats who fancied themselves cosmopolitan men of the world.[32] Raspe, by contrast, was middle class. Hamblyn observes that 'what really marked off Ferber, Raspe, Desmarest, and Whitehurst from the contemporary liberal spectator were their social and ideological positions: their outlooks had not been formed through close contact with the time-worn classical schedules of aristocratic travel; their interests were subsequently free from the gauze of aristocratic learning.'[33]

In the remainder of this essay I will be suggesting that in negotiating the shifting grounds upon which earth science was founded as a discipline during the eighteenth century, Raspe found himself caught between several competing *personae* of the natural philosopher, some of which were on the decline, some of which were newly emergent.

## Technologist and Alchemist

Raspe's dedication to severing what we understand as science from more humanistic pursuits is apparent in his geological writings. The prefaces to Raspe's scholarly works read as manifestos calling for the establishment of geology and mineralogy as disciplines restricted to empirical observation at a time when investigation of the earth was still an area of speculation inextricable from theology and human history,[34] as well as from the agendas of tourism and classical learning.[35] As an expression of his commitment to practicality, Raspe aligns himself with new technology and distances himself from devices tainted by association with alchemy and mysticism. Raspe presents himself as a dispenser of new methods of mineralogical analysis based on chemical classification, and of cutting-edge European mining techniques and technologies. In his preface to his translation of Born's mineralogical account of Transylvania and Hungary (1777), Raspe disparages old-fashioned methods, intoning that 'fire, crucibles, retorts, alembics, acids, and touchstones, are insufficient'[36] and extolling 'the modern invention of the magnetic needle,' which has 'made the art of surveying under ground actually more certain and more easy then it was before.'[37] Raspe disdains 'imposing quacks' and the 'divining rod,' writing, 'We are fully convinced that the latter has never answered any purpose but that of making dupes.'[38] At the same time, however, Raspe's language mystifies technology; in the preface to another work by Born on a new process of amalgamating ores (1791), Raspe speaks of his desire of 'putting the British Publick fully in possession of [the process's] secrets and mysteries.'[39]

Hamblyn argues that, by the mid-eighteenth century, 'the figure of the virtuoso had joined those of the connoisseur and the dilettante as representatives of learned amateurship of the worst (and thereby most profitably amusing) kinds.'[40] As an aspiring professional rather than a dabbling amateur, Raspe did not face mockery in any of these guises. More gallingly, given his self-presentation as a technical innovator, he was caricatured as an old-school magus. His self-styled futuristic mineralogical projects were often regarded as projections of an alchemical nature. In the eighteenth century, mineralogy as a field was tainted by association with alchemy. Minerals had to be stripped of their supposed alchemical properties and emblematic significance in order to be made fit objects for empirical study. Porter suggests that until the eighteenth century, 'the wonderful – e.g. combustible pyrites – the useful and the beautiful commanded more attention than synoptic interpretation of the mineral kingdom.'[41] The prominent role played by pyrite (known as fool's gold) in one of the most infamous incidents of Raspe's mining career may have encouraged subsequent generations to associate him with alchemy, as may have his seeming obsession with finding gold in the most unlikely places. In 1776 Raspe wrote to Captain Cook begging to accompany him on his voyage to Tahiti, asserting his conviction that 'Gold-dust' would be 'constantly found mixed with' volcanic sand washed up on beaches, and hence that his 'experiments will be rewarded with some Gold, and perhaps with some Diamonds.'[42] His request was declined. Although Raspe expressed the hard-headed interest in gold of an entrepreneur, not a mystic, statements like these no doubt fuelled the image of Raspe as a deluded, gold-obsessed monomaniac – an image that was already synonymous with the alchemist in the mid-eighteenth century.

The incident that sealed Raspe's reputation as a con man, and perhaps a conjurer, occurred in 1789, during his brief tenure as a prospector for the wealthy Scottish landowner Sir John Sinclair of Ulbster. This episode illustrates the kind of combustion that occurred when Raspe the professional man of science and Raspe the playful fiction-maker showed up together. Raspe may not have known how to turn base metals into gold, but he made fools of his employers with panache. The nineteenth-century accounts of what transpired during Raspe's month-long residence at Sir John's home, Thurso Castle, vary in the details, but the main thrust of the allegations posthumously levelled at Raspe is that he had pyrite – fool's gold – shipped in from Cornwall and buried on Sinclair's estate, which he then promptly dug up and presented to Sinclair as evidence of the riches his property contained, upon which he was given the funds and the go-ahead by Sinclair to commence mining in earnest.[43] 'But, in time the bubble burst,' as one nineteenth-century account of the events put it: Raspe vanished, leaving Sinclair out of pocket and feeling foolish.[44] Whether or not Raspe defrauded Sir John as alleged, he certainly would have been perfectly placed to exploit the aristocrat's naivety concerning the practical 'art' of mining.

The various nineteenth-century accounts of Raspe's stint at Thurso Castle do not quite add up. Although Raspe's deception of Sinclair is portrayed as 'fraud,' Raspe and Sir John are also depicted parting on good terms, and Sir John is cited as fondly recalling 'the amusement which the mineralogist had given them while a guest in their house.'[45] Was the bubble Raspe blew a get-rich-quick scheme or a whimsical prank? The question displays the historically particular criteria by which we categorize invention. In the nineteenth-century accounts, Raspe plays the role at once of performance artist and con artist, his inventions evading classification as either fraud or hoax. While today we tend to distinguish between artistic and fraudulent invention and associate each with different *personae*, the eighteenth-century projector's inventions admit no such distinctions.

The significance of these nineteenth-century accounts of Raspe's time at Thurso Castle lies less in their truth content and more the way in which the various authors – antiquarians, literary biographers, and novelists – mythologize the events. In their renditions, Raspe emerges as a romantic trickster, a foreigner who arrives in a small community, conjures up gold fever – and apparently gold – out of nowhere, and then mysteriously vanishes. At Thurso Raspe proved he could turn iron into gold, if only fleetingly, in the imaginations of his wishful clients. The accounts of Raspe's sojourn there, just five years before his death, retroactively bathe all his previous activities and incarnations in a mysterious light. Thus, in the nineteenth-century antiquary Robert Hunt's profile, Raspe is portrayed as an 'old conjurer working with all sorts of flames about him,' even while working in Matthew Boulton's industrial labs, which would seem to be the very forging ground of the persona of the modern industrial scientist.[46]

The depiction of Raspe as a magus receives its most full-blown treatment in Sir Walter Scott's fictionalization of Raspe's deception of Sinclair in his novel *The Antiquary* (1816). It is difficult to know whether Raspe, the champion of empiricist rigour and new technologies and ridiculer of quacks, would have been amused or mortified by Scott's immortalization of him in the ludicrous person of an alchemist. In Scott's novel the Raspe character goes by the dastardly sounding name of Herman Dousterswivel. Scott draws Dousterswivel's contours swiftly and mercilessly, as the opposite of Goethe's Faust; Scott's alchemist is not a long-suffering hero, but a comic villain. The various epithets by which the already meaningfully named Dousterswivel – he dowses for gold – is known effectively delineate his character. He is referred to most frequently as 'the adept,' a noun defined by Johnson as meaning 'he that is completely skilled in all the secrets of his art,' and a title, according to the *OED*, that was 'assumed by *alchemists* that professed to *have attained* the great secret.'[47] Other choice specimens in the long list of sneering monikers conferred upon Dousterswivel include 'the treasure-finder,' 'the charlatan,' 'the exorcist,' and the 'mendacious quack.'[48] He is also figured as a marvellous creature, hailed as a 'fairy' and a 'fiend'[49] and referred to by the Antiquary, with grim irony, as 'the swart spirit of the mine' who presides over 'the kingdom of subterranean darkness and airy hope.'[50] The characterization of Dousterswivel as more 'of a conjurer than a clerk' echoes Descartes's characterization of the German Jesuit polymath Athanasius Kircher as 'more charlatan than scholar.'[51] It is an echo that would have pained Raspe, who holds Kircher in special contempt in his scholarly works.[52] In marked contrast, the eminent Scottish geologist James Hutton is respectfully referred to as 'the learned doctor' and 'one of our first chemists.'[53]

Dousterswivel's character as a charlatan is fleshed out in descriptions of him as someone who 'mixed the terms of science with a strange jargon of mysticism,' which is actually an accurate description of Raspe's rhetorical style, which tends to disavow mysticism in one breath, only to evoke it in the next. We are told that when Dousterswivel 'is among fools and womankind, he exhibits himself as a perfect charlatan – talks of the *magisterium* – of sympathies and antipathies – of the cabal – of the divining rod.'[54] Such jargon establishes Dousterswivel's dubious provenance and his talk of the *magisterium*, or philosopher's stone, identifies him as an alchemist. Scott's portrait is a character assassination, and might have been the death of Raspe if he had not already met with his inauspicious end in 1794, the year in which *The Antiquary* is set.

## Scholar and Miner

While he openly embraced technology and disdained alchemy, Raspe trod more carefully in negotiating the *personae* of scholar and miner. By the late eighteenth century, the British community of natural philosophers had been transformed from an elite of Oxbridge-educated landowners to a more heterogeneous, broadly middle-class community. But although the world of late-eighteenth-century geology was more open than it had been a hundred years earlier, there remained a gap between affluent gentlemen-scholars and 'professional surveyors and mines consultants.'[55] Neither the scholarly nor the professional routes were easy. As Rachel Laudan argues, in Britain both the amateur geologist and the professional working with minerals were stymied by a lack of institutional support or opportunities for publishing their findings.[56] On the Continent, by contrast, 'a surprising number of [state industries] fostered the study of mineralogy and geology.'[57] The lack of institutional support in Britain was coupled with the pressure to specialize. Professionals working with minerals were specialists of necessity, having little leisure time for literary pursuits.[58] Among the more affluent ranks of the metropolitan intelligentsia, it became a mantra that 'to reduce Geology to a system demands a total devotion of time.'[59]

It was in this context that Raspe attempted to carve out a career for himself as a professional geologist in Britain. Given his limited funds, and the time he devoted to literary and other extra-mineralogical pursuits, it proved impossible to stay afloat. Had Raspe remained in Germany, he might have attended a mining academy. In Britain his research was dependent on the whims of patrons, or wealthier explorers who might allow him to tag along on their expeditions.

Faced with no institutional support for mining and a dearth of patrons, Raspe chose to capitalize on the lack of collaboration between the mining industry and scholarly geology in Britain by billing himself as the single-handed instigator of that dialogue. He styled himself as the intermediary for a cross-cultural exchange readily accessible by the advance purchase of one of his many volumes outlining the theory and practice of mineralogy in terms both the scholar and the miner could appreciate. Raspe's self-righteous rhetoric in his works, by which he depicts himself as the savior of modern mineralogy, is undercut by his pleas to his readers to place advance orders with his publishing house for future planned publications, admitting that he will not be able to complete them otherwise. The world of British mineralogy was not yet a hospitable environment for scientific popularizers. While Raspe may have been prescient in foreseeing the role popular publications might play in creating a discourse in which philosophers and practical men could make contact, and a culture in which science and industry could complement one another, his insight may have been more lucratively rewarded in Germany; in Britain, a lone proselytizer could only do so much without institutional support.[60]

The importance of bridging the gulf between scholar and miner is emphasized in all of Raspe's scholarly works. The preface to his final geological translation laments that miners 'have been scarce consulted at all, by philosophers who attempted to create and to dream mountains and worlds, and systems of mountains and worlds.'[61] In his preface to Swedish mineralogist Johann Ferber's natural history of Italy, Raspe lauds Ferber for 'improving science, for the scholar and the miner' and emphasizes that Ferber gathered information from 'the learned and the unlearned, from Philosophers, Chemists, Miners, and Smelters.'[62] Raspe's own status is indicated by the vague phrase 'friend to knowledge' and defined as a middleman between scholar and miner. Raspe deferentially addresses scholars as 'men of liberal minds, who consider not science as a job,' beseeching them not to regard 'the active friends of knowledge' – such as himself, presumably – 'as encroaching intruders upon their literary reputation.'[63] In the preface to another geological translation, Raspe argues that mineralogy, in its present state, does not 'fully answer the expectation of miners, of natural philosophers, and of friends to science,' and that 'to benefit the miner,' systems of classification 'should be explained in his own technical or provincial language, which is generally neglected; and to satisfy the natural philosopher, they should be established upon evident principles of chemistry.'[64] Such statements indicate Raspe's savvy grasp of the different linguistic registers spoken by his potential audiences, and of the importance of fluency in both languages. Raspe's words also indicate his debt to the work of Abraham Gottlob Werner (1749–1817), the German mineralogist who followed Linnaeus in his descriptive methods, proposing the identification of minerals according to external characteristics, because this was pragmatic for miners, and the classification of them by chemical characteristics.[65] But while Werner was a member of the mineralogical establishment in Germany, in Britain such statements were polemical, and paved the way for Raspe's advertisement of his own forthcoming works, which, of course, were designed precisely to fill the gap he had defined in the market.

Raspe assures his readership that his synopses of mineralogical works will offer 'something more satisfactory ... than hitherto has been given to the public.' He justifies his publication of these abstracts (as opposed to original works) by arguing that the dissemination of observations that render a 'true picture of the subterranean kingdom' will 'prove of nearly the same advantage to miners and philosophers.'[66] He explicitly addresses both audience sectors, playing upon the ambitions of one and the anxieties of the other, proposing to include a 'sketch of a new system of mineralogy for miners ... with a constant reference to the technical and provincial language of the miners and smelters,'[67] while also expressing the hope that his work will receive 'generous support' from 'gentlemen,' whom he hints may otherwise be duped by 'adventurous, unprincipled miners.'[68] This hint to his gentlemen-readers that they might be exploited without the benefit of his expertise plays upon confusion concerning the transferability of skills in the new arena of science. As Hamblyn points out, while, as connoisseurs, Raspe's gentlemen-readers were able to distinguish between true and forged works of art, the question remained open as to whether this cultivated discernment might just as easily detect the 'pretended Vesuvian precious stones' Raspe warned of.[69] As Hamblyn spells out the broader question at stake: 'Were the new kinds of scientific knowledge to be legislated in anything like the way in which the knowledge of taste had continually been?'[70]

## Connoisseur and Hack

Raspe was not only a beneficiary, but also a casualty of this doubt about how knowledge was to be legislated, for as neither amateur connoisseur nor professional geologist, he faced castigation as a hack. Although Raspe talked the talk of the professional geologist, disdaining the speculative whimsy of the dabbler and extolling empirical observation, he did not live up to his words. His diverse publications (including an essay on oil painting, an original romance ballad, and an edition of Leibniz) proved he was not a specialist. Moreover, even his scholarly magnum opus, his natural history of the terrestrial sphere, did not contain any original empirical observations, but was largely a digest of examples of terrestrial phenomena gleaned and inferred from classical sources, including mythology. For example, Raspe's case for the existence of floating islands depends almost entirely upon an ambiguous reference to a floating island in Homer's *Odyssey*. Despite the fact that many interpreters, by Raspe's own admission, do not interpret the reference to denote a literally floating island, Raspe justifies his opinion with recourse, not to empirical observation, but to the wisdom of the ancients. He argues that 'the point of view of the more ancient interpreters of Homer, and even the more ancient poets, since they are earlier in time,' should 'carry more weight.'[71] He thus defers to received opinion, validating the interpretation of Apollonius Rhodius and Aristarchus that the island truly 'moved back and forth,' as 'they are far more ancient that Dionysius and Eustathius.' For good measure, he also notes that 'a man who is worth many men to me, that delightful emulator of Homer, the glory of English letters, Alexander Pope,' also interprets the island to be literally floating, and cites a few lines from Pope's *Odyssey* to prove his point.[72] Raspe's professed stripped-down empiricism reads more like literary criticism.

Raspe's breadth was regarded as a mark of desperation, not a badge of his liberal humanism; to be a gentleman of wide tastes was to be a 'connoisseur,' but to be a professional of broad tastes was to be a 'hack.' There were cautionary examples of what might happen to professionals who spread themselves too thin. In 1759 the *Critical Review* identified the scientific popularizer 'Sir' John Hill as one casualty of the age, arguing that though Hill had been 'born a naturalist,' such was the 'trifling spirit of the age' that he had been compelled to become a hack, 'an universal scholar, a critic, polemic, casuist, physician, metaphysician, politician, essayist, poet, novelist, and astronomer, degenerating from, perhaps the best botanist Britain ever saw to one of the worst of her writers.'[73] The line between connoisseur and hack was a fine one, and one was more likely to emerge on the wrong side of that line by publishing in diverse fields. Hence, it is not surprising that throughout his life Raspe never publicly acknowledged himself as the author of *Munchausen*, his most flagrantly unscholarly work.

## 'Faithful Historian' and 'Idle Story Teller'

As one reads Raspe's geological and mineralogical works, a dichotomy emerges between that kind of philosophical work of which he approves and with which he wishes to be associated, and that which he values less highly and from which he distances himself. 'Diligence' is the key attribute of the first; 'invention' is the hallmark of the latter. Raspe suggests that he was spurred to publish his natural history of the terrestrial sphere by noticing that 'some ideas remain which my diligence could put in order,' and aligns himself with the philosophers who 'have favored the painstaking examination of nature to the momentary invention of some hypothesis.'[74]

Raspe repeatedly emphasizes the importance of vigilance in holding the worlds of fact and fiction at arm's length from one another. In the preface to his translation of Ferber's *Travels through Italy*, he notes disapprovingly how the volcano Vesuvius has 'struck the fancy, and engaged the curiosity, of philosophers and travellers.' Raspe maintains that such accounts are 'far from being satisfactory to Naturalists' because 'they indulged themselves either in sentimental and poetical flights, or in marvellous tales of wonders performed, or rather not performed.'[75]

As these examples suggest, science's identification with sober, self-effacing facticity was not yet taken for granted in the latter half of the eighteenth century. Rather, it had to be advertised and performed, a delicate balancing act in which Raspe's impassioned invective against speculation often shatters the very tone of scholarly sobriety it was intended to validate. Raspe insists that he himself is a paragon of empiricism, observing that his 'natural history of the earth' is 'planned less for speculation than for usefull practical Science. As it is established upon facts only I am conscious that it will agree with Nature better than the many philosophical Romances, which some learned men have hatched in their closets.'[76] And, indeed, at the opening of the very natural history to which Raspe refers here, he inveighs against going 'beyond the experience of our senses,' calling for the 'careful and diligent investigation of the surface of the globe.'

However, even when issuing stern warnings against the intrusion of the poetic imagination into science, Raspe cannot refrain from showing off his knowledge of the classics:

> What lies hidden deeper near the interior part of our globe is more safely left to the imagination of poets and men of more creative wit who may celebrate in letters, as it pleases them, the kingdoms of Klimmius, Tartarus, the Platonic states and the haunts of Aeolus, Vulcan and Pluto until such time as we are allowed to penetrate the greater depths and examine what was previously hid.[77]

Raspe may have followed his own advice to explore the interior of the globe through the literary imagination. In one of the most famous episodes of *Munchausen*, the Baron descends to the centre of Mount Etna, where he fraternizes with Vulcan and his beautiful wife Venus, an episode first appearing in November 1786.[78] Only the 1785 edition of *Munchausen* can be entirely and confidently attributed to Raspe; thus, it remains an open question as to whether this episode was penned by Raspe, 'safely' pursuing his speculations within another genre, or by another hand.

Whether or not he wrote this particular episode, Raspe made efforts to segregate his 'careful and diligent investigations' from his 'marvelous tales of wonders performed, or rather not performed,' confining each to different discourses and different *personae.* Raspe's double life as, on the one hand, publicity-hungry would-be empiricist geologist guarding his field against the intrusion of 'momentary invention' and, on the other, anonymous author of a fiction that was hailed on its first appearance as a work in which 'the marvellous has never been carried to a more whimsical and ludicrous extent,'[79] illustrates John Bender's point that in the mid eighteenth century, 'manifest fictionality underwrote the factuality of science.'[80] That is to say, Raspe's authorship of such an exaggeratedly fictional work confirms the expulsion of such inappropriate 'invention' from the realm of natural philosophy.

The most striking example of Raspe's coexistence within these contiguous discourses is his appearance in consecutive issues of the journal *The Critical Review,* in the first issue making a public appearance as Rudolf Raspe, diligent taxonomist, in the second skulking anonymously behind *Munchausen.* The first notice, appearing in the June 1786 'Monthly Catalogue' under the heading 'Miscellaneous,' reviews Raspe's catalogue of Tassie's casts of gems. While the review hails Tassie as 'an ingenious artist,' Raspe's catalogue of Tassie's mouldings seems more like a Linnaean taxonomy than an exhibition catalogue, laboriously indexing Tassie's 'twelve thousand different articles of ancient and modern engravings' according to precise subdivisions. As the review records, under the heading of ancient engravings are arranged

> Egyptian hieroglyphics, sacred animals, divinities, and priests; with Basilidian, Gnostic, and other abraxas, talismans, and amulets, oriental and barbarous engravings. Next to these are Greek and Roman originals, copies, and modern imitations. This class is distinguished into four subdivisions; the last of which is the historical age; and this is again subdivided into the history of Carthage, Greece, and Rome, with historical subjects unknown.

While the reviewer's highest praise is reserved for Tassie, Raspe receives commendation for his 'diligence,' approbation he would no doubt have appreciated.[81]

In the very next issue of the *Critical Review*'s 'Monthly Catalogue,' also under the heading 'Miscellaneous,' a notice appears of the anonymous *Gulliver Revived; or, the Singular Travels, Campaigns, Voyages, and Adventures of Baron Munikhouson.* Even as the public Raspe plays the diligent taxonomist, the anonymous Raspe masquerades as 'Gulliver Revived,' the title placing Raspe in a satiric, Swiftian relationship to natural philosophy. This was already the third edition of *Munchausen*; as the reviewer remarks, the *Critical Review* had given a favourable – and in fact the only – review of the first edition, which was little more than a pamphlet. This third edition, as the review notes, was 'much enlarged, and adorned with characteristic plates' and 'embellishements.' Like Tassie's Collection, Munchausen's marvellous narrative was already in the public domain, but it was not quite the same one; while the *Critical Review* notes Tassie's collection had received the 'general approbation' of polite society, Munchausen's narratives had garnered 'public attention' of an implicitly more vulgar sort.[82]

Raspe had compartmentalized his projects so effectively that his contemporaries, and even the succeeding generation, initially failed to make the connection between the diligent scholarly works to which Raspe signed his name and the anonymous, exuberantly silly *Munchausen* narratives. The earliest reference to Raspe's authorship of *Munchausen* thus discovered is from the 1830s, and it is not until the 1850s that it is revealed publicly.[83] For the eminent nineteenth-century geologist Charles Lyell, Raspe already had the status of an undiscovered gem. In the first edition of his *Principles of Geology* (1830), Lyell hailed Raspe's 'luminous exposition' of Hooke's ideas, expressing incredulity that Raspe's work 'should, for more than [a] half-century, have excited so little interest.'[84] Scientifically, 'Lyell considered Raspe's views as far beyond those of his age.'[85] But Raspe was revealed to be an even more rare specimen than Lyell originally suspected. In February 1830 the geologist John Hawkins wrote to Lyell: 'You will hardly believe that the Editor of the works of Leibnitz & Tassie's Catalogue of Gems, was the author of Baron Munchausen's surprising Adventures.'[86] Raspe, creator of the infamous relater of belief-defying narratives, was now the central character in a belief-defying narrative. Lyell's belated reply a little more than two years later must have satisfied Hawkins by expressing the incredulity he encouraged in his original disclosure. Lyell repeats the unlikely sequence of publications, as if to confirm the unthinkable. He writes, 'If I understand rightly your obliging letter of Feby 8th, 1830, your friend Raspe was Editor of the works of Leibnitz, Tassie's catalogue of gems, & Baron Munchausen's adventures ... Is it generally known that he wrote Baron M's adventures?'[87]

The incredulity expressed by Lyell suggests that by the nineteenth century, divisions between the cultural domains of art and science, and between polite and popular literature, were well established. As Stephen J. Weininger observes, 'Establishing a clear demarcation between fact and fiction was a matter of urgency in the nineteenth century.'[88] In both Hawkins's and Lyell's letters, what is deemed belief-defying is that the set containing the works of Leibniz and Tassie's catalogue of gems might also contain Baron Munchausen's adventures. That is, what evokes wonder is that any list of an author's *oeuvre* that begins 'Leibniz, Tassie ...' might end '... Munchausen.' For Hawkins and Lyell, the presence of Munchausen in such a catalogue does not add up. Like naturalists confronted with a singular specimen, Hawkins and Lyell find themselves confronted with a text that seems 'detached from all the other species of that genus to which it belongs.'[89] The wonder is not so much that Raspe wrote *Munchausen*; the wonder is that the *same* person produced all three texts. Not only is the inventor of Munchausen reincarnated as a belief-defying narrative; the diligent taxonomist is reconstituted as a Borgesian taxonomy that calls into question the line between fact and fiction.[90]

It is ironic that for all that he fastidiously distanced his fiction from his scholarly works, many of Raspe's scholarly writings nonetheless read today like the stuff of nonsense-poetry, or, indeed, episodes from *Mun-*

*chausen*, with their whimsical histories of floating islands, wishful visions of gold washing up on beaches, and impudent proposals that the Royal Society publish 'unphilosophical' rather than 'Philosophical Transactions.'[91] Raspe himself expresses the inextricability of profundity and silliness when he remarks, 'It is the fate of most truths that they are intertwined with whatever nonsense writers write and are not brought to light except by stage.'[92] Raspe understood his vocation as the disentangling of 'truth' from 'nonsense,' trusting that his faithful performance of this task would ensure, as he puts it, that he, as a 'faithful historian,' would not 'be looked upon as an idle story teller.'[93] But it was not to be: the tendency of readers to conflate Raspe with Munchausen, whose name literally defines idle storytelling, has virtually ensured that if Raspe is remembered at all, it is indeed as an idle storyteller.[94]

## Genius and 'Slave of the Book Sellers'

The incredulity that the revelation of Raspe's authorship of *Munchausen* elicited perhaps explains why he hid his authorship in the first place in a cultural climate increasingly intolerant of such intellectual promiscuity.[95] His authorship of *Munchausen* posthumously gained him the kind of fame he spent his life trying to outrun: notoriety. The nineteenth-century revelation of the true range of Raspe's abilities did not mean he was hailed as a genius; rather, he was held in some contempt, as a jack-of-all-trades, master of none. As Hawkins wrote to Lyell a few months after his first letter disclosing Raspe's authorship of *Munchausen*, 'Raspe was a man of extraordinary talents & information. I have never met with a man whose knowledge was so extensive or diversified, but he was a man of very irregular habits, which frequently brought him into great difficulties & made him the slave of the book sellers.'[96] Raspe here is commemorated not for his specialization or empirical rigour, but for the 'extensive' and 'diversified' nature of his knowledge. As Raspe's biographer puts it, trying to be kind, Raspe had 'a genius which was nothing if not versatile.'[97] This was not a conception of genius that was prized in the late eighteenth or early nineteenth centuries. Peter Galison has argued that the ideal of the eighteenth-century scientific genius was a 'genial author' whose most prized feature was his discernment in selecting the most excellent examples of nature to showcase for his readership.[98] The image conjured up by Hawkins of Raspe as 'the slave of the book sellers' is diametrically opposed to that of the genial, intervening genius.

Galison argues that by the early nineteenth century, the paradigm of the intervening genius had given way to a new model of objectivity in which 'it was only through superhuman self-restraint that the author could aspire to let nature "speak for itself."'[99] Rather than imposing his own vision upon nature, the nineteenth-century genius stepped into the shadows, allowing his discoveries to steal the limelight. But this new ideology of objectivity came too late for Raspe, whose advocacy (in words, if not deeds) of positivism might have been greeted more enthusiastically by nineteenth-century readers had it not been joined at the hip to his miscellaneous productions. As it was, the newly emerging contours of Raspe's persona seem to have been too messy for nineteenth-century encyclopedias, which took it upon themselves to clean up his act, expunging Raspe's scientific publications from the record, particularly as he became relegated to a mere footnote. Under the entry 'Munchausen' he was defined for posterity as an 'ingenious writer'; 'a German antiquary'; or 'a man of letters,' who 'was engaged in London in literary pursuits.'[100]

## Scientist and Artist

Poised at the watershed between two paradigms of the scientific persona – the intervening genius and the unobtrusive observer – Raspe failed to embody either during his lifetime: he chose to 'select and idealize,' not in his philosophical works, but in his fiction, where he thought it appropriate, and his self-effacing 'diligence' was only appreciated posthumously by Lyell and a few others.[101] Yet just as we seem to be moving inexorably towards the conclusion that Raspe's fiction was the fantasy bubbling up through the cracks in his scientific diligence, we should recall that the apparent conflict between mineralogy and *Munchausen* does not seem so clear-cut or inevitable when contextualized in terms of the changing relationship between the terms 'art' and 'science' during the eighteenth century.

Just as Raspe was caught between paradigms of the scientific persona, so too was he caught between competing understandings of art and science. On the one hand, in his geological works, Raspe was clearly involved in the modernizing realignment of the science/art distinction as a fact/fiction distinction, hence the importance of publishing *Munchausen* anonymously to avoid imputations that his scientific works were contaminated with the taint of 'invention.' But when it suited him Raspe chose to invoke an older understanding of the relationship between the arts and the sciences. For example, in the opening to his *Critical Essay on Oil-Painting*, Raspe distinguishes those 'most useful' 'arts' that take '*posi-*

*tive* truths and realities' as their 'objects and their pursuits,' and 'immediate *positive* advantage' as 'their result and reward,' from those 'sciences' that 'deal only in *ideal beings* in intellectual or sentimental objects, and in possibilities, which produce no other advantage but that of pleasing our fancy and of flushing our conceited pride.'[102] In the *Critical Essay*, Raspe understands 'sciences' as theoretical models and 'arts' as practical enterprises, ranging from 'polite' arts such as painting to 'mechanical' arts such as mining. Thus, Raspe will refer, for example, to 'the art of mining and its collateral science of mineralogy.'[103] It seems significant that Raspe, the self-styled modernizer, preserves the Aristotelian understanding of the distinction between arts and sciences rather than invoking the 'modern' understanding of these terms that was emerging during this period.[104] One can see why he might do this; when viewed in terms of the ancient distinction between arts and sciences, the apparent conflicts within Raspe's *oeuvre* disappear; all of the projects of which Raspe approves, and in which he took part, are arts, an expansive term, which, like projects as conceived by Defoe, encompasses all manner of supremely practical and self-interested endeavours not defined by preconceived distinctions between 'polite' and 'vulgar' culture.

*Munchausen* might seem to be the exception; on the one hand, it arguably pleases the fancy with its depiction of an 'ideal being' who transcends the limits of the possible. But on the other hand, *Munchausen* is an eminently pragmatic project; not only is the character of the Baron a can-do man, but the work is a can-do narrative concept and publishing venture: a fast write and a faster read, and infinitely adaptable – as its publishing history proves – to the topical issues of the day.[105] For Raspe, there may have been no anguished conflict between mining and *Munchausen*; he may have regarded both as 'arts' offering 'immediate positive advantage' as 'their result and reward.'

The trajectory of Raspe's reputation over the past two centuries has been a varied one, from renown to infamy, to posthumous modest scholarly acclaim, to languishing in the shadow of Munchausen, his larger-than-life creation. Raspe's projects both patrol and leap over the boundaries between competing ideologies, institutions, and *personae* of invention in eighteenth-century Britain. Raspe cast his persona into the crucible and we may marvel at the transformations that he wrought and that were wrought upon him: from technologist to alchemist, from scholar to miner, from connoisseur to hack, from genius to 'slave of the booksellers,' from 'faithful historian' to 'idle storyteller,' from scientist to artist, from notoriety to anonymity. Raspe's projects are significant, not merely as footnotes in the biography of an eccentric rogue, but as testaments to the strange history and alchemy of invention.

## NOTES

1 I adopt the German spelling of Raspe's first name here; his first name is more often given as 'Rudolph' in English works. Recent scholarship has established that Raspe was born in 1736, not 1737, as was previously thought. See Andrea Linnebach, ed., *Der Münchausen-Autor Rudolf Erich Raspe, Wissenschaft – Kunst – Abenteuer* (Kassel: euregioverlag, 2005), 28. On projection, see *OED* definition 9.b for projection, *n.* The entry cites uses of the term 'projection' in early translations of works by both Jung and Freud.

2 John Carswell, *The Romantic Rogue, Being the Singular Life and Adventures of Rudolph Erich Raspe Creator of Baron Munchausen* (New York: E.P. Dutton & Co., 1950), 185–8. See also Introduction, Rudolph Erich Raspe, *An Introduction to the Natural History of the Terrestrial Sphere Principally Concerning New Islands Born from the Sea and Hooke's Hypothesis of the Earth on the Origin of Mountains and Petrified Bodies to Be Further Established from Accurate Descriptions and Observations,* ed. Audrey Notvik Iverson and Albert V. Carozzi (New York: Hafner Publishing Co., 1970), lxxiii, and Haraldur Sigurdsson, *Melting the Earth: The History of Ideas on Volcanic Eruptions* (New York: Oxford University Press, 1999), 145.

3 See Headnote, Daniel Defoe, in *An Essay upon Projects,* ed. Joyce D. Kennedy, Michael Seidel, and Maximillian E. Novak (New York: AMS Press, 1999), xxiii, xlii. See also *OED,* definition 1.b for projector: 'In invidious use: A schemer; one who lives by his wits; a promoter of bubble companies; a speculator, a cheat.'

4 Rudolph Erich Raspe, *Singular Travels, Campaigns and Adventures of Baron Munchausen by R.E. Raspe and Others,* ed. John Carswell (London: The Cresset Press, 1948), xiv.

5 Raspe, *Natural History of the Terrestrial Sphere,* xxii.

6 Ibid., lxxii.

7 Carswell, *The Romantic Rogue,* 192.

8 See *OED,* 2nd ed., definition 2.a for projection, *n.*

9 Richard Hamblyn, 'Private Cabinets and Popular Geology: The British Audiences for Volcanoes in the Eighteenth Century,' in *Transports: Travel, Pleasure, and Imaginative Geography, 1600–1830,* ed. Chloe Chard and Helen Langdon (New Haven & London: Paul Mellon Centre for Studies in British Art; Yale Center for British Art, 1996), 200.

10 On the development of these discursive conventions, see Walter Moser, 'Experiment and Fiction,' in *Literature and Science as Modes of Expression*, ed. Stephen J. Weininger (Dordrecht, Boston: Kluwer Academic Publishers, 1989), 68. See also John Bender, 'Enlightenment Fiction and the Scientific Hypothesis,' in *Eighteenth-Century Genre and Culture; Serious Reflections on Occasional Forms: Essays in Honor of J. Paul Hunter*, ed. Cynthia Wall and Dennis Todd (Newark: University of Delaware Press, 2001), 236, 39.

11 The work is *Lectures and Discourses of Earthquakes and Subterraneous Eruptions*, written in 1668 and published posthumously in 1705. See Introduction, Raspe, *Natural History of the Terrestrial Sphere*, 20.

12 Sigurdsson, *Melting the Earth*, 140.

13 Albert V. Carozzi, 'Rudolf Erich Raspe and the Basalt Controversy,' *Studies in Romanticism* 8 (1969), 235.

14 Carswell, *The Romantic Rogue*, 32.

15 On Raspe's embezzlement see ibid., 63, 78–9, 80–5, 95.

16 Cited ibid., 84.

17 Cited in Robert Hunt, F.R.S., 'Rodolph Eric Raspe, Author of "the Travels of Baron Munchausen,"' *The Western Antiquary; or, Note-Book for Devon, Cornwall, and Somerset* 5.4 (1885), 74.

18 Cited in Carswell, *The Romantic Rogue*, 104.

19 Introduction, Raspe, *Singular Travels*, xxiv.

20 On Raspe's supposed work as an industrial spy for Boulton, see Carswell, *The Romantic Rogue*, 178, 196, 201; on his catalogue for Tassie see ibid., 209–11, 220–1, 232–5, 237, 240, 249.

21 Carswell, *The Romantic Rogue*, 255.

22 Ibid., 39.

23 The editors of the 1970 edition of Raspe's *Introduction to the Natural History of the Terrestrial Sphere* introduce Raspe as 'a romantic poet, antiquary, embezzler, spy, industrial chemist, and geologist' (xiii). Roy Porter's 1971 history of English geology notes that Raspe's diverse interests included 'sentimentalism, the picturesque,' and 'the promotion of industry.' Roy Porter, *The Making of Geology: Earth Science in Britain 1660–1815* (London, New York, and Melbourne: Cambridge University Press, 1977), 101. Most recently, Jennifer Uglow describes Raspe as 'geologist, gem expert, probable spy and anonymous author of *the Adventures of Baron Munchausen*.' Jennifer S. Uglow, *The Lunar Men: Five Friends Whose Curiosity Changed the World* (New York: Farrar, Straus, and Giroux, 2002), 262.

24 Douglas Lane Patey, 'Swift's Satire on "Science" and the Structure of *Gulliver's Travels*,' *ELH* 58.4 (1991), 811. See also Uglow, *The Lunar Men*, xx.

25 Patey, 'Swift's Satire on "Science,"' 12. On the semantics of the terms art and

science in the eighteenth century, see Raymond Williams, *Keywords: A Vocabulary of Culture and Society*, rev. ed. (New York: Oxford University Press, 1985), 42 and 277–8. See also Uglow, *The Lunar Men*, xx.

26 Porter, *The Making of Geology*, 108.

27 Ibid., 129.

28 Hamblyn, 'Private Cabinets and Popular Geology,' 188. The most prominent of the other participants in the debate over the origin of basalt rock were Hamilton and Desmarest.

29 Porter, *The Making of Geology*, 25.

30 Ibid., 132, 36.

31 Ibid., 140.

32 Ibid., 24, and also see Hamblyn, 'Private Cabinets and Popular Geology.'

33 Hamblyn, 'Private Cabinets and Popular Geology,' 195.

34 Porter, *The Making of Geology*, 107.

35 Hamblyn, 'Private Cabinets and Popular Geology,' 203.

36 Ignaz von Edler Born, *Travels through the Bannat of Temeswar, Transylvania, and Hungary, in the Year 1770. Described in a Series of Letters to Prof. Ferber, on the Mines and Mountains of These Different Countries, by Baron Inigo Born, ... To Which Is Added, John James Ferber's Mineralogical History of Bohemia. Translated from the German, with Some Explanatory Notes, ... By R. E. Raspe* (London: printed by J. Miller, for G. Kearsley, 1777), xxviii.

37 Ibid., xix.

38 Ibid., xxiv–xxv.

39 Ignaz von Edler Born, *Baron Inigo Born's New Process of Amalgamation of Gold and Silver Ores, and Other Metallic Mixtures, ... From the Baron's Account in German, Translated into English by R. E. Raspe with Twenty-Two Copper-Plates. To Which Are Added, a Supplement ... And an Address to the Subscribers ...* (London: printed for T. Cadell, 1791), xxii.

40 Hamblyn, 'Private Cabinets and Popular Geology,' 185.

41 Porter, *The Making of Geology*, 54.

42 Cited in Ruth P. Dawson, 'Rudolph Eric Raspe: The Geologist Captain Cook Refused,' *Studies in Eighteenth-Century Culture* 8 (1979), 286.

43 For accounts of these events, see Carswell, *The Romantic Rogue*, 220–30; Robert Chambers, *The Book of Days: A Miscellany of Popular Antiquities in Connection with the Calendar, Including Anecdote, Biography & History, Curiosities of Literature, and Oddities of Human Life and Character*, 2 vols. (London; Edinburgh: W. & R. Chambers, 1869), 85–6, and Hunt, 'Rodolph Eric Raspe,' 73–5. See also Introduction, Rudolph Erich Raspe, *The Surprising Adventures of Baron Munchausen*, ed. Thomas Seccombe (New York: Charles Scribner's Sons, 1895).

44 Chambers, *The Book of Days*, 86.

45 Cited in Hunt, 'Rodolph Eric Raspe,' 75.
46 Ibid., 74.
47 See Samuel Johnson, *A Dictionary of the English Language*, ed. E.M. Ginger, 3 vols., vol. 1 (Oakland, CA: Octavo CD-ROM, March 2005). See also *OED*, 2nd ed., definition for adept, *n.*
48 Walter Scott, in *The Antiquary*, ed. David Hewitt and David Punter (Harmondsworth: Penguin, 1999), 172.
49 Ibid.
50 Ibid., 96.
51 Ibid., 197. Descartes cited in Paula Findlen, 'The Janus Faces of Science in the Seventeenth Century: Athanasius Kircher and Isaac Newton,' in *Rethinking the Scientific Revolution*, ed. Margaret J. Osler (Cambridge, UK, New York: Cambridge University Press, 2000), 222.
52 See Johann Jakob Ferber, *Travels through Italy, in the Years 1771 and 1772. Described in a Series of Letters to Baron Born, on the Natural History, Particularly the Mountains and Volcanos of That Country, by John James Ferber, ... Translated from the German; with Explanatory Notes ... By R. E. Raspe* (London: printed for L. Davis, 1776), xii.
53 Scott, *The Antiquary*, 98.
54 Ibid., 100–1.
55 Porter, *The Making of Geology*, 136.
56 Rachel Laudan, *From Mineralogy to Geology: The Foundations of a Science, 1650–1830* (Chicago and London: University of Chicago Press, 1987), 56.
57 Ibid., 47.
58 Porter, *The Making of Geology*, 136.
59 From the Society's *Geological Inquiries* (1808), cited ibid., 147.
60 Ibid., 101. See also Laudan, *From Mineralogy to Geology*, 56.
61 Born, *Travels*, xxxiii.
62 Ferber, *Travels through Italy*, v–vi.
63 Ibid., viii.
64 Born, *Travels*, xxvii.
65 See Hankins, *Science and the Enlightenment* (Cambridge: Cambridge University Press, 1985), 153. See also Laudan, *From Mineralogy to Geology*, 27.
66 Born, *Travels*, xiv, xv.
67 Ibid., xxxvii.
68 Ibid., xxxvi.
69 Hamblyn, 'Private Cabinets and Popular Geology,' 192.
70 Ibid.
71 Raspe, *Natural History of the Terrestrial Sphere*, 48.
72 Ibid., 49.

73 *Critical Review*, 1759, 8: 272, cited in Porter, *The Making of Geology*, 100.

74 Raspe, *Natural History of the Terrestrial Sphere*, cxvi.

75 Ferber, *Travels through Italy*, xi-xii.

76 Cited in Dawson, 'The Geologist Captain Cook Refused,' 280.

77 Raspe, *Natural History of the Terrestrial Sphere*, 3.

78 I mention the month because four new editions came out in 1786 alone. See Erwin Wackermann, *Münchausiana: Bibliographie der Münchhausen-Ausgaben und Münchhausiaden Mit einem Beitrag zur Geschichte der frühen Ausgaben* (Stuttgart: Verlag Fritz Eggert, 1969), 162.

79 Cited in Introduction, Raspe, *Surprising Adventures*, vi.

80 Bender, 'Enlightenment Fiction and the Scientific Hypothesis,' 248.

81 A Society of Gentlemen, 'Monthly Catalogue,' *The Critical Review: or, Annals of Literature* 61 (June 1786), 479.

82 Ibid., 62 (July 1786), 79.

83 See *The Gentleman's Magazine January to June Inclusive. Being volume I in a new series* (London: John Henry and James Barker), January 1857, 2.

84 Cited in Carswell, *The Romantic Rogue*, 29.

85 Introduction, appendix, Raspe, *Natural History of the Terrestrial Sphere*, ci.

86 Cited ibid., cii.

87 Cited ibid., cv.

88 Stephen J. Weininger, 'Introduction,' in *Literature and Science as Modes of Expression*, ed. Frederick Amrine (Dordrecht, Boston: Kluwer Academic Publishers, 1989), xx.

89 The phrase is from Adam Smith's essay on the 'History of Astronomy' in which he likens the cognitive experience of wonder to the uncertainty experienced by a naturalist confronted by an unknown specimen. Adam Smith, 'The Principles Which Lead and Direct Philosophical Enquiries, Illustrated by the History of Astronomy,' in *Essays on Philosophical Subjects*, ed. W.P.D. Wightman and J.C. Bryce, *The Glasgow Edition of the Works and Correspondence of Adam Smith* (Indianapolis: Liberty Fund, 1982), 39–40.

90 I am thinking of the taxonomy Borges attributes to a 'certain Chinese encyclopedia' in his essay 'The Analytical Language of John Wilkins.' The taxonomy subdivides animals into fourteen categories, among them 'those that belong to the Emperor,' 'fabulous ones,' and 'those that are included in the present classification.' Jorge Luis Borges, 'The Analytical Language of John Wilkins,' in *Other Inquisitions 1937–1952* (New York: Washington Square Press, 1965), 108.

91 'Unphilosophical Transactions' was the title Raspe impudently proposed to the Royal Society's publisher, upon being expelled from its ranks. See Chambers, *The Book of Days*, 85.

92 Raspe, *Natural History of the Terrestrial Sphere,* cxv.

93 Rudolph Erich Raspe, *A Critical Essay on Oil-Painting; Proving That the Art of Painting in Oil Was Known before the Pretended Discovery of John and Hubert Van Eyck* (London: printed for the author by H. Goldney, and sold by T. Cadell, 1781), 6.

94 Ironically, it is not Raspe's scholarly works, but *Munchausen*'s precise descriptions that most strikingly adopt the matter-of-fact tone of the natural philosopher; as one critic puts it, 'When Raspe comes to write the stories he might be writing a report for Matthew Boulton.' Idris Parry, 'Munchausen Tells the Tale,' *PN Review* 15.1 (63) (1988), 14. The relationship between the narrative tone of *Munchausen* and Raspe's scholarly works is a fascinating subject beyond the scope of this essay.

95 Scholars often argue that Raspe's motivation for concealing his authorship was his burning desire to be recognized and remembered for his scientific works, rather than for a publication he would have regarded as a frivolity. See Raspe, *Singular Travels,* xxix; Sigurdsson, *Melting the Earth,* 144; and Raspe, *Natural History of the Terrestrial Sphere,* lxxiv. A more practical motivation for publishing *Munchausen* anonymously was self-protection from possible litigation, which the real Baron Munchausen – Hieronymous Karl Friedrich von Münchausen of Bodenwerder – did indeed embark upon, unsuccessfully, in response to the publication of Bürger's German translation of *Munchausen.* See Carswell, *The Romantic Rogue,* 191.

96 Cited in appendix, Raspe, *Natural History of the Terrestrial Sphere,* cvi, 30 May 1832.

97 Carswell, *The Romantic Rogue,* 27.

98 Peter Galison, 'Objectivity Is Romantic,' paper presented at the ACLS Philadelphia, 1999, 4.

99 Ibid.

100 Appendix, William Robertson, *The History of Scotland, During the Reign of Queen Mary and King James Vi. Till His Acession to the Crown of England. With a Review of the Scottish History Previous to That Period; and an Appendix Containing Original Letters* (1856), 85; Joseph Thomas, *Universal Pronouncing Dictionary of Biography and Mythology* (1870), 1869; *The American Cyclopaedia* (1873–6), 40.

101 Galison, 'Objectivity Is Romantic,' 3.

102 Raspe, *A Critical Essay,* 1.

103 Ibid., 5.

104 Patey describes how our current understanding of the terms art and science arises during the eighteenth century in the context of the debate between the Ancients and the Moderns. See Patey, 'Swift's Satire on "Science,"' 812.

105 For a bibliography of the eighteenth-century editions of *Munchausen*, see Raspe, *Singular Travels*. For a much fuller bibliography extending into the twentieth century and covering several languages, see Wackermann, *Münchausiana*.

chapter sixteen

# Measure for Measure: Projectors and the Manufacture of Enlightenment, 1770–1820

LARRY STEWART

Almost a century after Daniel Defoe had penned his *Essay on Projects,* the English MP Edmund Burke railed against the calculators who brought the world to the brink of ruin. For Burke in 1790, it was not France alone that then agitated him, but the whole reformist gang who relied on the manipulation of numbers, who made of the amalgam of trade, commerce, industry, and invention an elixir that corroded authority and tradition. France and its revolution was not his ultimate alarm. The Revolution was only a symptom (as another Irishman in another age would sense) of one 'rough beast ... slouching toward Bethlehem to be born.' As Burke saw it in 1790,

> France will be wholly governed by the agitators in corporations, by societies in the towns formed by the directors of assignats, and trustees for the sale of church lands, attornies, agents, money jobbers, speculators, and adventurers, composing an ignoble oligarchy founded on the destruction of the crown, the church, the nobility, and the people. Here end all the deceitful dreams and visions of the equality and rights of men. In 'the Serobian bog' of this base oligarchy they are all absorbed, sunk, and lost for ever.[1]

Burke's Bastille vision haunted him. But he worried more about the rubble that would remain in England if the calculators and reformers had their way. My intention is to explore the very calculus of innovation and promotion that, by the end of the Enlightenment, threatened to sweep away traditions and practices whose origins lay in time out of mind. For Burke there was much to fear. As custom was evaporated by commerce, tradition was submerged by the intellectual revolution of Enlightenment, overwhelming ties that once bound generations and societies together in their understanding of the world.

The particular concern here is with the complex relationship between science and commercial achievement in what has recently been called the Industrial Enlightenment.[2] But not all early-modern accomplishments in physical theory or technology enhanced the Enlightened ideals of progress. Profound problems lurked in the relationship between science and industry in the intensely commercial world of eighteenth-century empires. We must acknowledge that the industrial entrepreneurs and the seemingly vile projectors who then promoted technological invention were separated only by the most diaphanous of barriers. Even the mastery of natural and experimental philosophy provided an unconvincing veil to this relationship. In our mind-numbing age of Barings, Bre-Ex, Enron, Parmalat, and even Nortel and Arthur Anderson (to name only a very few that haunt us still) this may not seem so incongruous. Yet, in the eighteenth century, the adventures of accountants and assayers of ores were also tied to the exploitation and ownership of new ideas, to a version of intellectual property that had an unfortunate tendency to vanish all too often into the ubiquitous ether. In the notion of the ownership of ideas, there always lurked the aim of exploiting a market.[3] Just as ideas in natural philosophy and technology were juxtaposed in the Enlightenment, the same was also true of declarations of utility and claims to the generation of wealth.[4] In this context, much can be gleaned from the patent furies of the long eighteenth century, from their eruption in the 1690s, but most especially, from the aftermath of the South Sea Bubble, which tended to dampen overheated enthusiasm for grand schemes and wisps of fancy.[5]

## Projectors and the Power of Calculation

In the world of Isaac Newton and his immediate disciples, the growth of the fiscal state and its statutes gave agency to the power of calculation. Even though a Master of the Royal Mint and a moneyer, Newton did not alone invent the calculus of innovation and promotion. But his ghost stalked the century and its endless claims to the mastery of nature's powers. Then it was that moneyed men, the accountants, and the calculators compiled their own versions of private good and public wealth. Between the private and the public there was a great incongruity – in the gathering impetus of the industrial revolution and especially amidst the swarms of projectors who emerged from the smoke of the Midlands' manufactures. When, at the end of the century, Burke lamented the influence of the calculators, he was awake to the power generated by those such as James Watt and Matthew Boulton whose skill with numbers and instruments gave them advantage. The eighteenth century had become an age of measures, of longitude and the shape of the earth, of the weight of trade and the reach of empire, of meridians and metres.[6] Burke could not command this tide, as much as he might rail against it. In my view, Burke reflected an age of such economic transition that the appearance of untrammelled acceleration was bound to produce a high degree of concern. In this, he was very much of that age of projects – except that he gave clear voice to alarms previously inchoate and often misdirected. Of course, the immediate provocation was the project of the Revolution in France. But there was a much more general problem Burke managed to reveal – one that had surfaced notably in the immediate aftermath of the South Sea debacle of 1720.

It is to be remembered that Robert Walpole's administration had desperately needed to calm frightened nerves and uncertain fortunes. As John Carswell pointed out long ago, the passage of the Bubble Act was intended not merely to contain the wreckage but primarily to secure those paper fortunes that had been created in the furious run on funds. It was therefore imperative to control the enthusiasm for endless rival schemes before a complete collapse occurred.[7] Walpole and the Hanoverians needed to worry about the credibility of projects and the future of family fortunes. But what was it then that Burke owed to the aftermath of these events? The calculators had manufactured a crisis. The fundamental issue was that the purveyors of stock in insurance, lotteries, and a myriad technological schemes (especially for raising water) were selling the future and not the present. This was the essence that lured investors to the calculations of projectors, and it has been so since. Hence, the Bubble Act set out, *inter alia*, to prohibit the use of 'Patents to raise Subscriptions' by way of joint stocks, especially those insurance schemes openly competing with the South Sea Company – in which Newton himself speculated.

It was no accident that some of Newton's first generation of disciples were intimately connected with commercial ventures. But their immediate response to the Bubble chaos was remarkably astute and long-lasting. Reflecting on the fancies that overwhelmed so many, especially in the mechanical schemes he knew best, the Newtonian John Theophilus Desaguliers meditated in 1744 about the necessity of caution, 'for there are several Persons who have Money, that are ready to supply boasting

Engineers with it, in hopes of great Returns; and especially if the Project has the Sanction of an Act of Parliament to support it – and then the Bubble becomes compleat, and ends in Ruin.'[8] It was the calculations of some self-styled engineers that laid the foundations for fraud or, if we are to believe our Newtonians, provided the best preservative against them. Hence, the engineer Henry Beighton, editor of the *Ladies Diary*, argued in 1721,

> We generally see, those who pretend to be Engineers, have only guess'd, and the Chance is, they sometimes succeed; else they have made them like others that have done pretty well. But he who has skill enough in Geometry, to reduce the Physico-Mechanical Part to Numbers, when the Quantity of Weight or Motion is given, and the Force designed to move it, can bring forth all the Proportions, in a Numerical Calculation, so as it may be impossible to Err.[9]

Calculation was everything, it seems, simultaneously the basis of many a scam and the only hope for preservation. In such a world, the instruments of measure and precision would inevitably be much in demand. Amateurs and merchants, as much as philosophers, frequented instrument-makers' shops, where the apparatus of measurement and experiment could increasingly be bought off the shelf.[10] The rage of experimentation induced enough interest to keep makers at work, and the increasing need for accurate measures in commerce and manufactures meant an explosion of demand in the international instrument trade. At the same time, the literate efforts to disseminate technical information failed dismally even when the monopolies of guilds were readily breached.[11] Even the brilliant images of technology, revealed most famously in the *Encyclopédie*, were without much impact on the everyday lives of the eighteenth-century artisan or trader.[12] The highly skilled craftsman who might make a machine work owed little to this literary tradition, however long-standing it then was. Yet, instrumental precision was increasingly a skill that philosophers, as well as engineers, promoted. Of necessity, the capacity to measure and calculate would transform the mechanical capacity to do work. It would also change the way in which projects were perceived.

## Measuring Certainty

The desire for precision and accuracy was one response to the uncertainty of investment in many a commercial or mechanical venture. After all, the growth of global trade and European empires magnified the eighteenth-century desire for some kind of certain and objective measure of success.[13] At the same time, the exploration of natural laws and the computation of forces were invariably tied. In an attempt to assess the credibility of many a mechanical claim, it was crucial to be able to establish a firm foundation for the mechanical measures of strength and work. It was not merely a matter of knowing the capability for the work of a man, or a horse or a machine.[14] Throughout much of the century, throughout much of Europe, it was increasingly necessary to determine how a quantitative measure of work might actually be determined, if at all. This could be the difference between the success and failure of many a mechanical project. It was precisely for this reason, in the first two decades after the South Sea debacle, that Desaguliers and increasing numbers of his competitors would demonstrate mechanical contrivances and promote natural laws as a means of assessing claims to invention. This was to be fundamental. Credibility depended on the ability to compute problems of impact, momentum, and the effects of friction, for example, as these were the very complexities that might lead many a machine to fail and investments to evaporate.

Inevitably then, in industry, a very close relationship was established between the accountant and the artisan. And we need to understand that the effort to establish a link between early-modern science and engineering was part of this same search for certainty. This was not, as some have claimed, 'mere window dressing, intended to attach science's authority to engineering ingenuity.'[15] Such an assertion misses a major initiative of the eighteenth century. In fact, the search for a scientific language was broadly based throughout a century overwhelmed by nature's bounty. In such a world, natural philosophy provided more than a sign of cultivation and of friendship, linking men of means with men of knowledge. Increasingly, nature's bounty was transformed by ideals of usefulness that revived the spectre of Baconian dreams.[16] To propose trials in natural utility opened the gates for those who would turn all objects into goods. The result was obviously problematic, especially when claims for mechanical achievement spawned proposals for investment. Thus, the evaluation of these claims and the management of mechanical enterprises were profoundly important to those who invested and to those whose reputations and livelihood rested on the triumph of their new machines.

On what authority, then, could confidence rest? James Watt's separate steam condenser may or may not have owed much to Joseph Black's notion of latent heat; Josiah Wedgwood's innovations in glazes and pottery may have owed little to chemical knowledge within the broader disputes over the theory of phlogiston, but none of this suggests that theoretical claims were 'no more than a way of rationalizing what was physically observable.' On the contrary, as Wedgwood otherwise revealed, the close attention to computation and measurement in any manufactory was essential to its success.[17] This was likewise the case within the very laboratories in which both Watt and Wedgwood spent so much time. And the same was true in the precise measures undertaken in the laboratories of their contemporaries Joseph Priestley and Antoine Lavoisier, among others, out of which a revolution in chemistry would emerge at the end of the century.[18] Knowledge was, in several ways, a confidence game.

## The Calculation of Power

Manufactures magnified the Enlightenment ideal of improvement. One way in which this happened was through the skills of artisans, no skill being more important than the capacity for calculation. This was no simple matter. It involved more than the teaching of straightforward arithmetical skill, for which many schools could be found in many a seaport. It also required agreement over the means by which precise computation and physical principles could be linked. It might be recalled that this was also the century of definition, not only in the likes of John Harris's *Lexicon Technicum* or Samuel Johnson's *Dictionary*, but also in the very catalogues of discoveries and taxonomies in biology, geology, and chemistry with which the century was overrun. Definitions and calculations together amplified the power of manufactures. As John Heilbron once pointed out, the development by mid-century of increasing numbers of tables concerned the transformations not only of monies, but also the translation of the myriad weights and measures that confounded trade, both national and international.[19] Here, commerce incessantly encountered the impediment of customary traditions. Trade demanded measurement, excise the use of rates, and navigation star catalogues and logarithms. Here, I propose, were the beginnings of a revolution in public understanding of both mathematical and physical operations – operations reflected in groups such as the Spitalfields Mathematical Society in London's industrial and immigrant east end from 1717 until its demise in 1845.[20] This was one antecedent to the mechanics institutes of the nineteenth century. And it was also but one example of many that fixed their gaze on the power of mathematics in a mechanical world. We need to turn to those trained as artisans to understand the measurement of the physical powers that would lay the foundation for industrialism.

Industrial projects made precise calculation more necessary than ever. The emergence of large-scale manufacture in the eighteenth century had a significant impact on the way in which physical concepts were defined and understood. For example, James Watt made a major contribution to the establishment of a universal measure of force by comparing his engines to the power of a number of horses. But this was hardly a new idea in the engine trade. As early as 1698, Defoe's contemporary Thomas Savery had laid claim in *The Miner's Friend* to an engine equal to the capacity of two horses constantly working together (or of ten or twelve necessary to allow continuous operation to happen). Indeed, the calculation of the power of the later Newcomen engine was defined in 1721 by the engineer Henry Beighton as the result of a variety of factors such as that of the piston and that of the pump drawing water, along with the amount of water and the depth from which it was drawn.[21] Yet, the horse-power analogy surely stood as a useful one, as many engines for raising weights or water had commonly required horses, as in mills, mines, or working cranes. Similarly, almost a century later in his calculation of a steam engine in 1775, the engineer John Smeaton measured its capacity to raise water over a day to a height of 53 feet to be the equal of 400 horses in total. But was this definition or calculation? Did all artisans or mechanics understand, even intuitively, the relationship between force and work, between power and energy? Not likely. And yet such computations were the means by which machines were to be marketed.

Patentees such as James Watt made claims and, for the payment of a fee, received monopolies. But claims did not always have the hard evidence of calculation to support them. The problem of computation, of course, required some sense not only of what an engine was capable of on a consistent basis, but also of what work a horse might accomplish in a particular amount of time. It is, therefore, of some interest that the early applications of the Watt steam engine ran into intense scepticism not only from competitors but from purchasers. One notable case occurred during the early phase of construction of an engine for the waterworks at Richmond, near London, in 1778. Doubts arose from the common comparisons made with the use of the numerous horses long employed to work the pumps. The overseer complained to Boulton and Watt,

> Our first setting out was from a Principle of saving from our Mode of Work-.

> ing by Horses – but really hitherto we have met with nothing but delay, disappointment & Vexation. We were always told a Mere Boy might conduct the Engine – being upon so Simple a Plan but from some Very Unfair Tampering with our Man We now Employ him at 18s/pr Week, a house to live in and Coals, and another Man to Attend the Services at 13s/pr Week besides.

The anticipated savings from horses had not materialized and the outlay for an engineer or carpenter constructing the unmanageable engine began to trouble the investors.[22] Construction skills were very much in demand and generated even more new costs. In any case, horsepower was clearly becoming the measure of significance because it was actual horse power that was a primary avenue of saving and that could be estimated in a large variety of ways, although none of these methods was clearly satisfactory.

The concept of horsepower was readily subject to considerable interpretation. There is little hint at how Watt arrived at his definition, in 1782, of the capacity of a horse to pull 180 pounds through 60 yards per minute. Leaving aside this precise proposition, it is clear that, during the 1780s, Watt was routinely calculating the capacities of engines at either a horse-power standard of 32,400 pounds raised one foot high per minute – or 33,000 by 1783. The specific nature of these evaluations need not concern us here. What is apparent is that Watt was soon describing his engines to prospective buyers in horsepower.[23] During the 1790s, such assessments were made of the engines built by interlopers in the steam-engine trade like J.C. Curwin at Workington or even in cases of replacing engines that had been working in the mills for some time. Thus, in 1796, while his father at the Spa in Bath sought a cure for endless illnesses, James Watt, Junior, calculated for one prospective buyer that a recommended 20 horsepower engine 'will turn Two Thousand spindles if the Millwork is tolerably constructed and will then burn 24 Cwt per 12 hours. But to do the work of your present mill, [he told the proprietor, Richard Paley] say to turn 1400 Spindles, it will not burn more than 18 Cwt per 12 hours, whereas we are told that the Engine you now have burns 45 Cwt in the same space of time.' Watt junior then followed with a calculation that the savings of coal based on annual fuel estimates would pay for itself 'in less than three years.'[24] It is significant that in his own account of the derivation of the standard, James Watt acknowledged a certain amount of fudging and a tradition derived from pre-existing mechanical practice. Boulton and Watt

> felt the necessity of adopting some mode of describing the power, which should be easily understood by the persons who were likely to use them. Horses being the power then generally employed to move the machinery in the great breweries and distilleries of the metropolis, where these engines first came into demand, the power of a mill-horse was considered by them to afford an obvious and concise standard of comparison, and one sufficiently definite for the purpose in view. A horse going at the rate of 2½ miles an hour raises a weight of 150 lbs. by a rope passing over a pulley, which is equal to the raising 33,000 pounds one foot high in a minute. This was considered the horse's power; but in calculating the size of the engines, it was judged advisable to make a very ample allowance for the probable case of their not being kept in the best order.[25]

The proposition of horse power, whatever its real meaning, followed a custom of a lengthy lineage and of the common experience of mechanics in mills and mines.

To give one example of such issues, the heterogeneous circumstances for which engines were intended obviously might alter the consumption of energy in a range of mechanical ways. The Soho firm reminded the proprietors of a grain mill in Essex that a small engine of four horse would not meet their needs efficiently as 'the taxes upon a 4 Horse Engine (such as Friction, loss of Heat & Steam, Friction of ye Mill work) are so great in proportion to the power that we are persuaded that such very small Engines will not answer' and that it would actually be more profitable for them to erect an engine of twice the horsepower.[26] Writing from Clifton, near Bristol, James Watt, Jr, suggested in 1797 that the increasing want of power in most of the mines represented an opportunity that those who had attempted to evade the Watt patent could not possibly meet. The year earlier, from Newcastle, he had reported on trials of a winding engine for a colliery at Benwell. Rainy weather had interfered with the pit gear, but 'The Engine however acted its part perfectly and gave great satisfaction to a large Company who attended with great patience,' immediately resulting in six orders for 20-horsepower engines.[27] The horsepower standard furthermore made it possible to determine the value of an engine for the purpose of a premium to be paid to Boulton-Watt so long as their patent remained in force. Various means had been made to establish a yardstick of the translation of energy into power, such as savings of coal, the size of the cylinder, the stroke of the cylinder, or even the depth of water. This had made discussions with proprietors a potentially complex matter, as in the Boroughs waterworks in

Southwark, where the engine premium followed the size and stroke of *both* the engine cylinder and the pump. Ultimately, however, a fixed rate per horsepower was established for all new engines and this greatly simplified the negotiations.[28] Definition was essential to new mechanical contrivances.

It may thus appear that Watt's successful measure was to some extent arbitrary. Nonetheless, it did provide the requisite basis for a comparative computation between engines of diverse types. Whether Watt's calculation was a matter of convenience 'to facilitate office work,' as was once suggested, or was a means of 'facilitating and regularizing his own work,' it was a brilliant attempt at standardizing the measure of, and the difference between, engines of various types.[29] In standardization, commercial possibilities abounded. Horsepower became, in one sense, a marketing device. This is readily apparent when the concept took hold in the 1790s. Thus, while James Watt, Jr, in his discussions with his father, could describe engines in terms of the diameter of the cylinder, length of stroke of the piston, load, consumption of coal per day, and even the cost of coal per ton, ultimately orders for their engines were to be placed in terms of horsepower.[30] The notion became a kind of shorthand or proxy measure by which to describe, but not to evade or disguise, the technical considerations involved in comparing engines. Out of such calculations an industrial Enlightenment would arise.

## Computing Success

To create a definable concept that made sense of work and efficiency was not enough. Measures needed to appear to be more than ethereal.[31] Above all else, these measures needed to make comparative computations of mechanical power meaningful in the marketplace. James Watt and Matthew Boulton's marketing depended on their ability to establish a measurable definition for engine power. In this they succeeded, but they were hardly alone. Many reformist natural philosophers and engineers were great promoters of new systems. No systems, were they to be widely accepted, could appear subservient to local considerations. As Heilbron points out, Enlightenment weights and measures could have no localized or customary foundation, as in the foot of a king or the length of an arm. This turned out to be one of the major factors in the promotion of metric reform by French revolutionaries.[32] But, even more important, the commonplace 'skill in calculation' of the 'inferior orders of common men' could not be exceeded. As one of Burke's own contemporaries insisted, it could not result in 'the perplexing of all dealings, and the benefitting of knaves and cheats.'[33] But calculating limits did not defeat the enthusiasm of projectors or entrepreneurs. They may even have encouraged them.

It was thus not so much the calculation of momentum that proved crucial, but rather the momentum of calculation. Moreover, measures of nature's powers needed to be transparent. This was not so easy. Calculation itself might be taught, but philosophical concepts like power, work, energy, load, duty, friction, or momentum were not so readily grasped. This was exceedingly important, since, as Watt knew as well as anyone, the patents for invention upon which a manufacturer might rely were hardly enforceable. The very legal process of patent application made the measure of mechanical success even more complex – and yet seemingly necessary. Until well into the eighteenth century, the patent process for inventions was merely an exercise in legal futility. There was no real means of technical evaluation. Even by the 1740s, when mechanical diagrams began to appear in patent applications, these did not improve upon the verbal claims of patentees. It was clearly impossible to duplicate an invention based on these schematics. Similarly, the engravings of the 'encyclopedic ideology' of the mid-century were dismally inadequate in the transfer of technological skill.[34]

Under such circumstances, the desire for quantitative analyses of mechanical achievement became even more urgent. Precisely for these reasons, there was a remarkable effort made at the end of the eighteenth century to establish universal measures for physical principles like force and work. For example, the Derby clockmaker John Whitehurst, immensely skilful in instrument design, set out to establish a universal measure of length by way of experiments with a pendulum. The exercise did not escape the attention of many philosophers, and much of the rest of the century was overtaken by the enthusiasm to try finding ways of translating measures across the wide range found on the Continent and in Britain. Thus, Whitehurst's contemporary James Watt sought to establish a decimal translation for 'the Philosophical pound' and Sir George Schuckburgh Evelyn sought a universal 'Standard of Weight and Measures.'[35] Similarly, the experience of Watt and Wedgwood suggests the industrial workshop had some influence on the experimental evaluation of physical principles long before the nineteenth century.[36] The explorations in their private laboratories could well have had enormous public significance, especially if such measures had ever been adopted. But in the absence of universal definitions, compounded by the lack of any means of assessment, patent descriptions in the eighteenth century necessarily revealed very little indeed. A policy of secrecy was hardly the only impediment.

## Patenting Fraud

Enlightenment urgency over definition and calculation was one response to the failure to accurately, or even meaningfully, describe how many mechanical devices worked. This proved critical in the process of patent making. From Defoe to Desaguliers, it was well understood that patents were often vehicles for investment and frequent ruin. Whence, it could be argued, followed the Bubble Act and endless efforts to draw the Royal Society and the Royal Society of Arts into schemes of mechanical promotion.[37] Even such a well-known principle as the debilitating effect of friction did not inhibit, but rather actually encouraged, creative minds. Antifriction claims, in conjunction with proposals for improved efficiency in energy production and translation from natural sources to machine power, were continual themes in patent applications.[38] The multiple discussions, in many mechanical texts and lectures, about the impediment of friction and the difficulties of its elimination were virtually a challenge to inventors. Consequently, even at the end of a century when the problem was well known, various pieces of apparatus were contrived to reduce friction in machine design. Thus, in 1795, the improved construction of cranes was proposed, likely by the Rev. Edmund Cartwright, F.R.S., through the introduction of a worm gear to balance the effect of friction on the teeth of the wheels.[39] In the same year, James White received 40 guineas from the Society for the Encouragement of Arts, Manufactures and Commerce for his invention of a crane for wharfs that reduced friction by pulleys so that the human operator could shift his position and diminish the resistance of the load. Likewise, in 1800, James Fussell, an iron manufacturer in Somerset, and James Douglas, an engineer in Surrey, received a patent for a device using chains, wheels, and rollers 'for the Purpose of lessening Friction in raising, lowering, driving, and conducting Heavy Bodies.' Also, in 1801, Joseph Gaston, or John Baptiste, styled Count de Thiville, then of Piccadilly, received a patent for 'certain new Methods of giving an independent moving Power to all Machines, by Means of Hydraulic Engines and also of constructing and employing separately several of their Parts, such as Wheels, Pistons, and Apparatus for reducing Friction, upon new Principles.' Notably, Gaston claimed his invention was based on 'hydrostatical principles not at all but little known till now, which contrary to the laws of statics in solid bodies, give to incompressible fluids the property of rising almost spontaneously above the place from which they have fallen.'[40] It is impossible now to tell whether any of these ever met with expectations in any degree. Statical forces remained as much undefeated as of continual concern. This bespeaks a fundamental paucity of understanding, not only on the part of inventors, but critically on the part of the law officers of the Crown who granted the patents.

In circumstances where physical principles and mechanical innovation collided, it is understandable that inventors would be drawn to explore the fundamental limits of energy and power. How else were limits to be described but mathematically? But the results were ultimately counterintuitive, especially if we were to adopt the Enlightenment mantle of rationalism and mathematical precision. In the 1790s, the engineer Joseph Bramah obtained a patent for 'certain new Methods of producing and applying a more considerable Degree of Power to all kinds of Mechanical Apparatus, and other Machinery requiring Motion and Force, than *by any Means at present practised* for that Purpose.' Hinting that devices were not intended to infringe on any existing invention or patent was a common defensive manoeuvre. Thus, in 1799, Matthew Murray of Leeds claimed 'Improvements in the Steam-Engine, for the purpose of saving Fuel, lessening the expense of erecting Steam-Engines, and producing a more steady Motion therein *than by any Means at present practised*'; John Luccock, of Morley, near Leeds, proposed an invention for a machine 'upon Hydrostatic Principles, to produce a very considerable Mechanical Power, applicable to all the Purposes of a Steam-Engine, without the Aid of Fire, Steam, or a Water-wheel'; and James Lambie of Paisley, a mechanic, received a patent 'for his Invention of a Mode of applying additional Power to all Kinds of Machinery, by which the Force of Man is greatly increased, and he is enabled to do much more than he could otherwise perform.'[41]

In such a world, piracy might be rife, but evasion was easier. This induced greater attention to the legal protections and limitations of the patent process.[42] But the response of the public could also be gauged in the sheer rage for invention that caused plenty of patents to be taken out by the misguided and manipulative as well as the mechanically adept.[43] Any proposals relating to basic physical principles, as well as mechanical improvements, were too difficult to assess. The increasing use of engine specifications notwithstanding, it is apparent that the Crown officers had little to guide them in technical matters. This was not simply the consequence of a technological or economic determinism at work.[44] Thus, it seems that we might usefully examine the force of the unrelenting tide of patents for invention as one *cultural* measure of the Industrial Revolution. More than reflecting a penchant for profit, this suggests a passion to explore physical principles and natural limits. Projectors would have a field day if patents were the only impediment.

## Power, Patents, and Projects

When James Boswell had visited the manufactory at Soho and was shown round by Matthew Boulton in 1776, he had the occasion to note Boulton's (now often repeated) boast: 'I sell here, Sir, what all the world desires to have – POWER.' This has since become part of the romance of industrialism, endlessly repeated by commentators and historians keen to demonstrate a Britain on the verge of technological superiority.[45] Boswell's diary may well reflect the state of play in a Britain then otherwise preoccupied with the tortuous constitutional crisis in America. But it also may reflect the more evasive nuances in the measures of power and energy that were often capricious and arbitrary. The production of mechanical power was by no means certain. The fact that Boulton and Watt succeeded in their designs, even when engine makers were not always sufficiently able to make them perform, reveals that the attractions of power often overcame fears of risk and disappointment.

Natural laws were so ill defined, misunderstood, and subject to variance (even in their computation and measurement) that they actually inspired the swarm of patents for perpetual motion that existed alongside the workable industrial innovations of the likes of Watt and Smeaton. Hence, even at the end of the century and well into the nineteenth, patents for perpetual motion were sought and granted by the Crown in parallel with some of the most effectual mechanical achievements in the new technologies of power. By the middle of the nineteenth century this had become such a scandal that the engineer Henry Dircks laid part of the blame at the 'The ready means afforded for making such plans public, through the medium of cheap popular scientific journals and magazines, and likewise the facilities afforded for patenting inventions.' Indeed, Dircks pointed to the York engineer Thomas Mead, who claimed that 'it was not for any particular apparatus for which the letters patent were granted!'[46] Take, for example, Robert Davidson's patent in 1791 for a magical and mysterious 'Engine, in place of a steam-engine, to be worked without fire, wind, or water, and with or without a horse.' Likewise, continuous motion was a repeated source of inspiration, as in 1800 and 1801, when William Johnson of Bromley, in Kent, received a patent 'for a machine with new means of obtaining power in mechanical operations, of the nature of a self moving power or perpetual motion,' or in 1809 with William Pleasants's 'Self-mover, or machine which can keep itself in motion,' or George Linton's 1821 miraculous patent for 'Propelling machinery without the aid of steam, water, wind, air or fire.' One would surely like to know more about the entrepreneurial moths drawn to such flames. These mechanical projects were suitably endless and, most important, were not to be shamed even by the most spectacular achievements of the pre-eminent industrialists in the last two decades of the eighteenth century – even when the reputation of Watt-Boulton was at its height and sustained by the very Crown that simultaneously granted patents to ethereal imaginations.[47]

The intellectual evaluation of the process of work and the determination of physical labour was the key to industrial innovation. It was also the key to the way in which patents failed and projects promoted the impossible. Hence, the translation of energy into work, and the abridgment of human labour by devices that permitted not simply the expansion of work but the measurement of concepts such as power, load, efficiency, duty, or horsepower as physical and not merely arbitrary or simply customary, equally belonged to the mechanic inventor as to the philosopher. For this reason, the public world of the apostles of Newtonian natural and experimental philosophy of the Enlightenment, their innumerable lectures and demonstration devices, and the public reaction to seemingly endless claims for patents were indispensable to the comprehension of technical achievement. This was something that, during the eighteenth century, the law officers of the Crown proved incompetent to establish, even in the adjudication of the legal, if not the technical, foundations of new patents for the abridgment of labour.

Power was not only sold. It seemed to be projected out of thin air. As Andrew Ure argued in his *Philosophy of Manufactures* in 1835, new manufactures meant coming to terms with the deployment of labour, the design of factories as well as machines, and the marketing of the results.[48] The implementation of new definitions and calculations thus improved immensely upon marketing once dependent merely on promise and description. In such a climate, industrialism was not only triumphant, but it was patents ironically that made projectors smile. Virtually anything seemed possible in the generation of energy and power – even when it was not. There lay the subtle paradox behind Matthew Boulton's boast. Along with it, by the 1790s, lay a not so subtle confirmation of Burke's alarm over calculators.

## NOTES

An earlier version of part of this essay was published in *Economia e Energia. Secc. XIII–XVIII.* Istituto Internazionale di Storia Economica 'F. Datini' Prato (2003).

1 Edmund Burke, *Reflections on the Revolution in France*, ed. J.C.D. Clark (Stanford: Stanford University Press, 2001), 364.

2 See Joel Mokyr, *The Gifts of Athena: Historical Origins of the Knowledge Economy* (Princeton: Princeton University Press, 2002); and Margaret C. Jacob and Larry Stewart, *Practical Matter: Newton's Science in the Service of Industry and Empire 1687–1851* (Cambridge, MA: Harvard University Press, 2004).

3 Cf. Rob Iliffe, '"In the Warehouse": Privacy, Property and Priority in the Early Royal Society,' *History of Science* 30 (March 1992), 29–68; Eric Robinson, 'James Watt and the Law of Patents,' *Technology and Culture* 13 (1972), 115–39; and Nigel Stirk, 'Intellectual Property and the Role of Manufacturers: Definitions from the Late Eighteenth Century,' *Journal of Historical Geography* 27 (2001), 475–92.

4 Ulrich Wengenroth, 'Science, Technology, and Industry,' in *From Natural Philosophy to the Science: Writing the History of Nineteenth-Century Science*, ed. David Cahan (Chicago: University of Chicago Press, 2003), 223.

5 The purpose of legislation amidst the panic of the Bubble has caused much misreading of the Bubble Act, the main intent of which seems to have been to protect the interests of the South Sea Company. Cf. Ron Harris, 'The Bubble Act: Its Passage and Its Effects on Business Organization,' *Journal of Economic History* 54 (1994), 610–27.

6 J.L. Heilbron, 'The Measure of Enlightenment,' in *The Quantifying Spirit in the 18th Century*, ed. Tore Frängsmyr, J.L. Heilbron, and Robin Rider, eds. (Berkeley: University of California Press, 1990), 231. See also Ken Alder, 'A Revolution to Measure: The Political Economy of the Metric System in France,' in *The Values of Precision*, ed. M. Norton Wise, (Princeton: Princeton University Press, 1995), 29–71; Mary Terrall, *The Man Who Flattened the Earth: Maupertuis and the Sciences in the Enlightenment* (Chicago: University of Chicago Press, 2002); and Ken Alder, *The Measure of All Things: The Seven-Year Odyssey and Hidden Error That Transformed the World* (New York: Free Press, 2002).

7 John Carswell, *The South Sea Bubble* (London: Cresset Press, 1960), 139. See

also Malcolm Balen, *A Very English Deceit: The Secret History of the South Sea Bubble and the First Great Financial Scandal* (London: Fourth Estate, 2002), 148.

8 Desaguliers, *A Course of Experimental Philosophy*, vol. 2 (London, 1744), 8. On Desaguliers and projectors see L. Stewart, *The Rise of Public Science* (Cambridge: Cambridge University Press, 1992).

9 Quoted in A.E. Musson and Eric Robinson, *Science and Technology in the Industrial Revoluton* (Toronto: University of Toronto Press, 1969), 49.

10 James A. Bennett, 'Shopping for Instruments in Paris and London,' in *Mechants & Marvels. Commerce, Science, and Art in Early Modern Europe*, ed. Pamela H. Smith and Paula Findlen (New York: Routledge, 2002), 370–95.

11 Cf. Simon Schaffer, 'Golden Means: Assay Instruments and the Geography of Precision in the Guinea Trade,' in *Instruments, Travel and Science: Itineraries of Precision from the Seventeenth to the Twentieth Century*, ed. Marie-Noelle Bourguet, Christian Licoppe, and H. Otto Sibum (London: Routledge, 2002), 20–50; and L. Stewart, 'Science, Instruments, and Guilds in Early-Modern Britain,' in *Early Science and Medicine* 10 (2005), 392–410.

12 Svante Lindqvist, 'Labs in the Woods: The Quantification of Technology during the Late Enlightenment,' in *The Quantifying Spirit*, ed. Frängsmyr, Heilbron, and Rider, 292.

13 Cf. Lindqvist, 'Labs in the Woods,' 298. On the background to mathematical measurement see Alfred W. Crosby, *The Measure of Reality: Quantification and Western Society, 1250–1600* (Cambridge: Cambridge University Press, 1997).

14 Lindqvist, 'Labs in the Woods,' 307.

15 Cf. Wengenroth, 'Science, Technology, and Industry,' 228

16 L. Stewart, 'Global Pillage: Science, Commerce, and Empire,' in *The Cambridge History of Science: Eighteenth-Century Science*, ed. Roy Porter (Cambridge: Cambridge University Press, 2003), 825–44.

17 See Neil McKendrick, 'Josiah Wedgwood and Factory Discipline,' *The Historical Journal* 4 (1961), 30–55; McKendrick, 'The Role of Science in the Industrial Revolution: A Study of Josiah Wedgwood as a Scientist and Industrial Chemist,' in *Changing Perspectives in the History of Science: Essays in Honour of Joseph Needham*, ed. Mikulas Teich and Robert Young (London: Heinemann, 1973), 274–319; and Philip Mickleburgh, 'Accounting and Management in the British Industrial Revolution: A Re-examination of the Wedgwood Archives' (MSc. dissertation, University of Manchester Institute of Science and Technology, 1998).

18 See Jan Golinski, '"The Nicety of Experiment": Precision of Measurement and Precision of Reasoning in Late Eighteenth-Century Chemistry,' in *The Values of Precision*, ed. Wise, 72–134.

19 Heilbron, 'The Measure of Enlightenment,' 211.

20 L. Stewart and Paul Weindling, 'Philosophical Threads: Natural Philosophy and Public Experiment among the Weavers of Spitalfields,' *British Journal for the History of Science* 28 (1995), 37–62.

21 Henry Beighton, *The Ladies Diary*, 1721.

22 Clement Smith, Richmond Water Works, to Boulton and Watt, 29 August 1778. Birmingham Central Library, Matthew Boulton Papers, 254.

23 H.W. Dickinson and Rhys Jenkins, *James Watt and the Steam Engine* (London: Encore Editions, 1989), 353–5; Richard L. Hills, *Power from Steam: A History of the Stationary Steam Engine* (Cambridge: Cambridge University Press, 1989), 88ff.; Cf. Uglow, *The Lunar Men*, 376.

24 James Watt, Jr, 'Observations on the Engines of J.C. Curwin, Esq.,' 6 November 1798, Birmingham Central Library, James Watt Correspondence, box 19, B4; Watt, Jr, to Richard Paley, 27 July 1796, ibid., box 38.

25 Quoted in Dickinson and Jenkins, *James Watt and the Steam Engine*. 355–6.

26 Boulton-Watt to James Cooper, 23 November 1792, Birmingham Central Library, Matthew Boulton Papers, 254.

27 James Watt, Jr, to Matthew Robinson Boulton, 11 February 1796, Birmingham Central Library, James Watt Correspondence, box 38/3.

28 Boulton-Watt to Thomas Griffith, 14 July 1795; Griffith to Boulton-Watt, 22 July 1795, Birmingham Central Library, box 31/1; and James Watt, Jr, to Matthew Robinson Boulton, 25 May 1796, box 31/3, ibid.; Dickinson and Jenkins, *James Watt and the Steam Engine*, 344–5. See, for example, the queries of James Watt's assistant, John Southern, for an engine to power hammers at a forge in Derby. John Southern to John Spencer, 9 February 1798, Birmingham Central Library, box 19.

29 Cf. Dickinson and Jenkins, *James Watt and the Steam Engine*, 5, 354.

30 James Watt, Jr, to his father, 24 September 1795, Birmingham Central Library, James Watt Correspondence, parcel E.

31 On the role of technologies as mediators, between concepts and things, see M. Norton Wise, 'Mediations: Enlightenment Balancing Acts, or the Technologies of Rationalism,' in *World Changes: Thomas Kuhn and the Nature of Science*, ed. Paul Horwich (Cambridge, MA: MIT Press, 1993), 207–56. See also Wise and Crosbie Smith, 'Work and Waste: Political Economy and Natural Philosophy in Nineteenth Century Britain (II),' *History of Science* 27 (1989), 391–449, esp. 418ff.

32 Cf. Ken Alder, *The Measure of All Things: The Seven-Year Odyssey and Hidden Error That Transformed the World* (London: The Free Press, 2002).

33 John Riggs Miller, MP, quoted in Heilbron, 'The Measure of Enlightenment,' 209.

34 Lindqvist, 'Labs in the Woods,' 305.

35 See Uglow, *The Lunar Men*, 357; Heilbron, 'The Measure of Enlightenment,' 220–1; and Sir George Shuckburgh Evelyn, 'An Account of some Endeavours to ascertain a Standard of Weight and Measure,' *Philosophical Transactions of the Royal Society* 88 (1793), 133–82.

36 Cf. Wengenroth, 'Science, Technology, and Industry,' 249; H. Otto Sibum, 'Experimentalists in the Republic of Letters,' *Science in Context* 16 (2003), 89–120; Sibum, 'Reworking the Mechanical Value of Heat: Instruments of Precision and Gestures of Accuracy in Early Victorian England,' *Studies in the History and Philosophy of Science* 26 (1995), 73–106; and Simon Schaffer, 'Accurate Measurement Is an English Science,' in Wise, *The Values of Precision*, 135–72.

37 David Philip Miller, 'The Usefulness of Natural Philosophy: The Royal Society and the Culture of Practical Utility in the Later Eighteenth Century,' *British Journal for the History of Science* 32 (1999), 185–201.

38 Christine MacLeod, *Inventing the Industrial Revolution* (Cambridge: Cambridge University Press, 1988), 178; on the background to the theoretical problem see also Simon Schaffer, 'The Show That Never Ends: Perpetual Motion in the Early Eighteenth Century,' *British Journal for the History of Science* 28 (1992), 157–89.

39 On Cartwright and patents, see MacLeod, *Inventing the Industrial Revolution*, 71, 142, 167–9; and H.I. Dutton, *The Patent System and Inventive Activity During the Industrial Revolution, 1750–1832* (Manchester: Manchester University Press, 1984), 157, 180.

40 *The Repertory of Arts and Manufactures* 3 (1795), 312–16, 318, 113–18; 12 (1800), 303–8; 14 (1801), 289–300. See Henry Dircks, C.E., *Perpetuum Mobile: Or, Search for Self-Motive Power during the 17th, 18th, and 19th Centuries. With an Introductory Essay* (London: E. & F.N. Spon, 1861), 82.

41 See *Repertory of Arts and Manufactures* 6 (1797), 289–96; 8 (1799), 73–96, 309–14, 371–3; my italics. Cf. MacLeod, *Inventing the Industrial Revolution*, 105–6, 143, 220; and Dutton, *The Patent System and Inventive Activity*, 141, 161, 183.

42 Dutton, *The Patent System and Inventive Activity*, 36–7, 180.

43 Cf. MacLeod, *Inventing the Industrial Revolution*, 147.

44 Maxine Berg, *The Age of Manufactures 1700–1820: Industry, Innovation and Work in Britain*, 2nd ed. (London and New York: Routledge, 1994), 170–1.

45 Geoffrey Cumberledge, ed., *Boswell's Life of Johnson* (London, New York: Oxford University Press, repr. 1953), 704 (22 March 1776). For the latest expression of the British achievement in the face of endless visitors to Soho, see Roy Porter, *Enlightenment: Britain and the Creation of the Modern World* (London: Allen Lane, 2000), 432.

46 Dircks, *Perpetuum Mobile*, 13, 67; Christine MacLeod, Jennifer Tann, James Andrew, and Jeremy Stein, 'Evaluating Inventive Activity: The Cost of

Nineteenth-Century UK Patents and the Fallibility of Renewal Data,' *Economic History Review* 56 (2003), 537–62, esp. 550–2.

47 See William Nicholson, *A Journal of Natural Philosophy, Chemistry, and the Arts; Illustrated with Engravings*, 5 vols. (London, 1797–1802), 1:375–80. 'On the mechanical Projects for according a Perpetual Motion,' *The Repertory of Arts and Manufactures* 12 (1800), 303–8, 431; 14 (1801), n. 11. Bennet Woodcroft, *Alphabetical Index of Patentees of Inventions* (London: Evelyn, Adams & Mackay, 1854, repr. 1969).

48 Cf. Wengenroth, 'Science, Technology, and Industry,' 225.

# Contributors

**Albert Boime** is Professor of Art History in the Art Department of UCLA. In addition to his multi-volume, ongoing *Social History of Modern Art*, he has written widely on the subject of the Academy (*The Academy and French Painting in the Nineteenth Century*, 1971, and *Thomas Couture and the Eclectic Vision*, 1980) and on the relationship between art and revolutionary politics.

**Carole Fabricant** teaches in the English Department at the University of California, Riverside, and has published widely on eighteenth-century topics such as colonialism, Anglo–Irish relations, travel literature, and the politics of landscape design. She is the author of *Swift's Landscape* and is currently working on editions of *Swift's Miscellaneous Prose* and *Swift's Irish Writings.*

**Martin Gierl** is currently Privatdozent at the University of Göttingen. He has been working on the intellectual history of seventeenth and eighteenth centuries, especially Pietism and German Enlightenment. After completing a study on the organization of science and the Academies of Sciences in the late nineteenth century, his current project is about Johann Christoph Gatterer and the history of historiography now. His main interest is to understand how science and, more generally, history worked as a process of institutionalization.

**Paul Hammond** is Professor of Seventeenth-Century English Literature at the University of Leeds, and a Fellow of the British Academy. He is the author of *Dryden and the Traces of Classical Rome* (1999), *Figuring Sex between Men from Shakespeare to Rochester* (2002), and *The Making of Restoration*

*Poetry* (2006), and co-editor of the Longman Annotated English Poets edition of *The Poems of John Dryden*, 5 vols. (1995–2005).

**David Boyd Haycock** read 'Modern History' at St John's College, Oxford, and has a doctorate from the University of London. Since 1998 he has held research fellowships at Wolfson College, Oxford, De Montfort University, UCLA, and the London School of Economics. He is presently Curator of Seventeenth Century Maritime and Imperial Studies at the National Maritime Museum in Greenwich, London. His most recent book is *Mortal Coil: A Short History of Living Longer* (2008).

**Sarah Tindal Kareem** is Assistant Professor in the Department of English at UCLA. She is completing a book about the role of wonder in eighteenth-century fiction.

**Margery Kingsley** is Professor of English and Chair of the Department of English and Foreign Languages at Cameron University. She is the author of *Transforming the Word: Prophecy, Poetry, and Politics in England, 1650–1742.*

**Kimberly Latta** teaches Milton, Restoration, and Eighteenth-Century Literature at the University of Pittsburgh. She is currently completing a book, provisionally titled 'Married to the Market: Towards a Cultural History of the Imagination.'

**Maximillian E. Novak** is Distinguished Professor of English and Comparative Literature, Emeritus, at UCLA. He is the author of books on Defoe, Congreve, and eighteenth-century literature; as well as the editor of several volumes of the *California Edition of the Works of John Dryden* and several collections of essays, including *Enchanted Ground: Reimagining John Dryden* (2004) with Jayne Lewis.

**Alison F. O'Byrne** is a Lecturer in English at the University of Portsmouth. She has held an Ahmanson-Getty Postdoctoral Fellowship with the Center for Seventeenth- and Eighteenth-Century Studies at UCLA and a Teaching Fellowship at the University of York. She is currently completing a book entitled 'The Art of Walking in London: Representing the Eighteenth-Century City.'

**Steven C.A. Pincus** is Professor of History and International and Area

Studies and chair of the Council on European Studies at Yale University. His forthcoming 'First Modern Revolution' is a reinterpretation of the Revolution of 1688–9 in England.

**Manuel Schonhorn**, Professor Emeritus of Literature and History at Southern Illinois University–Carbondale, has written essays on and edited Defoe's writing on piracy (1972, 1999), along with *Defoe's Politics* (1991). In addition, he has written essays on Fielding, Pope, Defoe, Hemingway, and Twain. His current project is 'Dumezil, the Professions, and the Birth of the Artist.'

**Larry Stewart** is Professor of History at the University of Saskatchewan. He is author of *The Rise of Public Science* (1992) and, most recently, with Margaret Jacob, of *Practical Matter: Newton's Science in the Service of Industry and Empire, 1687–1851* (2004).

**Elliott Visconsi** is Assistant Professor of English at Yale University. He works on literature, law, and political thought in the early modern period, with special emphasis on the Restoration. His first book, *Lines of Equity: Literature and the Origins of Law in Later Stuart England*, was published in April 2008.

**William Weber**, Professor of History at California State University, Long Beach, wrote *Music and the Middle Class, 1830–1848* (1975/2003), *The Rise of Musical Classics in Eighteenth-Century England* (1992), and *Great Transformation of Musical Taste: Concert Programming from Haydn to Brahms* (2008). Recently a visiting professor at the Royal College of Music in London, he has helped arrange several conferences at the William Andrews Clark Memorial Library, UCLA. An article related to that in this volume, 'Canonicity and Collegiality: "Other" Composers, 1790–1850,' will ap-pear in *Common Knowledge* 14 (Winter 2008).

**Howard D. Weinbrot** is Ricardo Quintana Professor of English, and William Freeman Vilas Research Professor in the College of Letters and Science at the University of Wisconsin, Madison. His latest books are *Menippean Satire Reconsidered: From Antiquity to the Eighteenth Century* and *Aspects of Samuel Johnson: Essays on His Arts, Mind and Afterlife.* He is currently at work on 'Hearts of Darkness: Swift, Johnson, Burke, and Confrontations with Evil in the Eighteenth Century.'

# Index